T0297383

Algorithmic Information Dynamics

Biological systems are extensively studied as interactions forming complex networks. Reconstructing causal knowledge from, and principles of, these networks from noisy and incomplete data is a challenge in the field of systems biology. Based on an online course hosted by the Santa Fe Institute Complexity Explorer, this book introduces the field of Algorithmic Information Dynamics, a model-driven approach to the study and manipulation of dynamical systems to solve general inverse problems. It draws tools from network and systems biology as well as information theory, complexity science, and dynamical systems to study natural and artificial phenomena in software space. It consists of a theoretical and methodological framework to guide an exploration and generate computable candidate models able to explain complex phenomena in particular adaptive systems, making the book valuable for graduate students and researchers in a wide number of fields in science from physics to cell biology to cognitive sciences.

Hector Zenil is a senior researcher at The Alan Turing Institute, British Library, researcher at the Machine Learning Group, Department of Chemical Engineering and Biotechnology, University of Cambridge, and the leader of the Algorithmic Dynamics Lab at the Karolinska Institute in Sweden. Previous positions include Computer Science faculty member at the University of Oxford, NASA Payload team member for the Mars Gravity Biosatellite at the Massachusetts Institute of Technology, and researcher at the Evolutionary and Behavioural Theory Lab at the University of Sheffield. He helped develop the factual answering Artificial Intelligence engine behind Siri and Alexa at Wolfram Research. He has published over 120 peer-reviewed papers, edited six books, is Editor of the journal *Complex Systems*, and the author of *Methods and Applications of Algorithmic Complexity* (2022).

Narsis A. Kiani is a senior researcher at the Department of Oncology-Pathology and the leader of the Algorithmic Dynamics Lab at the Center for Molecular Medicine, Karolinska Institute in Stockholm, Sweden. She was Vinnova-Marie Curie fellow at the Karolinska Institute (2014) and held a postdoctoral position at BioQuant, Heidelberg University. She is the co-editor of the *Uncertainty, Computational Techniques and Decision Intelligence* book series that focuses on modern advances and innovations in computational intelligence and decision sciences. She is also co-editor of the book *Networks of Networks in Biology: Concepts, Tools and Applications* (2021).

Jesper Tegnér is Professor of Bioscience and Computer Science at King Abdullah University of Science and Technology and Adjunct Chaired Strategic Professor of Computational Medicine at the Karolinska Institute. He was awarded a research fellowship position from the Alfred P. Sloan Foundation (USA). In February 2002, Dr Tegnér was recruited to the first Chaired Full Professorship in Computational Biology (Department of Physics) in Sweden four years after completing his PhD. He has held the positions of Strategic Professor Computational Medicine at the Center for Molecular Medicine, Director for the Unit of Computational Medicine, Department of Medicine, Solna Karolinska Institute and Karolinska University Hospital since January 2010.

Algorithmic Information Dynamics

A Computational Approach to Causality with Applications to Living Systems

HECTOR ZENIL
Department of Chemical Engineering and Biotechnology, University of Cambridge

NARSIS A. KIANI
Karolinska Institute, Stockholm

JESPER TEGNÉR
King Abdullah University of Science and Technology, Saudi Arabia

Shaftesbury Road, Cambridge CB2 8EA, United Kingdom

One Liberty Plaza, 20th Floor, New York, NY 10006, USA

477 Williamstown Road, Port Melbourne, VIC 3207, Australia

314–321, 3rd Floor, Plot 3, Splendor Forum, Jasola District Centre, New Delhi – 110025, India

103 Penang Road, #05–06/07, Visioncrest Commercial, Singapore 238467

Cambridge University Press is part of Cambridge University Press & Assessment, a department of the University of Cambridge.

We share the University's mission to contribute to society through the pursuit of education, learning and research at the highest international levels of excellence.

www.cambridge.org
Information on this title: www.cambridge.org/9781108497664

DOI: 10.1017/9781108596619

First published 2023

A catalogue record for this publication is available from the British Library.

Library of Congress Cataloging-in-Publication Data
Names: Zenil, Hector, author. | Kiani, Narsis A., author. | Tegnér, Jesper N., author.
Title: Algorithmic information dynamics : a computational approach to causality with applications to living systems / Hector Zenil, Narsis A. Kiani, Jesper Tegnér.
Description: First edition. | New York : Cambridge University Press, [2023] | Includes bibliographical references and index.
Identifiers: LCCN 2022048120 (print) | LCCN 2022048121 (ebook) | ISBN 9781108497664 (hardback) | ISBN 9781108596619 (epub)
Subjects: LCSH: Computational complexity.
Classification: LCC QA267.7 .Z46 2023 (print) | LCC QA267.7 (ebook) | DDC 003/.85–dc23/eng20230123
LC record available at https://lccn.loc.gov/2022048120
LC ebook record available at https://lccn.loc.gov/2022048121

ISBN 978-1-108-49766-4 Hardback

Contents

Preface

This book is based on the Massive Open Online Course (MOOC) on Algorithmic Information Dynamics, supported by the Foundational Questions Institute (FQXi) and hosted by the Complexity Explorer. The course and textbook introduce the field of *Algorithmic Information Dynamics* (AID). AID is a type of model-driven approach to the study and manipulation of dynamical systems. A type of digital calculus that studies dynamical systems in software space.

The material we have elected to cover derives from work undertaken over the last 15 years. Though its arrangement suggests a series of discrete topics, in fact what we attempt here is the exploration of avenues to build bridges between them, mirroring the kinds of research programmes we ourselves have been engaged in.

In Chapter 1, we outline the broader intellectual landscape within which the course is situated. To paraphrase a famous line, 'No book is an island'. The course was conceived and designed to be very different from others covering related topics. On the one hand, both the course and textbook are highly eclectic in nature, being research-orientated and focused largely on topics and questions at the forefront of scientific inquiry from what we believe to be an unconventional angle. For example, some of contemporary science's main challenges arguably involve topics such as causation, the corresponding causal networks, and the network dynamics that effectively produce regularities in data or observations. Questions such as system identification, control, and reprogrammability are therefore germane. These topics can be approached from a generative equation-based (physics) perspective, or a statistical machine-learning perspective driven by the data. Here, we develop a fundamental information-based approach to address these challenges. To this end, we assess the extent to which algorithmic information theory can provide a framework within which to understand regularities in data and observations. In an important sense this reflects our vision of a more horizontal science, which proceeds by cross-pollinating seemingly disparate areas that together push the boundaries of the individual fields involved. On the other hand, our ambition from the outset was to design the course to be self-contained, and this has meant having to include a large preliminary section, with individual chapters that could conceivably be expanded into one or several free-standing courses. However, this has not been our choice, as to do so would in our view disrupt the flow between disparate areas and mask their deep interconnectedness, which is precisely what we think deserves careful investigation. Our course and textbook are thus quite different from others in that they treat emerging areas of interest in depth while both broadening in scope and disrupting every one of the established fields drawn upon. In doing so we are able to offer, in return as it were, some new insights into the many concepts and tools lent us by the said fields. This is the case with, for example, graph and network theory, where we make specific contributions by defining native measures of algorithmic complexity, or with dynamical systems, where we provide new tools to characterise systems, and help reconstruct and explore dynamics from disordered data, asking, for example, how informative eigenvalues are in the context of graph spectra. Likewise, with respect to classical Shannon information theory, we contribute a new measure that joins forces with estimations of algorithmic complexity to make each measure stronger than it would be on its own. And finally, to causation, where our mechanistic/algorithmic approach is intended to make its greatest contribution, we bring a fresh and novel viewpoint, moving beyond a solely statistical approach to data analysis towards model inference and program synthesis.

The Introduction will provide the background to the many concepts involved in the study of AID and, in particular, prior approaches to the challenge of causality or causation that we have built upon.

We will briefly introduce graphs and networks from the mathematical perspective, as well as in terms of the way in which they serve to represent interactions in biology. Networks will be a fundamental object of study throughout the book.

In Part I, we will have the difficult task of providing a general overview of information theory, computability theory, and algorithmic complexity, the three topics, apart from dynamical systems, which are most essential to an understanding of AID.

We will also walk you through another fundamental area, that of dynamical systems, one of the main disciplines needed to understand algorithmic information dynamics, where we will be able to study the evolution of systems over time.

In Part II, covering theory and methods, we will begin by introducing the measures on which AID directly stands, that is, the so-called Coding Theorem Method or CTM, and the so-called Block Decomposition Method or BDM, as well as the concept of sequence models.

At the end of the second part (Chapter 8), we will introduce Algorithmic Information Dynamics, its associated algorithmic causal calculus, and its direct application to understanding and reconstructing dynamical systems. The seminal concept of reprogrammability will close this part of the book.

While we would have preferred the process of learning and communication to be less sequential, we humans have few options other than to present written material in a consecutive order, even when ontologically unwarranted and pedagogically less than optimal.

However, you may elect not to move sequentially through the book, especially if you are comfortable and knowledgeable in these areas, in which case you can go directly to those topics that are of particular interest to you, even if that means going directly to Part II. Figure 1 should help you understand the dependencies.

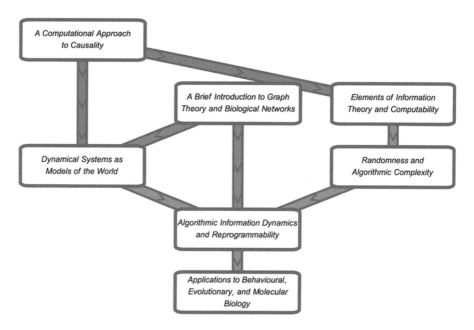

Figure 1 Module dependencies of the course and of the new field of algorithmic information dynamics. Orange links signify conceptual/motivational dependency. Pink links signify weak dependency. Blue links indicate strong dependency.

Part III of the textbook pulls everything together, from networks to evolving systems, from perturbation analysis to complexity and computability, from information theory to causation and applications to genetics, molecular biology, cognition, and evolution. Because of the level of multidisciplinarity of our work, we will include a summary some chapters that explains how an observer from the field under review in that chapter would regard the fields covered in each of the remaining chapters.

Finally, in Chapters 11 and 12 we will explore the various applications of algorithmic information dynamics, in particular to behavioural, evolutionary, and molecular biology, to static and evolving genetic networks, and in the new area of algorithmic machine learning, providing pointers toward future research.

Acknowledgements

We are very grateful to our friends and collaborators whose work has enriched our own, including Fernando Soler, Jean-Paul Delahaye, Santiago Hernández-Orozco, Felipe S. Abrahão, Joost J. Joosten, Stephen Wolfram, Gregory Chaitin, Peter Minary, James Marshall, Nicolas Gauvrit, and Jürgen Riedel. And, of course, thanks should go to our many MOOC students.

We want to thank the Algorithmic Nature Group and LABORES. Special thanks to our course alumni, in particular Bart Wauters, who very kindly read our first drafts and provided feedback; to several other alumni who ended up becoming close collaborators; and to our MOOC TA, friend and collaborator, Dr Alyssa Adams.

Finally, we want to acknowledge the institutions that have hosted or supported us while we carried out the research that forms the bases of this course: the Karolinska Institute, the Centre for Molecular Medicine, SciLifeLab, the University of Oxford, the King Abdullah University of Science and Technology (KAUST), The Alan Turing Institute, and Oxford Immune Algorithmics. We also acknowledge the institutions that offered their sponsorship and support: Oxford Immune Algorithmics, Wolfram Research, the Foundational Questions Institute (FQXi), the Santa Fe Institute, the John Templeton Foundation, the Swedish Research Council (Vetenskapsrådet), and Vinnova.

Introduction: Living Systems, Causality, and Information

While it is clear that life and living systems are fascinating and complex, the reader may still ask why we talk about living systems in the context of causality and information. You now have a text titled *Algorithmic Information Dynamics – A Computational Approach to Causality and Living Systems* in front of you. This may appear an unlikely constellation of terms. Computation is not something you readily associate with life or living things. Causality and information are also usually treated in very different domains of science and research. Life and causality appear not to be that close either. Now, let us disentangle some of the reasoning behind our assembling what appear to be an ill-assorted melange of terms.

First, some words on living systems. Fundamentally, from a physics standpoint it is not evident how we are to understand life and the emergence of autonomous living systems on the basis of fundamental equations such as those of Schrödinger and Maxwell. To formulate the problem in these terms may not yield a productive research program despite the fact that living systems must obey the fundamental laws of physics. Asking how to construct or generate models of living systems from these equations does not appear to be a pragmatic way forward considering the currently impoverished state of our fundamental understanding of living systems. Historically, the quest for an understanding of living systems, in contrast to what can be referred to as dead systems, has fuelled development of different scientific and non-scientific views and ontologies. Scientifically, we can broadly discern a path originating in Aristotle's teleological biology, including his analysis of four different types of causes, leading to different types of explanations of phenomena in the world. Aristotle had realised that biological systems appeared to be fundamentally different from rocks and stones. The next conceptual shift came with the publication in 1859 of Darwin's *On the Origin of Species*. It was Darwin's genius was natural selection as a sufficient driver and producer of biological diversity, a claim that has since been supported by ample empirical evidence. An element of randomness, altering a code, resulting in modified phenotypes (organisms) is in this sense sufficient to generate diversity. Now, at this juncture one can ask several questions: about the source of randomness, about the kind of code involved. The next chapter in this history – addressing the issue of the code – was written in the middle of the last century, when the structure of DNA was deciphered and proved to be a vehicle to store and transfer genetic information, which turned out to be fundamentally a discrete computational representation. Following the complete sequencing of the human genome and the realisation of the sheer complexity in genome organisation and magnitude of the various kinds of biomolecules, we saw the renaissance of systems biology. Making the transition between reading the discrete genomic code and assessing the impact on the phenotype, the Darwinian organism, proved to be an exceedingly complex business. Systems biology can therefore be succinctly characterised as an integrative approach, albeit one grounded in biomolecules and data, that tries to grapple with the complexity of living systems in this specific sense. The problem can be deconstructed into several steps, each tremendously challenging. First, how does a code produce a diversity of different cells? Next, how does the collective behaviour of cells eventuate in tissues and organs, and ultimately in a functional living system capable of surviving in and acting on its environment? In parallel with these selected historical stepping stones, there has been an unprecedented development of algorithms, architecture, and hardware since 1936, when Turing and others offered a pioneering conception of computation. Of particular relevance for the understanding of living systems is the mathematical exploration of necessary and sufficient conditions for self-reproducing machines by John von Neumann in the 1950s using cellular automata. Fundamentally, a living system has to somehow sustain and rebuild itself over time. The genius of von Neumann was first to ask how this happened, and secondly to develop a formalism with which to address the question in a precise manner. Even though von Neumann was a leading expert in quantum mechanics, it's important to realise that he understood that breathing life, as it were, into such equations to produce a self-reproducing system, was neither practical nor conceptually attractive. Instead, his approach opened up to ask: What is the minimal program sufficient to generate a self-reproducing system? Subsequently, there have been numerous research programs targeting topics such as complex systems, self-organisation, and non-linear dynamics – in computer science, chemistry, and mathematics. Thus we can safely assume that notions such as information, complex systems, and causality cannot be ignored in the analysis of living systems, and indeed they should not be sealed off from such notions.

Let us be careful about the kinds of questions we can ask about living systems, and precise in specifying the particular questions we wish to address. This sets the stage for understanding the architecture of the current book. In principle, we can ask four different types of questions:

A: Historical: What is the origin of life? This can be elaborated on in a number of ways. When did it originate? How we define life affects this temporal question, as well as the question of how life developed from simpler components. Did life originate on earth or in space? Addressing the order of events and the primacy of different types of molecules is still fertile ground for arguments and experiments. Is our world an RNA world or not?

B: Temporal progression of species: How did life develop? What are the drivers of evolution beyond or behind natural evolution? Current issues include elucidating molecular mechanisms as well as understanding species dynamics over space and time in response to different environmental conditions.

C: Causation and the mechanisms by which information is transferred across generations: With the discovery of DNA, we now have a blueprint for how this works. The problem is solved in the sense that DNA is the carrier, whereas massive amounts of work has been carried out since the mid-twentieth century on exactly how this works in biochemical detail. A hot topic generating a lot of interest is the notion of transgenerational transfer as a consequence of acquired epigenetic changes.

D: Dynamical, i.e., the stability and sustainability of living systems: This amounts to asking why living systems are not a molecular house of cards that falls apart given the second law of thermodynamics. What are the mechanisms that are sufficient to generate systems that are alive, robust, self-reproducing, and flexible enough to learn and adapt from their local environments?

First, we would like to underline that these are very different types of questions, and that each can be addressed using different techniques originating in different professional fields. Secondly, it is interesting that Schrödinger effectively addressed C and D in his classic book *What Is Life?* published in 1944. As a physicist, he speculated, prior to the discovery of DNA, that for problem C, the answer had to be some kind of repetitive crystal structure transmitting the information (the code). A decade later, DNA was established as just such a structure, which interestingly turned out to be discrete. Furthermore, problem D is conceptually similar to an earlier foundational problem in physics. Physicists had been intrigued by how the atom could be a stable structure, a challenge which required the development of quantum mechanics before it could be addressed. In brief, why does the periodic table look the way it does? Who can interact with whom, and what are the fundamental rules determining the reactivity and stability of the different elements? Thus, it is rather natural to ask a similarly fundamental question – as Schrödinger did – about what accounts for the stability of biomolecules defining molecular machines and, by extension, for the persistence of living systems – functional metabolic circuits, cells, tissues, organs, and eventually organisms surviving in an environment.

In our book, our initial entry point is essentially problem D, viewing living systems as complex systems. We search for a conceptual and practical approach to assess what they are, how they work, and how such systems can extract regularities from their environment. Extracting regularities is a consequence of the fact that the world is not random; there are causal regularities out there. Without such regularities it is difficult to conceive of the existence of living systems, let alone Darwinian evolution. Indeed, the existence of causal regularities in the world appears necessary for the existence of living systems, as how would they survive otherwise? Furthermore, such systems benefit from having the ability to detect regularities and exploit them in their actions. Naturally, since living systems are part of such a world of regularities, their very construction reflects causal dependencies. This outlook – which motivates our work – therefore leads us to consider topics such as causality, complex systems, and information from a fundamental point of view, and to specifically address how they are intertwined, this intertwining being precisely the challenge that the existence of living systems poses.

Let us briefly digress and consider the notions of complex systems, networks, and causality, which will eventually set the stage for using the concept of information as an organising principle upon which to base our analysis. On the one hand we have learned, ever since the pioneering studies of Poincaré, that non-linear dynamical systems can exhibit chaotic behaviour as well as produce regular repeating patterns, which, however, are not predictable from a microscopic point of view. Poincaré was also interested in, or rather challenged by the Swedish King to tackle, the question of stability. Given Newtonian dynamics, are the planetary orbits stable or not? The famous mathematician Mittag-Leffler headed the prize committee and this was the inception of the famous Mittag-Leffler Institute for Mathematics in Stockholm. In response to the challenge and building on his ongoing work, Poincaré invented the field of dynamical systems. In the 100 years since, stunning progress has been made in understanding phenomena such as sensitivity to initial conditions, bifurcations, and attractors using some of Poincaré's tools. The key point for our purposes is that systems, which from an equation point of view appear simple, can indeed exhibit very complicated chaotic dynamical behaviour. The flip side of the coin is that very complicated systems of equations can effectively display low-dimensional dynamical behaviour, which can be captured in reduced, simpler mathematical descriptions. These yin-yang, complex-simple facets of non-linear systems have been extremely useful stepping stones toward further exploration. Now we have a mathematical language to describe and analyse such dual phenomena in various complex systems. For example, modern machine learning techniques such as deep convolutional networks detect low-dimensional representation in data, akin to techniques such as principal component analysis (PCA), and matrix decomposition algorithms more generally. Exploration of the sensitivity and complexity of chemical systems has been pioneered by Prigogine in his Nobel Prize work (1977) on non-equilibrium systems, offering one example of how to ground complex non-linear experimental systems using such language as non-linear dynamics. These systems are, as a rule, simple in the sense that their dimensionality is limited. The number of equations is either rather small, say one or two orders of magnitude (non-linear dynamics in applied mathematics), or alternatively, the systems are homogeneous (e.g., spin glass models in physics), thus enabling effective low-dimensional mean-field descriptions. In contrast, the amounts and complexity of data produced in the sciences and society is staggering and continuously increasing. In response to this development, since the turn of the century we have witnessed increased research activity on networks, i.e., graphs. They have become fundamental as efficient representations of data. We will therefore carefully discuss networks with and without (non-linear) dynamics. We will do so without necessarily restricting ourselves to small networks since we need tools to disentangle causality in large non-linear networked systems. Here, at the intersection between structure (networks) and dynamics (the equations describing the unfolding of the mechanism behind a graph), is a key frontier where challenges to understanding causality converge. For example, why does this neural network succeed or fail at a classification task, what are the causes that produce the data originating from a networked system? Advancing from observations of data to a causal understanding of what goes on behind the scenes, coming to grips with the generative processes producing observed phenomena, remains a fundamental challenge in science. Paraphrasing Shakespeare, if the world is a stage, then what is the play, or what is the program or code generating the dynamics we observe playing out on the stage? In other words, how can we find the low-dimensional effective description of a system from observations, i.e., from collecting data? The amount of data makes the stage huge, and generally speaking we do not know the size of an effective low-dimensional description. This is how causality enters the picture. Which state variables are sufficient to capture the dynamics of the system, including different kinds of interventions (causes) and the resulting responses (effects) from this low-dimensional system? In science, we like to ask what effects arise if X occurs, where X could be a natural cause or a human intervention into a system. These are questions we usually ask about the world, i.e., questions about how things work. When dealing with living systems, we like to ask causal questions concerning how such complex systems remain stable and persist (or not) over time. In essence, the existence of regularities in the world, including living systems, implies the existence of manifolds in a geometrical mathematical sense, thus bringing us back full circle to Poincaré's original insights. Here, we'd like to further remark that living systems, such as the human brain, for example, could be viewed as devices evolved to decipher the causal structure around us. Hence causality is not only an ontological question having to do with how the world works; the challenge of causality also has an epistemological component, as it appears to be central to how living systems acquire useful knowledge about the world, facilitating their survival. Thus, what appears to be purposeful behaviour is akin to a teleological aspect of living systems, teleology being Aristotle's attempt to address the issue. Regardless, our focus is to fundamentally ask whether – given observational data – we can reconstruct the generative (causal) program producing a string of events. This is relevant for understanding complex systems, including living systems, where as a rule we do not have prior access to the generative equations. We note that by building powerful systems able to decode such generative models from data, we may gain insight into how living systems acquire and act upon their internal representations of regularities in the world.

In summary, there is a dual sense in which causality is central to understanding living systems: understanding their organisation, which evidently is a composite of complex non-linear systems, and understanding the machinery that enables these systems to extract causal regularities from the world. Now, returning to the topic of stability and persistence as a consequence of the organisation of living systems and the ability to extract causality from the world, it is clear that we need strong fundamental tools to probe causality in such complex systems. Information, in a precise technical sense, provides us with language to assess organisational features of systems and their transformations, without being committed to a particular physical ontology. For example, living (and other) systems can be viewed essentially as information processing entities, thus requiring a language which does not necessarily refer to the underlying physical entities, such as information theory. Classical Shannon information theory is an excellent tool to describe communication channels and transformations, in essence counting bits or symbols given a probability distribution. There are also deep connections with the second law of thermodynamics, for example see [1], as regards to how there is – on average, over time – an increase of entropy, amounting to an increase in the number of micro-states compatible with macroscopic behaviour. There is an immense literature on various applications of Shannon entropy in physics, biology, and language research. Yet, as we shall see in this book, there are cases where a counting (Shannon) approach is not sufficient to detect differences in structure in systems, i.e., to capture differences in regularities, and therefore by extension, causality. To count bits or states, we need to know what macroscopic property we are interested in. This is a bias, akin to the challenge of optimisation in machine learning; we need to know which function to minimise or maximise. This is not an issue in cases where we know which feature we are searching for. However, it becomes a fundamental restriction when we require a powerful tool to detect regularities in the absence of a prior assumption of an objective function. This, in essence, is what has motivated us to search for a more unbiased quantitative language to describe complex systems and their transformations. Algorithmic information theory (AIT), with roots in precise fundamental theorems in mathematics, is exactly such a framework. The seminal work of Kolmogorov, Chaitin, Solomonoff, Levin and Martin-Löf proved to be sufficiently powerful to enable an unbiased characterisation of randomness and complexity, in effect solving Hume's inference problem definitively. The minor catch is that quantities in AIT are not computable in a Church–Turing sense. This has been a stumbling block with regard to applications of AIT, for obvious reasons. Recent developments spearheaded by our teams and colleagues have demonstrated that these quantities can be numerically approximated beyond the limitations of lossless compression (which we also proved to be almost exclusively related to Shannon entropy rather than to algorithmic complexity), thus providing upper bounds to the complexity of objects.

In this book, we develop a comprehensive account of these concepts from AIT and their temporal, i.e., dynamical, extensions, thus relating them to dynamical systems and causality, using networks as a conceptual bridge in several cases. Our title, *Algorithmic Information Dynamics – a Computational approach to Causality and Living Systems*, refers exactly to these new developments and how they are practically (computationally) and conceptually related, intertwined, and key to probing living systems from the dual perspectives of external and internal causality. We strongly believe that by facilitating an understanding of all these themes and their interconnections, we can hope to make new breakthroughs across several areas of science and, in particular, in the science of living systems.

Part I

Preliminaries

1 A Computational Approach to Causality

Chapter Summary

As discussed in the previous chapter, distinguishing cause from effect is essential in science, in particular for understanding complex and living systems. In this chapter, we will provide a general overview of approaches bridging computation and causality. We begin by surveying concepts and outlining the current state of the field to convey its richness, its multiple difficulties, and its challenges. In particular, we survey how causality has been addressed from different conceptual standpoints. Interestingly, two of these strands reflect the tension that has characterised AI research since the 1956 Dartmouth workshop considered the inception of AI. The first modern take on causality derives from a discrete logical and symbolic mode of thinking. Central in this paradigm is the belief that to construct intelligent reasoning, i.e., causally competent systems, we need symbolic representation and computations thereof. The second framework that we discuss in this chapter essentially displaces causality as represented in terms of discrete symbolic elements in favour of statistical continuous distributions. It has pre-empted progress in the study of causality and replaced it with the study of correlations and associations, plunging the field of causality into a spell of wintry aridity. Correlations in data and observations assumed saliency during this period spanning several decades and still prevail in all areas of science. The third line of thinking flips the problem on its head. Instead of saying 'let us assume a statistical model and ask how our data could or could not be accounted for by it', a machine learning approach reformats the problem as a learning task. Given the data, it says, 'let us discern an implicit model capturing the regularities that characterise it'. There is also a major line of research that merges statistical and learning approaches to finding patterns in data. Bayesian approaches, including network representation, are examples of this. Yet, as we will see, both the statistical and the learning approach often confound causes and effects in data, a shortcoming inherited from traditional statistics. This is not entirely surprising, as many of these approaches were not primarily designed for elucidating causality from data without recourse to probability distributions. The fourth paradigm that we review originates in what could be considered a scientific computing and dynamical systems mode of thinking. In practice, it uses a model which is generative in a mechanistic sense (e.g., differential equations), not a merely statistical model, and it asks whether such a generative model is sufficient to account for the observations. The advantage is that, here, candidate models are made explicit and can be tested against data and predictions. Yet the sufficiency of a model does not imply its necessity, as there could well be other casual models (explanations) for the same set of observations. This chapter sets the stage for the major ways scientists have dealt with causality in the quantitative sciences, exploring their advantages and inherent limitations. It paves the way for critically inquiring into what we need to accomplish in order to come to grips with causality. The chapter ends with a set of multiple choice and discussion questions to establish key takeaway points.

1.1   The Challenge of Causality

Below we will review several concepts – as outlined above – that are relevant to the objective of this book, which is to explain how the theories of computability, dynamical systems, and algorithmic complexity can help address the challenge of causal discovery, specifically by helping to generate mechanistic models underlying data and observations, that is, models that one can run on a computer, for example, in order to make educated guesses about the cause of a system's behaviour, its likely future behaviour, and how it may be manipulated.

Because there are many things we cannot understand or predict, we tend to believe that we are dominated by chance. However, science tells us that much of what we believe to be random is in fact highly determined by events in our world, events that precede and determine other events. Science, or at least classical physics, tells us that everything happens for a reason. What turns out to be difficult is to find or identify these reasons. Later in this book, we will explain that determinism and predictability are not the same thing and do not imply each other. In particular, we will see that determinism does not imply predictability.

Causality is a property of objects and systems that have been produced by a cause, as opposed to having occurred or appeared by chance. Causality is the property that distinguishes science from magic, and is the driver of the scientific method as it attempts to discover the causes of what happens in the natural world. The ultimate aim of science is to understand the world through simplified causal models and thereby to predict and change events. These events can be of any kind, for example, producing a new drug to treat a disease, or designing a new electric plant to generate more energy. Most of the time, the connection between cause and effect is anything but trivial. One of the most popular examples is the so-called butterfly effect, the idea that a small perturbation in one part of the world can cause a cascade of unpredictable effects on the other side of the world.

This type of butterfly phenomenon is referred to as *non-linearity*. Non-linearity means that effects feed back into themselves, becoming further magnified in complex and unpredictable ways. Later on in this book we will explore some of the technical aspects of complex systems.

At the core of the challenge of causality is something more mundane, and that is the fact that we rarely witness a process unfolding in real time. We usually start studying something when it has already happened, whether it is how dinosaurs became extinct, how multi-cellular life emerged on earth, or how a virus spreads. And when we do happen to witness a process of interest unfolding, or perform experiments to make it happen, we may attempt to isolate the system of interest from other causes, but in fact we rarely succeed. In the real world, we must necessarily deal with incomplete data from limited observations, and confront systems interacting with other systems at all scales, forcing us to settle for educated guesses about causation rather than strive for complete certainty. This is why we are forced to use tools such as statistics to study probable causes of natural phenomena. Usually, systems interacting with each other appear to us as *noise* as they cannot be distinguished from the data of interest, thus impeding our understanding of events.

Science has managed to come up with partial solutions to some of these challenges. Scientists perform experiments to force events to happen in real time so as to watch them unfold firsthand. Researchers have also come up with tools that we like to group under different rubrics, such as logic, statistics, probability, and information theory, to mention a few. But all of these have the same purpose even if they approach the problem in different ways, namely, to empower scientists with suitable means to characterise the relationship between observations, causes, and effects. To the action of discovering the direct causes of natural phenomena, we give the name science, and all subfields of science, from data science and logic to astrophysics and biology, are devoted to this activity.

In the next section, we will discuss causality and models of the world, in particular mechanistic models.

1.2   What Is a Mechanistic Model?

Central to causality is the notion of a 'source' whence a process arises, a generating mechanism underlying an observation. We often talk about primordial causes as first principles. We also think in terms of what we call a mechanistic model, a model that can be followed from cause to effect step by step, as in an algorithm.

What has come to be known as a Rube Goldberg machine after its creator, a cartoonist, is the perfect illustration of a mechanistic model, because in this rather silly, convoluted mechanism, every process corresponds to an action and a consequence, all linked in a long chain of causes and effects (see Fig. 1.1). In this cartoon, a so-called self-operating napkin is activated by the action of a person using a spoon to eat soup. It is comical because the mechanism for activating the napkin comprises a series of risibly sophisticated tasks, only to achieve the rather mundane end result of moving the napkin.

Figure 1.1 Self-operating napkin cartoon by Rube Goldberg illustrating a long and rather silly chain of intermediate cause-effect events with a trivial end result.

A Rube Goldberg machine, or Goldberg machine for short, illustrates another important aspect of the challenge of causality, particularly in relation to the area of modelling, which is that if we were not able to witness the processes leading to the movement of the napkin, we would not be able to rule out a Goldberg machine as the generating mechanism. The funniest thing about a Goldberg machine is that it is very transparent – we see its inner workings. But if we were only looking at the napkin, we wouldn't know whether it is the hand of a person that moves the napkin or a series of comically sophisticated actions. In other words, a Goldberg machine also illustrates that it is generally impossible to know the underlying cause of an event.

While we would like to say that a Rube Goldberg machine is among the least likely explanations for what happens in the world, we truly do not know, and we have to find arguments in favour of or against such explanations. Perhaps what strikes us as funny in this cartoon is not that far from what actually happens inside the human mind, for example, or at a molecular level, even if it does not involve gears, pulleys, and parrots. Clearly, a causal model like this one is far from random because the mechanism constrains the degrees of freedom of movement so that movement can only happen in a constrained space and can only be activated by a previously activated movement. The more control from original causes, the less random the effect when movement is limited in space and controlled by a prior cause.

For many centuries a geocentric model of the universe reigned and was based on what we call epicycles (Fig. 1.2). This model was used to explain the movement of the stars and planets in the sky, with the earth occupying the centre of the universe. Planets traced an additional movement around a small circle called an epicycle, which, in turn, moved along a larger circle called a deferent. Both circles rotated clockwise and were roughly parallel to the plane of the sun's orbit. This explained the apparent retrograde motion of the five planets known at the time, but one can see how complex the epicycles model was.

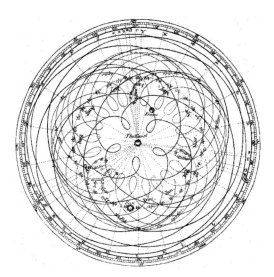

Figure 1.2 Intricate planetary motions as observed in the night sky leading to Ptolemy's epicycles model.

Even though the epicycles model of the universe was, for most purposes, incorrect, it was able to explain and describe, with virtually the same accuracy as heliocentric models, the movements of objects in the sky (see Fig. 1.3).

One of the most difficult movements to explain was that of the interior planets (closer to the sun than the earth), because these planets appeared to move in one direction and then suddenly switched to the opposite direction depending on the relative positions of the planet and the earth in their revolution around the sun. This retrograde movement was difficult to explain without placing the earth outside the centre of the solar system, so the epicycles were needed if the heliocentric model was to be causal and mechanistic.

Moreover, both the epicycles and the later heliocentric models had about the same predictive power for certain features, meaning that it was possible to run the models into the future to simulate the movements of the sun, planets, and stars as they appeared in the sky and make predictions (see Fig. 1.4). For example, they explained changes in the apparent distances of the planets from the earth.

So both the geocentric and heliocentric interpretations involved mechanistic models, with the epicycles being a concomitant of the mechanistic geocentric model and Kepler's laws of planetary motion, for instance, being associated with the mechanistic heliocentric model of Copernicus. Clearly, the effectiveness of both the geocentric and heliocentric models are an indication that generating mechanisms and predictions are not necessarily related. Nevertheless, mechanistic models are very important because they are not passive descriptions of an active process; they can actually be built, run, and followed step by step. This is why they are called mechanistic.

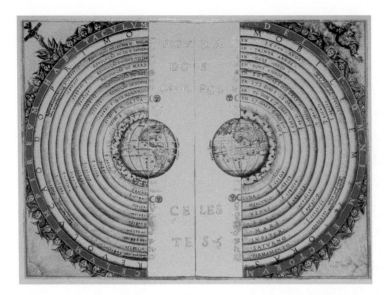

Figure 1.3 Heavenly bodies – An illustration of the Ptolemaic geocentric system by Portuguese cosmographer and cartographer Bartolomeu Velho, 1568 (Bibliothèque Nationale, Paris).

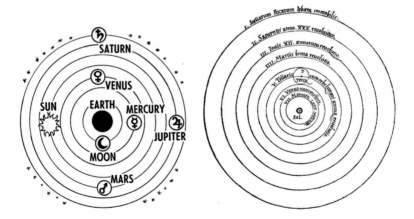

Figure 1.4 An illustration of the (left) geocentric and (right) heliocentric models.
Source: Archives of Pearson Scott Foresman, donated to the Wikimedia Foundation.

A mechanistic model may reveal more information by virtue of being mechanistic as opposed to being, say, merely descriptive. For instance, in the heliocentric model, if the sun's cycle requires 2 gears 20 times larger than the gear for the earth's moon, the mechanical model would suggest that the diameter of the sun is 2 multiplied by 20 times the diameter of the moon and would also shed light on the moon's distance from the sun, especially as observed during solar eclipses, when the moon and sun appear to be about the same size. Geocentric models would have required more convoluted explanations to reveal just the right distances at all times. And this makes mechanistic models more important than descriptive models.

1.3 The First Approach: Logic and Symbolic Reasoning

When humans started thinking about the world, they could not explain most of it and they tended to attribute everything to magic or the will of capricious gods. This is exactly the kind of explanation that is not mechanistic. Non-mechanistic models can come in many forms. Many cultures have attached importance to astronomical events, connecting them to human and terrestrial affairs, and to a certain extent we continue to do so. Every time they face something inexplicable some people are still apt to attribute it to extraterrestrial intelligences or paranormal phenomena. But the first record of progress toward finding tools to separate fact from myth and understand the natural world is to be found in Greek philosophy, in the development of a form of formal logic in arithmetic and geometry that remains the basis of modern science.

Aristotle was the first to deal with the principles of formal logic in a systematic way. The history of logic is the history of valid inference. In fact 'logos' means 'reason' in Greek, among other things. Logic as a formal way of reasoning was reinvigorated in the mid-nineteenth century, at the beginning of a revolutionary period when the subject developed into a rigorous and formal discipline that took as its paradigm the method of proof used in Greek mathematics.

This copy of Euclid's Elements (Fig. 1.5) was published in 1573 but it was written around 300 BC, long before the Bible itself. The concept of mathematical proof reached maturity at this point, and was systematically used thereafter.

The early thirteenth century witnessed a recovery of Greek philosophy after its eclipse in the Middle Ages. And to return to mechanistic models, a philosopher named William of Occam was exercised by the problem of finding arguments to rule out

Figure 1.5 Euclidis – Elementorum libri XV Paris, Hieronymum de Marnef & Guillaume Cavelat, 1573 (second edition after the 1557 ed.); in-8, 350, (2)pp. THOMAS-STANFORD, Early Editions of Euclid's Elements, no. 32. Mentioned in T. L. Heath's translation. Private collection H. Zenil.

explanations along the lines of a Rube Goldberg machine, that is, possible but overly complicated explanations for even the simplest phenomena.

Today we know his argument as 'Occam's razor', also known as the 'law of parsimony'. Occam's razor establishes that when presented with competing models, one should select the one that makes the fewest assumptions. So, in the case of the self-activated napkin drawn by Rube Goldberg, we would simply discard all the mechanisms involved if we had not seen them in operation firsthand, and we would instead favour simpler models as possible explanations. For example, that the person themselves or someone else was manipulating the napkin and not an overly complicated machine.

In science, Occam's razor is used as a guiding principle in the development and selection of theoretical models, but we will also see that there is good evidence in favour of Occam's razor. One of my own contributions to the topic is to show that Occam's razor is not only formalised by a concept central to this book, the concept of algorithmic complexity, but also that numerical approximations to algorithmic complexity provide strong evidence in its favour. We will see this later, formally and in detail; we will see how algorithmic complexity is connected with the kind of simplicity advocated by Occam.

The development of modern 'symbolic' and 'mathematical' logic is owed to authors such as George Boole, Gottlob Frege, Bertrand Russell, Giuseppe Peano, David Hilbert, and Kurt Gödel. One of the most basic kinds of mathematical logic is that of propositions that can be assigned a truth value, either true or false (Fig. 1.6).

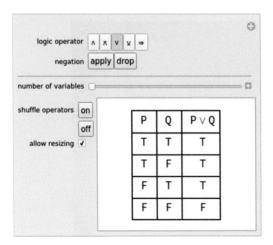

Figure 1.6 Charles Sanders Peirce appears to be one of the earliest logicians (1893) to devise a truth table matrix. While truth tables are relatively modern, their underlying principles are similar to those of the Aristotelian syllogism (a subset of the current Boolean operators), which Frege, Boole, and others would much later build upon to formalise what is today known as first-order predicate logic. A program designed to illustrate the concept is available at: http://demonstrations.wolfram.com/TruthTables/.

This type of logic is equivalent to a type of algebra known as Boolean logic, after George Boole. Boolean logic turned out to be especially important for computer science because it fits nicely with the binary system, in which each bit has a value of either 1 or 0, comparable to True or False. Boolean logic will be important later on in the book as we will be using a type of system called a Boolean network. Boolean networks are networks with propositional formulae, such as the ones on a truth table, that process information according to the way in which the network is connected.

In the truth table shown in Fig. 1.6, for example, there is a formula meaning P and Q and written P AND Q, where P and Q are propositions that can be False or True, represented by 'F' or 'T', which can also simply be 0 and 1. The connective 'AND' is called a Boolean operator and assigns a value to two propositions, telling us whether they can be true or false together. For example, for the operator AND, only when the two propositions P and Q are True can the final formula be True.

Say the proposition P tells us that 'the lights are on' while proposition Q tells us that 'there are empty seats'. Both statements may be True or False, but if we say that 'The lights are on AND there are empty seats' such an assertion can only be True if both P and Q are true at the same time. This is different from the Boolean operator OR, where if we claim that 'The lights are on OR there are empty seats' then the proposition denoted by P OR Q can be True if either of the two propositions is True, that is, even if one is False, for example, if the lights are off but there are empty seats (see Fig. 1.7).

These kinds of formulae can be arbitrarily long and so they become less trivial and can implement more complicated structures when combined as, for example, is shown in Fig. 1.8.

It can now be clearly seen how this simple logic can be deeply connected to causality. For example, the formula P OR Q can only be True if either P, Q, or both are True, and thus the observation that P OR Q is True or False tells us something about the truth

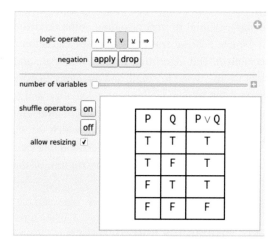

Figure 1.7 The truth table for the Boolean operator OR. Source: http://demonstrations.wolfram .com/TruthTables/.

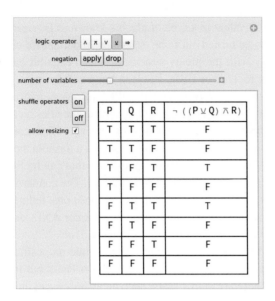

Figure 1.8 The truth table for a formula with combined logic operators. Source: http://demonstrations.wolfram.com/TruthTables/.

value of the causes, that is, the truth values of P AND Q individually when the truth value of the formula is True or False.

However, in general, mathematics and logic are more interested in truth, which is only indirectly connected to causality, because a proof is not a generating mechanism as it cannot be constructive. That is, it is impossible to write a program to effectively carry out the details of a mathematical proof, as in the case of Peano's famous diagonalisation methods. But in the twentieth century, truth and proof in logic were uncoupled in a very fundamental way, and this somehow broke the direct connection to causality.

Nevertheless, as we will see in this book when covering aspects of the theory of computation, this rupture was essential to understanding some fundamental properties of the ways in which we can achieve truth and discover causation by means of pure reasoning, logic, mechanism, and algorithm. Indeed, that logic and proof were uncoupled did not mean that we regressed to magic and astrology to explain complex phenomena. On the contrary, it led to new methods of calculation and to the way in which we approach science today – through the lens of computation. And even more importantly, it led to the concept of universal computation, which we will explore in detail in the following chapters. But it is important to understand that causality is very difficult to calculate, for the same reasons that led to the rupture of the connection between truth and proof by formal logic.

1.4   The Second Approach: Probability and Statistics

Following what we think of as a first revolution in the formal approach to the problem of causation, the advent of logical inference, there was a second revolution, so to speak,

Figure 1.9 The current statistical approach to science assumes that processes are mostly random, so when a deviation, such as a black swan, is observed in a long sequence of other uninteresting events (white swans), it generates great surprise. However, if one assumes that a sequence of events may not be randomly arranged and has been, for example, carefully sorted, then black swan events can be expected, provided the underlying mechanism allows for them or even promotes them.

consisting of an attempt to quantify the idea of chance, of things simply happening without a particular reason, as opposed to being produced or having a cause. There are all sorts of interesting epistemological challenges entailed in such an enterprise.

For example, science, and particularly classical mechanics, establishes that everything has a cause, and that there is no such thing as chance. This may suggest that something appearing to happen by 'chance' is precisely that, an appearance.

Italian and French mathematicians such as Cardano, Pascal, Fermat, and Borel were the first to make attempts to characterise chance in a mathematical way, and we owe them many of the concepts and ideas still in use today in probability and statistics. Further progress was made by German-speaking mathematicians such as Jacob Bernoulli and, much later, Richard von Mises, laying the foundations of modern probability theory.

But it was Andrey Kolmogorov who formulated the definitive version of what we know today as probability theory in its current formal mode, called an 'axiomatisation' in mathematics. Interestingly, both von Mises and Kolmogorov felt they had arrived at a limited and weak characterisation of randomness by means of probability distributions. However, while some crucial concepts necessary to an effective characterisation of randomness were unavailable to von Mises, Kolmogorov combined ideas from the rising field of computation to arrive at what we know today as the mathematical definition of randomness, which is at the centre of this book and the techniques introduced and studied here.

The aim of classical probability and traditional statistics is to help solve the problem of causal inference by calculating probability distributions. A key criticism of the way statistics approaches causality has to do with its dependence on assumptions and expectations based on probability distributions. Indeed, without knowing the generating cause of the appearance of, for example, five thousand white swans parading in single file, a statistician would approach the problem of guessing the colour of the next swan to appear by simply calculating a distribution based on how many times white swans have been spotted versus swans of other colours (Fig. 1.9).

Viewed through the lens of traditional statistics, the sudden appearance of a black swan in the long parade of white swans would be a great surprise and would be considered an oddity – and may in fact be so. However, if there was a swan owner releasing the swans in a specific order who had decided to release all the white ones first followed by the black ones, we would have made the erroneous assumption that swans were

appearing at random and thus that a black swan was an oddity when we saw one for
he first time. In other words, using, or attempting to devise a model of a generating
mechanism not based on an assumption of randomness has the advantage of leading
to better explanations, and affords the means to make more accurate predictions. Of
course it is one thing to attempt and another to achieve such an objective, but statistics is
not designed to generate candidate models but rather to describe data probabilistically.

Its limitations may include confounding causes and effects, that is, either difficulty
isolating a common cause in certain tangled situations, as happens frequently in science
and is among the main considerations in designing experiments, or else a lack of clarity
as regards what is cause and what effect. For example, a certain study may claim that
children who watch a lot of TV are the most violent. Clearly, TV makes children more
violent, it says. But this effect could easily be produced by some other cause. Perhaps
it could be that violent children watch more TV than less violent ones because they are
more likely to be reclusive.

In cases such as this, what one needs is to introduce what is called a 'control experiment' to sort out whether the children were already violent and became more addicted
to TV as a result of their violent personalities. Another problem is that one cannot reset
the same child as violent and then non-violent or vice versa, so the experiment has
to be performed on an already existing population that will necessarily be influenced
by other indirect causes that cannot be completely isolated. The purpose of control
experiments, however, is to control the most obvious bias or confounding cases, so as
to be able to draw more meaningful conclusions. Control experiments are always of
the form 'what would have happened if something else had or had not happened, or
what would happen if we do or do not apply or remove a certain other influence'.

Pierre-Simon Laplace was the first to use what are called uniform priors when faced
with a complete lack of knowledge, that is, a distribution that assumes that all events
are equally likely. He introduced a principle known as the 'principle of insufficient
reason,' also known as the 'principle of indifference' (Fig. 1.10).

The 'principle of insufficient reason' is similar to Occam's razor in that it is a guiding
principle with no strong evidence in favour or against. The principle states that if there
are n possible causes indistinguishable except perhaps by their names, then possible
causes should each be assigned a probability equal to $1/n$, that is, equal probability,
and none should be discarded or ruled out.

While the principle of insufficient reason is a reasonable principle to follow, we
will challenge some of its assumptions because while it may be desirable to assign all
possible causes non-zero probability, as suggested by another principle, the Principle
of Multiple Explanations, which establishes that if several theories are consistent with
the observed data we should retain them all, there are strong reasons to assign different
rather than equal probabilities to different explanations. In fact, the 'principle of indifference' seems to contradict Occam's razor, which suggests that we not assign equal
probability to overly complicated causes unless there is a good reason for doing so.

All these assumptions made in traditional statistics reveal that there is a highly subjective component in the way that classical probability deals with causality, particularly
in the absence of data and knowledge about the generating source, which is pretty much

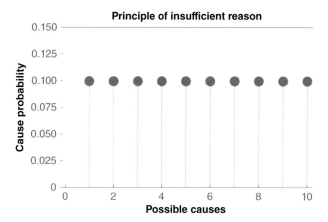

Figure 1.10 Principle of insufficient reason: when no other information is consistent with the data, or the data is not observed, this principle suggests that we assume a distribution where all events are equally possible, so if 10 events are possible causes of an effect, all them can, independently, be responsible for the said effect with probability 1/10.

the general case. Nonetheless, all these methods are widely accepted and used in the field of machine learning, for instance.

We will challenge the use of uniform priors, that is, the use of uniform distributions as a first assumption when undertaking the study of causal systems, as opposed to random systems, also called stochastic systems. We have shown that challenging the common use of these uniform priors affords interesting insights – such as the acceleration in the convergence of biological evolution – in comparison to assuming uniform random mutations, something that we will cover in the last chapter, which will return to all this in greater detail.

The idea that 'probability' should be interpreted as the 'subjective degree of belief in a proposition' was proposed by John Maynard Keynes in the early 1920s, but even today, methods such as Shannon entropy are taken far more seriously than they should be. Entropy, in the sense in which Claude Shannon introduced it, is usually presented as a measure of surprise in the context of communication.

As we will see, Shannon entropy is a measure of degree of uncertainty, which reflects one's own lack of knowledge rather than any objective indeterminacy in phenomena. So, contrary to what's generally claimed, we will demonstrate that Shannon entropy is not a syntactic measure at all, but a highly semantic one – though this does not mean that it is necessarily better or worse. Shannon entropy is interesting because it introduced logic and computation as descriptions and operations of information. We will explain in greater detail later why some of these ways to characterise randomness are often, if not always, very fragile. We will illustrate this with examples, but it is nevertheless important that you have a grasp of these concepts.

A fundamental concept in statistics is that of correlation, not a minor concept but actually the heart of statistics in a sense. Statistics is all about finding statistical patterns in the form of regularity. As we will see, anything more sophisticated than that will be missed by statistical approaches to inferring causal mechanisms in data.

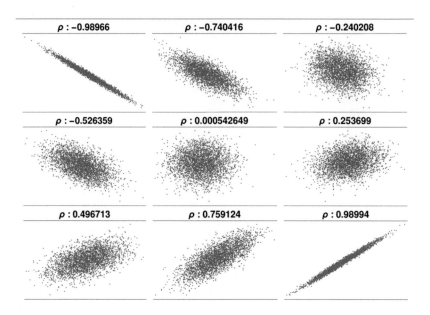

Figure 1.11 Correlation values for different scatter plots. Source: Justin Matejka and George Fitzmaurice, ACM CHI 2017. Same stats, different graphs: Generating datasets with varied appearance and identical statistics through simulated annealing. (www.autodesk.com/research/publications/same-stats-different-graphs).

A statistical regularity can be, for example, the tendency of some data points to lie on a plane, or of a time series to display periods. These are typical plots of positive and negative correlation. Think of two processes from which we obtain data, wishing to ascertain whether they correlate and/or are causally connected by correlation, say, a time series. A time series is a collection of data points sorted by time. Think of the x-axis as recording the values of one time series and the y-axis those of the other one. Then one can test whether the data points get aligned, meaning that they distribute in a similar fashion.

Correlation test values used to measure the strength of association between variables, usually denoted by the Greek letter ρ (rho), are typically given between -1 and 1; when ρ is close to 1 or -1 the data is positively or negatively correlated, respectively (Fig. 1.11).

The plots in Fig. 1.11 are called scatter plots. There are several ways to measure correlation, but they are all very similar and consist of measuring distances among data points. One of the most popular measures is called Pearson's correlation, which gauges the correlation among data point values. Another popular measure is Kendall's or Spearman's correlation. These measures rank correlation when only the order matters.

Because of the limitations mentioned above, traditional statistics often leads, with high probability, to spurious models from false negatives and false positives. A false positive or false negative is a regularity in the data that appears real but is only an artefact giving the wrong impression regarding its cause or effect. Figure 1.12 shows examples of false negatives, meaning that the correlation test, quantified by rho, suggests that there is no correlation between axis x and y. But if we look at the plots themselves, we

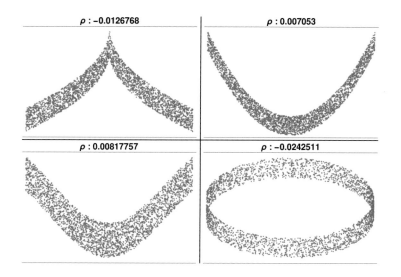

Figure 1.12 False negatives. Source: Justin Matejka and George Fitzmaurice, ACM CHI 2017. Same stats, different graphs: Generating datasets with varied appearance and identical statistics through simulated annealing. (www.autodesk.com/research/publications/same-stats-different-graphs).

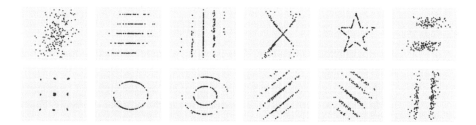

Figure 1.13 Scatter plots with equal correlation values but different structures. Source: Justin Matejka and George Fitzmaurice, ACM CHI 2017. Same stats, different graphs: Generating datasets with varied appearance and identical statistics through simulated annealing. (www.autodesk.com/research/publications/same-stats-different-graphs).

see structure immediately, suggesting that something interesting is happening in the way the data points are distributed along the axes. However, rho is almost 0 in these four plots, suggesting that, on the contrary, nothing interesting is happening.

These are other examples where not only is the correlation 0 but all plots have exactly the same rho value, despite there clearly being lots of different structures (see Fig. 1.13).

The Stanford Encyclopedia of Philosophy refers to attempts to analyse causation in terms of statistical patterns as 'regularity theories' of causation [2]. Statistical regularities are only a subset of the possible properties that a phenomenon may display. A statistical approach offers an explanation for the distribution of data but leaves to the scientist the arduous task of formulating an interpretation in order to come up with a model underlying the data. Traditionally, what a scientist does is to fit a curve. Then the equation of the curve is taken as the generating model, both to explain the distribution of data and to make predictions. In the typical case of positive correlation, for example,

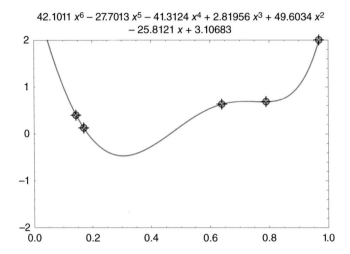

$$42.1011\ x^6 - 27.7013\ x^5 - 41.3124\ x^4 + 2.81956\ x^3 + 49.6034\ x^2$$
$$- 25.8121\ x + 3.10683$$

Figure 1.14 Polynomial fit. One can always find a polynomial of degree equal to or greater than the number of data points that fits the data points perfectly, but most likely this is only necessary when such data points have no apparent common cause and are algorithmically random. When at least one data point can be explained by a combination of the others, then the polynomial equal to or greater than the number of original data points is most certainly over-fitting the data.

it is not difficult to fit a line. This is called a linear fit because the function is linear as fitted by a line.

However, one can always force a curve to pass through any number of data points using a polynomial of degree proportional to the number of points. One can see how, by increasing the degree of the polynomial, one can make the curve go through or very close to the data points (see Fig. 1.14).

All these limitations of traditional statistical approaches to causality can be summarised in what is one of the most common adages in the field: 'association is not causation', or 'correlation does not imply causation', though lack of correlation does not mean lack of causation. In other words, that you can fit a curve to a set of data points does not necessarily mean that the curve actually has anything to do with said data points.

In [3], the degree to which regression and correlation can be misleading was made clear. All the scatter plots in Fig. 1.15 have identical values, that is, mean, standard deviation, and Pearson's correlation to 2 decimal places (x mean $= 54.26$, y mean $= 47.83$, x SD $= 16.76$, y SD $= 26.93$, Pearson's $R = -0.06$), yet clearly they were generated in different ways. They started from some points scattered randomly with little displacement then using an annealing technique they pushed every dot to a target image (a dinosaur), keeping all summary statistics the same. Clearly, once the data points reach the dinosaur a lot of computation has been invested and the result is a long chain of highly directed cause and effect, as opposed to more random configurations. Yet all these forms are very different and represent shapes that do not imply

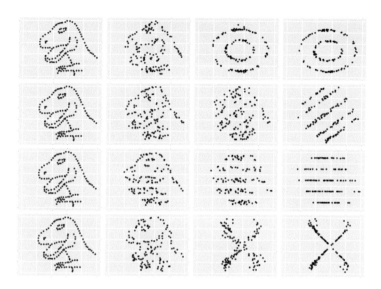

Figure 1.15 Deceptive stats, graphs with the same statistical correlation values forming all sorts of shapes, from a dinosaur to various other shapes that tools like correlation are blind to. Credit: Justin Matejka and George Fitzmaurice, ACM CHI 2017, Source: www .autodeskresearch.com/publications/samestats.

randomness. Some versions of traditional statistics, including Shannon entropy, may distinguish certain trivial cases, but will fail in many ways, as illustrated in these plots. What is missing is a method other than statistics that may be used to infer the underlying algorithmic probability and linked chains of cause and effect. In the chapters to follow we will propose that some of this can be tackled using computation and algorithmic probability.

Another example that can serve to illustrate the limitations of statistical analysis can be found in Fig. 1.16. According to the statistical description of the phenomenon, car arrivals at any point on an open road follow what is known as a Poisson distribution. This is because distributions can have characteristic shapes when they are plotted. However, what the statistical approach describes is the effect of the mechanistic cause and not the generating mechanism. It may provide clues to the causes that are then left to the scientist to interpret, but the statistics in and of themselves do not provide a model.

The reason why a Poisson distribution is produced is that slower drivers cause faster drivers to cluster in their wake, producing knots of cars running together on the highways and arriving at gas stations at about the same time. But it is not the Poisson distribution that causes the cars to cluster, nor does it suggest how or why this happens. In contrast, a mechanistic approach attempts to provide a causal model that may help design ways to manipulate the effects, as it points out the exact submechanisms that can be changed to achieve a different end. Probability and statistics have led a revolution in the study of causality, but they have, in some fundamental way, exhausted their potential

Figure 1.16 Times between cars in common traffic on a road can be described with statistics and an associated probability distribution (Poisson), but the cause can only be suggested or advanced externally. Taking a different approach, one can look for causes directly. In this case, the cause is that slower cars lead to clustering. Source: JArrevillaga/Wikimedia, licensed under the CC BY-SA 3.0 licence.

to engender further progress in modern science. This book is all about trying to provide an alternative and complement to traditional statistics and classical probability. We will see how what we call algorithmic information dynamics (AID) provides interesting tools to help reveal and deal with mechanistic causes.

It is a common observation that some events are less predictable than others. When are we justified in regarding a given phenomenon as random?

The most common approach in classical probability theory is to frame events within a set-theoretic framework. Events have an outcome that belongs to a set of possible outcomes. Traditionally, events are assumed to be repeatable, and such a repetition is called a trial. Outcomes are direct observables. Let us denote the probability P of an event A_n as $P(A_n)$. For example, in the roll of a die, events $\{1\}, \{2\}, \{3\}, \{4\}, \{5\}, \{6\}$ are elements of a space of possible outcomes of the experiment. Two events A and B are considered mutually exclusive, or disjoint, if the occurrence of A rules out the occurrence of B. In the dice example, if a roll turns up 'three dots', this event rules out the event 'six dots'.

1.5   The Third Approach: Perturbation Analysis and Machine Learning

In the previous sections we saw how much progress had been made by formalising the concepts of causality and chance with mathematics, in particular with logic, probability, and statistics. Another revolution brings these ideas together and takes probability and statistics to their limits.

How informative an experiment may be in determining the strength of a cause depends on a number of factors, such as whether there are control experiments to discard or rule out certain causes, as we have seen before, but also how good the measurements are, how well placed the observer that makes them, both at the input and output levels of a system, and how distinguishable individual events are.

We have seen how correlation is central to statistics, but also how limited it can be. One way to make the most of statistics is by performing systematic perturbations according to a probabilistic calculus introduced by Judea Pearl and colleagues, with a view to finding possible causes that can easily be ruled out upon further inspection. These ideas are at the forefront of the practice of probabilistic causal discovery. Perturbation analysis allows us to update beliefs under conditions of uncertainty based on previous knowledge derived from, e.g., performing perturbations or observing a stream of data. It also accommodates the concept of a control experiment. For example, if skin cancer is related to sun exposure and failure to wear sunscreen, then, using Bayes' theorem, a person's exposure to sunlight and failure to wear sunscreen can be used to assess the probability that they will develop skin cancer more accurately than would be possible if one knew nothing about their degree of exposure.

Central to this kind of calculation is the work of Thomas Bayes (circa 1701–1761). In its simplest form, Bayes' theorem establishes a relationship between the probabilities of two events A and B, $P(A)$ and $P(B) \neq 0$, and the conditional probabilities $P(A|B)$ and $P(B|A)$. Bayes' theorem establishes that:

$$P(A|B) = \frac{P(B|A)P(A)}{P(B)}. \tag{1.1}$$

Bayes' theorem as a rule provides some kind of 'backwards probability' where $P(A)$ is traditionally called the *prior* (or initial degree of belief in A) and $P(A|B)$ is the constructed *posterior*.

As in the case of the experiments with time series mentioned in previous sections, we can ask whether two time series resulting from different observations are causally connected to each other (see Fig. 1.17). Now, traditional statistics would typically suggest that the behaviour of two time series, let us call them X and Z, are causally connected because they are correlated, but there are several other cases that are not decidable following a simple correlation test.

A first possibility is that the time series simply show similar behaviour without being at all causally connected, despite the possible positive result of a correlation test. Another possibility is that they are causally connected, but correlation does not tell us whether it is a case of X affecting Z, or vice versa. Yet another possibility is that the two have some common cause upstream in their causal paths and that X and Z are therefore not directly causally connected but connected through a third cause Y that is concealed from the observer. So how are we to test all or some of these hypotheses?

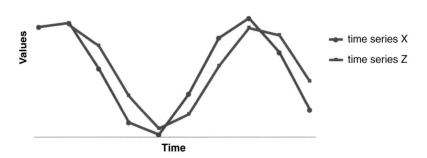

Figure 1.17 Time series as an example of a sequence of observations to be tested for a causal relationship.

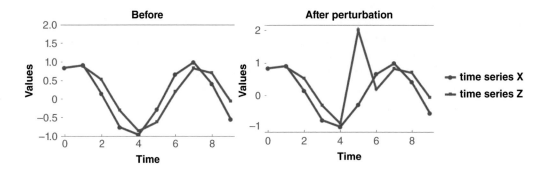

Figure 1.18 Time series X and Z before and after perturbation.

One way to do so is to perform a perturbation on one time series and see how the perturbation spreads to the other time series. Let us perturb a data point in the time series Z; let us say that we multiply by -2 the data point in position 5. We can see that nothing has happened to the values of time series X; it looks exactly the same as before (see Fig. 1.18).

So if we perturb the values in the time series Z, at least for this data point we can see that X remains the same. This suggests that there is no causal influence of Z on X.

However, if the perturbation is applied to a value of X, Z changes and follows the direction of the new value, suggesting that the perturbation of X has a causal influence on Z (see Fig. 1.19). From behind the scenes, we can reveal that Z is the moving average of X, which means that each value of Z takes two values of X to calculate, and so is a function of X. The results of these perturbations produce evidence in favour of a causal relationship between these processes if we did not know that they were related by the function we just described.

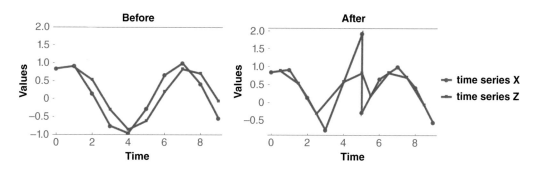

Figure 1.19 Time series X and Z before and after another perturbation.

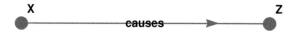

Figure 1.20 Causal relationship between series X and Z.

This suggests that it is X which causally precedes Z. So we can say that this single perturbation suggests the causal relationship shown in Fig. 1.20.

It is important to consider that performing interventions is not always simple or possible in reality, but their simulation or emulation can offer great insights. This kind of simulation of 'impossible' situations is what Judea Pearl identifies a 'counterfactual', i.e., the opposite of a possible fact. Simulation and emulation will be at the heart of our AID approach. Think of drugs as a way to perturb a biological system. Some drugs may have adverse consequences, so interventions in the way of prescriptions are highly regulated and one cannot easily do experiments relating to diseases on humans. But assuming that one may perform such interventions, in cases where an intervention does not lead to a change, it produces evidence against a causal connection between the events.

Moreover, the underlying argument is that after an intervention, correlation may or may not occur, and thus one still relies on classical statistics to make the final calls if an inference engine that uses probability theory, regression, and correlation is not replaced with something that does not. So while perturbation analysis does help to rule out some cases, it still inherits the pitfalls of statistics and correlation analysis, for the simple reason that we have not yet changed the tools; we have merely performed more experiments on the data. And this is what AID is about, taking probability and statistics out of the core of candidate models, even if these models must still rely on some statistics or on a probability distribution – which we will never be completely rid of.

As mentioned above, there are three possible types of causal relationships that can be represented in what is known as a directed acyclic graph, that is, a graph that has arrows implying a cause and effect relationship but has no loops, because a loop would make a cause into the cause of itself, or an effect into its own cause, some-thing that is not allowed because it would be incommensurate with causality (see

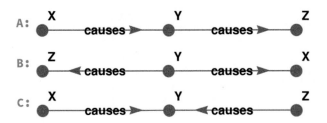

Figure 1.21 Direct acyclic graphs of all possible causal relationships among three unlabelled variables with no self-loops, i.e., no effect is the cause of itself.

Fig. 1.21). We will cover all this graph jargon in the next chapter. In these graphs, nodes are events and events are linked to each other if there is a direct cause-and-effect relationship.

In the first case, labelled A in orange, the event X is the cause of event Y, and Y is the cause of event Z, and so X is said to be an indirect cause of Z. In general, we are of course always more interested in direct causes, because almost anything can be an indirect cause of anything else. In the second case, B, an event Y is a direct cause of both Z and X. Finally, in case C, the event Y has 2 causes, X and Z. With an interventionist calculus such as the one performed on the time series above, one may rule out some cases but not all of them, but more importantly, the perturbation analysis offers the means to start constructing a model explaining the system and data rather than merely describing it in terms of simpler correlations.

In our approach to causality, we incorporate the ideas of an interventionist calculus and perturbation analysis in AID, and we replace traditional probability theory and classical statistics and correlation by a fully model-driven approach that is fed by data but is not merely data-driven. The idea is to systematically produce the most likely generating models that explain the observed behaviour. In the case of our two time series experiments, the time series X is produced by the mathematical function $f(x) = \sin(x)$, and thus $\sin(x)$ is the generating mechanism of time series X. On the other hand, the generating mechanism of Z is MovAvg($f(x)$), and clearly MovAvg($f(x)$) depends on $f(x)$ which is $\sin(x)$, but $\sin(X)$ does not depend on MovAvg($f(x)$), so how could we have guessed these two functions that statistics alone cannot systematically find? In other words, we need some sort of method to infer the function or algorithm behind the data. Having found MovAvg($f(x)$) and then $f(x) = \sin(x)$, we could not only establish the actual causal relationship, we could also produce the observation and any number of other data points in the future with perfect accuracy.

Some of this perturbation analysis and the graphical approach to causality was based on what are called Bayesian networks, involving the calculation of conditional probability distributions with the additional feature of asking 'what happens with some variable representing a possible event if this other variable representing another event is observed, perturbed, or simulated'. These ideas on an interventionist calculus are fundamental to causality and are incorporated into AID.

1.5.1   Perturbation Analysis and Interventionist Calculus

Judea Pearl's main contribution was to come up with the idea of perturbations, coun-terfactuals, a calculus, and graphical representation of cause and effect. We built upon Pearl's idea of perturbing systems. But as we have established, Pearl could not move away completely from traditional statistics and probability theory because he had to resort to the very tools he was trying to leave behind to compare the original and the perturbed or simulated system. This means that anyone using Pearl's do-calculus would be forced to resort to regression or correlation in the absence of better tools. In AID we take the next step toward removing probability from candidate causal models. Once a system is perturbed or simulated, one uses AID to compare the result to the origi-nal system, that is, one examines the differences in the candidate models explaining the original and the new model. AID incorporates perturbations and counterfactuals naturally. In AID, we apply all possible perturbations to a system, including those that can be seen as counterfactuals, and then we look at how the sets of candidate models change. In contrast, perturbing a system or estimating a counterfactual effect in Pearl's calculus in practice involves comparing distributions, which means that we fall into the same regression/correlation trap.

Candidate models in AID are computable models that explain both the original data and the new (perturbed or simulated) data. Then one can see how much the underlying model has changed as a result of the perturbation and decide which changes impact the dynamics of the original system, something that Pearl's calculus is unable to do without an inference engine that does not offer the do-calculus, or at least not without having to resort to probability distributions, which require traditional statistics.

There is a relationship between causal graphs, entropy, and algorithmic complexity that we explore throughout the book. Algorithmic complexity is a generalisation of entropy and has absolutely no problem dealing with Pearl's causal graphs; they are simply not necessary because we are no longer dealing with a language based on probability distributions (which are usually difficult if not impossible to access in the best case).

Solomonoff was aware that the theory of algorithmic probability, on which AID is heavily reliant, was the ultimate optimal theory of inference. The AI community also agreed that Solomonoff's theory was the answer to causal inference, but because it was uncomputable and there was not enough computational power back in the 60s – and indeed until recently – it was difficult to really explore the field numerically, and most researchers simply turned their backs on it. It is easy to see how Marvin Minsky could say that the way forward, possibly the only or the most important way forward, was algorithmic probability. He was speaking of AI at a panel discussion on The Limits of Understanding at the World Science Festival, NYC, on 14 December 2014, just one year before he passed away. Marvin Minsky is widely considered to be the founding father of artificial intelligence. His astonishing claim describes what turns out to be precisely the objective of our research programme and the main purpose of this book. To quote his closing statement to the panel:

It seems to me that the most important discovery since Gödel was the discovery by Chaitin, Solomonoff, and Kolmogorov of the concept called Algorithmic Probability, which is a fundamental new theory of how to make predictions given a collection of experiences and this is a beautiful theory, everybody should learn it, but it's got one problem, that is, that you cannot actually calculate what this theory predicts because it is too hard, it requires an infinite amount of work. However, it should be possible to make practical approximations to the Chaitin, Kolmogorov, Solomonoff theory that would make better predictions than anything we have today. Everybody should learn all about that and spend the rest of their lives working on it.

1.6   The Fourth Approach: Model-Driven Inference, Dynamical Systems, and Computation

1.6.1   Introducing Computation in Causal Analysis

Causality and computation have been linked since the inception by way of the concept of calculation. Figure 1.22 shows an ancient Greek calculator used to predict celestial events such as eclipses. It is known as the Antikythera Mechanism after the island on which it was found. It is the first known 'analogue computer', believed to have been made circa 87 BC, that is, more than 2,000 years ago. Possibly built by Archimedes, this device is very similar to the mechanical artifact shown above that simulate the planetary movements of the solar system. These cogwheels would turn in a precise fashion to make a prediction based on the epicyles model that we encountered above. Figure 1.23 shows a computer program designed to showcase the way in which the mechanism worked and what it was able to predict.

Not everything is known about this ancient device. For example, it is not known what additional components were needed to simulate and display planetary motions. Turning the handle causes a sequence of interlocking gears to rotate, moving the Sun

Figure 1.22 The Antikythera mechanism. Courtesy: National Archaeological Museum, Athens.

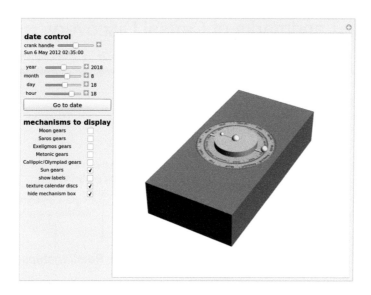

Figure 1.23 A computer simulation showing the presumed inner workings of the Antikythera mechanism. Source: http://demonstrations.wolfram.com/AntikytheraMechanism/. Contributed by Adam P. Goucher.

and Moon markers around a calendar disc. The reason that researchers have cracked this device with partial success, inhabiting the mind and knowledge of the people who designed it 2,000 years ago, is that it is a mechanical device disclosing the model that it mechanistically simulates.

As we have seen before, mechanistic models such as this one does not necessarily represent the actual mechanisms of the movements of the celestial bodies that they are attempting to simulate, but they do capture some of their regularities. What is most interesting is that while the mechanism was not the cause, the mechanism itself was the cause of the numerical calculations indicating the future position of the celestial bodies. This is literally a computer simulation performed more than 2,000 years ago, illustrating how seriously the concept of causation was approached even back then – with the use of a mechanical calculator or (analogue) computer.

What we will see in this textbook is that modern computer simulations are not very different from this one. Simulations can be used to make optimal predictions and even to try to figure out the most likely mechanistic models of natural processes. This simple artefact actually illustrates the kinds of concepts that bring together computation and simulation to serve the purpose of finding possible causes for natural or astronomical phenomena, just as they set out to do on the Greek island of Antikythera.

Figure 1.24 shows various landmark approaches to the problem of causal discovery. One can see how until very recently most approaches were based largely on data, with almost no model production. This was the statistics-led approach of the last few decades that is still most widely used today in the practice of research and science. In mathematics, however, we have dealt with models all the time. These are mostly

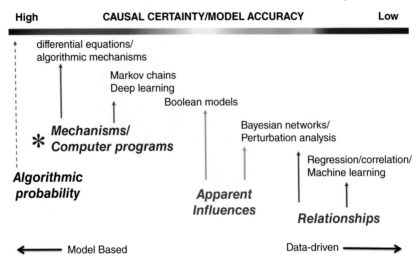

Landmarks of Causal Discovery

High   **CAUSAL CERTAINTY/MODEL ACCURACY**   **Low**

differential equations/
algorithmic mechanisms

Markov chains
Deep learning
Boolean models

Mechanisms/
Computer programs

Bayesian networks/
Perturbation analysis

Regression/correlation/
Machine learning

Algorithmic
probability

Apparent
Influences

Relationships

⟵——— Model Based   Data-driven ———⟶

Figure 1.24 Landmarks of causal discovery indicating the place of model-driven approaches, with algorithmic information dynamics based on algorithmic probability being the ultimate theory of optimal model inference.

theoretical models in the area of dynamical systems, which we will cover in detail in a future chapter using differential equations, one of the cornerstones of the field, used in producing numerical and computer simulations. One can also see how perturbation analysis and Bayesian networks move things in the direction of models.

Statistical machine learning (which we will refer to as ML) cannot find or generate mechanistic models from data. Sometimes the work of scientists is facilitated by AI and ML approaches, but traditionally neither AI nor ML can produce models by themselves unless they are empowered by an inference engine of a symbolic type. Thus, if the fundamental distinction between the statistical analysis of patterns and classification versus mechanistic generative models is not taken into consideration, confusion may occur. One example of this comes from biomarker discovery research in the medical domain. Since a number of biomolecules exhibit correlated behaviour they will appear as clusters in data analysis. There are therefore several distinct subsets of biomarkers that are equally effective as predictors of a given phenotype, such as disease. Machine learning and data-driven analytical techniques are excellent tools for finding such correlated patterns in large data sets. Yet, such sets of biomarkers are not equivalent to the mechanisms of disease. The generative model of a disease is not necessarily the same as a set of features correlated with the disease. The lack of appreciation of this (academic) distinction has caused a significant amount of confusion, for example, through neglect of the lack of repeatability in biomarkers.

Trends and methods in these areas, including deep learning and deep neural networks, are black-box approaches to data classification – and even prediction – that work amazingly well but provide little to no understanding of generating mechanisms. As a consequence, they also fail to be scalable to domains for which they are not trained. Supervised learning typically requires that tons of data be trained before anything interesting can be done, and training is needed every time such methods must tackle (even slightly) different data.

It seems reasonable to expect ML and AI to get deeper into model-driven approaches, leaving traditional statistics behind by incorporating algorithmic principles. This means promoting fundamental science rather than throwing more computational resources at data-related problems, as is characteristic of current ML approaches. This is exactly what we are doing with AID, and we will encourage you to test it on your own data for your own purposes, and tell you how to do so in the last chapter involving real-world applications.

Our aim is to go all the way and start from the opposite end of the spectrum in the landmarks of causal discovery, at the most model-driven end, and from there move to the data-driven side when necessary, in a feedback loop, so that a model is generated and improved upon based on the data observed, and once a candidate model is produced or chosen it is tested against new data until we are left with only a handful of likely mechanistic models that we can study and exploit. At this extreme end of the spectrum of model production, generation, and selection, is a theory called algorithmic probability, which is going to be at the centre of our methods and applications in this book.

This book is all about establishing a strong bridge between data- and model-based approaches, between observations and perturbations, and it offers a causal calculus based on the theory of algorithmic probability as the optimal and ultimate theory of induction. To this end, we connect different areas of science such as perturbation analysis, complex networks, information theory, dynamical systems, algorithmic complexity, and even machine learning, where we think our approaches will help better understand fundamental concepts such as deep learning, and will contribute to refining tools and making them more powerful in the quest to uncover causes.

1.6.2 Computational Mechanics

Our approach can be viewed as complementing the area of computational mechanics (traditionally based on, for example, Markov processes and Bayesian inference) with an inference engine powered by algorithmic probability and an empirical estimation of the so-called universal distribution (the distribution associated with algorithmic probability). Similar to program synthesis in inductive inference, albeit mostly of a theoretical nature, there are approaches (such as AIXI) that combine algorithmic probability with decision theory, replacing the prior with the universal distribution, something that AID also suggests. In actual deployments, however, many of these approaches circumvent uncomputability and intractability by relying heavily on popular lossless compression algorithms such as Lempel–Ziv–Welch (LZW), minimum description length approaches, Monte Carlo search, and Markov processes, thereby effectively adopting weaker models of computation. Another set of approaches are based upon

Levin's search and other variations, based on a dovetailing algorithm interleaving computer programs one step at a time, from shortest to longest, with each program assigned a fraction of time proportional to its probability during each iteration.

Some methods used by AID can be linked to resource-bounded algorithmic (Kolmogorov–Chaitin) complexity, which imposes an upper bound on program length, which in turn effectively requires the adoption of a linear finite automaton computational model that can be continuously improved upon by throwing more computational resources at it, which we circumvent by allowing improvements while increasing the running time (though in practice each calculation is restricted to a resource-bounded calculation).

1.7   Modelling from Observation

1.7.1   Systems as Black or Grey Boxes

The real world always behaves in more complicated ways than do theoretical models, and often when studying an object we have to treat it as a black box because we cannot see behind or inside it, whether it is the brain or a computer. Take as an example an aeroplane, a man-made artefact for every part of which we have accurate blueprints. We can understand the basics of how it flies but hardly anyone understands in detail the way in which millions of components work together in an aeroplane such as an Airbus A380 or a Boeing 747, with their roughly six million parts (see Fig. 1.25). Instead, manufacturers specialise in different parts and different aspects of their assembly.

Thus even engineers at Boeing and Airbus are compelled to see aeroplanes as black boxes. And this with something that humans have designed and manufactured. The situation is even more complicated with things that have evolved naturally and with which we may have less of a shared history and no involvement at all in their design, such as biological organisms. And even natural phenomena such as the weather or fluid dynamics turn out to be extremely difficult to model and predict. This is because we always have a very partial point of view, we are overwhelmed by the number of

Figure 1.25 Pictures taken by H. Zenil at the main Boeing factory in Everett, WA.

interactions involved, we cannot easily isolate them from one another and at all possible scales. This is why we have to always deal with apparently noisy and incomplete data and why we need to learn to model complex systems with our best tools.

One way to do so is to try to understand a system by simulating it. In this process of modelling, the first step is to understand the value and also the limitations of performing an observation. Let me show you an extremely oversimplified case consisting of an unknown system behind a black box. What typically happens is that the observer is at some point of the system that can be identified as its output.

One can also see how the input can actually be an induced input, an experiment, or a perturbation of the system. It is like throwing a stone at something to see how it reacts. What one can throw at black-box functions are inputs in a certain sequence.

However, behind a black box there could be a function that may appear to be the function we identify, but that in fact produces the outputs in a much more convoluted way, say by adding a random number and subtracting the same random number and thus behaving like the identify function while not being the simplest identify function, thus dissembling the actual operating mechanism.

So it is relevant to ask what made us think that the function behind a black box is actually the most simple version of the identify function, that is, that there was not instead a Rube Goldberg machine pretending to calculate the identify function but doing so in some incredibly, risibly, sophisticated way. Indeed we cannot ever be completely certain that a function is the function that appears to be without actually opening the black box, or that it is a mathematical function at all. It only appears to be so in the range given and for the type of perturbations or stones thrown. This is similar to the black swan problem in statistics, where one can only see the output but not where the swans are coming from, that is, the generating mechanism, and may be misleading at thinking that black swans are impossible to see until we do.

It is also worth noting that though we can follow an order in the input sequence, observations could have been carried out at random without loss of generality as long as the sample is congruent with the sought confidence level of the function behind. However, things are not always that straightforward, and as soon as we get into slightly more complicated cases, it may be more difficult to establish a relationship between the sequence of perturbations. One can call the sequence of observations a sample of the behaviour of the function.

Obviously all this becomes much more difficult in the real world because we usually have no idea of the magnitudes of the possible inputs, nor is there necessarily a privileged order. But we can see what the input for a biological organism may be. For example, to eat or drink can be an input, just as to learn or to read may be considered an input for the mind.

1.7.2   Noise versus Sampling

How informative can a single observation or a collection of observations be when it comes to producing a reasonable model with some degree of confidence? In other words, what type and how many observations should we perform to decide whether the function behind a black box is the one we have hypothesised? How many input and output pairs is it sufficient to gather in order to infer an underlying function? Is this a property of the observer or the observed? We will see that the quantity and quality of an experiment depends both on the conditions of the experiment and the capabilities of the observer.

Let us look at this sigmoid function that is usually a representation of how a neuron works, because there is a short interval, called a threshold, where after it is reached, the output behaviour of the system radically changes. There are two main factors allowing or preventing us from gathering enough information about the system to correctly infer the function. One is how much noise there is in the environment, and the other is how precise our measurements are. Notice that these two factors may not be independent and one may condition the other. For example, poor measurement capabilities may look like noise, and noise may look like inaccuracies of measurement. This is why, traditionally, tools are calibrated in a controlled experiment to assess how good they are before being used in more complicated cases.

In Fig. 1.26 we see some large boxes going up and down the sigmoid function represented by a white line. These boxes represent how off a measurement can be from the actual function value for an input on the X axis in case of noise or measurement

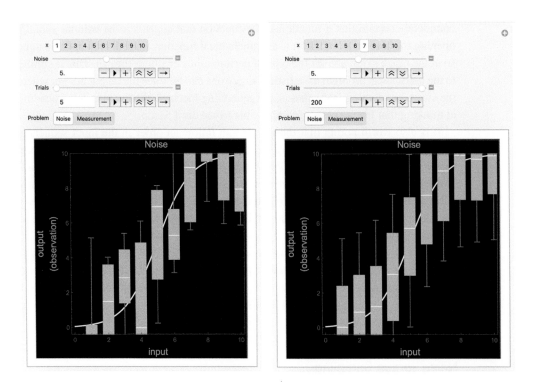

Figure 1.26 The problem of measurement from sampling as an observer in the face of noise (at source or as a result of measuring limitations). Increasing the number of observations (right) reduces uncertainty (box size) and increases accuracy (median).

inaccuracies, and what the measurements or observations on the Y axis would look like. If these boxes are too large, values will start overlapping, preventing us from making educated guesses. The smaller the error, the better and faster we can guess at the generating function.

Notice that when the error bars in yellow are too large, the mean indicated by a small horizontal white line inside the yellow bars will converge to the true value of the function, illustrating how the larger number of observations and measurements increases our chances of finding the generating mechanism. Even though individual measurements indicated by the red dots on the Y axis are highly misleading, this shows that increasing the number of samples increases the accuracy of the prediction.

This is related to what is known as the 'law of large numbers', a principle of probability according to which the average of the results obtained from a large number of trials should be close to the expected value and will tend to become closer as more trials are performed. A trial consists of a repetition of the same input value several times, so that even if the underlying system is completely deterministic, and perhaps even if the error bars are fixed, the result may be variant output values, but their average will converge to the true value. A trial is also often called a replicate, because the idea is to replicate the same experiment.

Another interesting observation is that the error bars may be of different lengths and even depend on the place they occupy along the function. In this case the error bars are very similar and do not depend on the function. The kind of noise they introduce is then called additive and linear, but more complicated cases exist as well.

The relationship between noise and the number of samples is thus proportional; the greater the noise the larger the number of samples needed, and the smaller in magnitude the noise, the fewer the samples needed. We will see how the field of information theory can help us make all these choices.

What is most interesting in these examples is that no matter how oversimplified, they illustrate how everything may be studied in a similar fashion. There is always an input and an output in a system of interest, even in areas such as biology and cognition, as we will see in the last chapter.

For example, an input can be a drug and the output the development of a disease, an input can be heating a protein and the output the way in which it folds, an input can be the accumulation of clouds and the output is whether it rains or not. Notice also that inputs are usually outputs from other systems, and outputs are usually inputs for other systems. So every aspect of these examples is related to causality.

In the next section we will see how we may study these systems by introducing computation into the study of causation.

1.8 Causality as Computation

In the previous section we saw how functions could be concealed behind or inside black boxes, and how we could infer them by performing experiments such as perturbing them and then recording the corresponding observations. Functions can also be seen as computer programs. As we have said before, the identify function, for example, can have an infinite number of implementations. For example, the simplest version $f(x) = x$ will look exactly the same as $f(x) = x + a - a$ from the perspective of an observer feeding in the input and observing the output, and versions of the same function can be very complicated. If one only cares about the output, then functions may suffice as the object of study, but if we care about mechanisms, then other types of time-dependent equations and computer programs are better representations. Representations that have a time variable are traditionally the object of study of an area called dynamical systems. A cellular automaton is a type of discrete dynamical system that we will be using throughout the book.

An intuitive way to explain what a cellular automaton is is to look at the way they work, using what we call their space-time diagrams. In a space-time diagram of one of the simplest instances of a cellular automaton, called an elementary cellular automaton (ECA) as introduced by Stephen Wolfram, time runs from top to bottom, starting from an initial tape placed on the top. Cellular automata run on a discrete space that looks like a Sudoku grid. One may start the system with a black cell and then apply a set of local rules from top to bottom, row by row. The set of rules is called the rule icon. The rule icon of a cellular automaton dictates how each cell changes over time. You can see that the rule that corresponds to the initial black cell surrounded by white cells is a black cell in the second row, and so on.

This particular example illustrating how a cellular automaton works is slightly misleading because each row should be updated in parallel. In contrast, in this example, the cells highlighted in red show the local rule that was applied at each step sequentially.

A cellular automaton has the advantage of being a highly visual computer program whose evolution can be inspected and followed step-by-step in real time. In other words, it is a sort of transparent computer program that allows us to understand every single one of its components visually, tracking it down at the lowest scale. It is a deterministic system and hence causal, so it will play an important role in our computational approach to the exploration of causality.

Notice that time can be seen as flowing from top to bottom while space runs sideways. This is why the diagram is called a space-time diagram, and it provides a 2-dimensional view of the evolution of a 1-dimensional discrete dynamical system. Notice that one can also start from a less simple initial configuration rather than from the black cell, for example, a random initial configuration. Yet the rules are applied in exactly the same fashion. A minor technicality is what happens at the borders of the system. Each rule is defined every three cells, but at the beginning and end on both right and left extremes one cannot evaluate the last two cells without a third neighbour. In these cases, convention dictates that we take the missing cells from the opposite side, as seen in the example in Fig. 1.27.

Cellular automata can be seen as part of a larger ensemble of means to simulate aspects of causal reality using what is known as an agent-based approach, given that individual cells can be seen as agents interacting with other agents according to specific rules. Agent-based models and cellular automata are bottom-up approaches for constructing models of causal phenomena. Cellular automata can be studied as black,

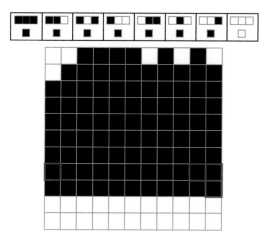

Figure 1.27 Illustration of the application of local rules in the evolution of a cellular automaton. Cells at the boundary take cells on the opposite extreme to continue applying rules. This is called a cylinder configuration. Adapted from http://demonstrations.wolfram.com/ CellularAutomataEvaluation/. Based on code by Paul-Jean Letourneau.

grey, and transparent boxes for the purpose of testing old and designing new tools. Traditionally, cellular automata are completely transparent boxes, as already noted, as their evolution is very visual, but as we have also seen, this is not always the case in the real world.

So if you are an ideal observer you would probably witness the whole space-time diagram running, and if sufficient observations were possible the rule icon could be cracked, that is, the computer code and generating mechanism for a particular cellular automaton. Thus a cellular automaton is an ideal simplification on which ideas related to causality can be studied and tested by, for example, concealing the cellular automaton's evolution to see what an observer looking at the last output can infer about it (see Fig. 1.28).

We can perform some sort of organised attack on this particular cellular automaton's black box to see if we can crack the code. The strategy consists in giving it an ordered set of initial conditions based on a binary enumeration and then seeing what cellular automaton it may be by looking at the output, just as we did with mathematical functions.

It looks as if for any input, as marked in red in the example, the output is always blank, so the rule icon must be something that takes any input and then writes a white cell if this is an elementary cellular automaton (Fig. 1.29).

This rule is called Rule 0 because the output is all white cells and, taken as if they were zeros in binary, they would represent the zero in decimal. However, not all rules are so simple. Figure 1.30 demonstrates the so-called ECA rule 30, also in binary, according to Wolfram's enumeration scheme for ECA. And its evolution, depicted here from 1 to 70 steps, looks quite random, and has even been used as a random number generator in the past. Trying to crack the code by only observing the last row at each

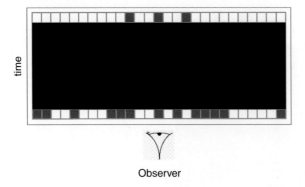

time

Observer

Figure 1.28 A typical example in the real world is an observer looking at the output of a system, with such a system typically being a black box about which little, if anything, can be known or extracted first hand, because either the process that is the output is already complete or it is the result of multi-scale top-down and bottom-up causation to which we have no easy access. We have to find the best tools to infer what the black box may be doing just by looking at the output for different natural or induced inputs.

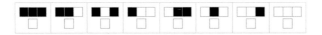

Figure 1.29 All local rules write a white cell for any input.

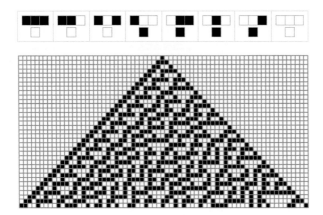

Figure 1.30 Evolution diagram of the ECA rule 30. The icon on top is the rule determining how each cell will be coloured according to the previous three cells evolving from top to bottom and starting from a single black cell.

time is much more difficult than in the previous examples, and one has to perform the observations in the right place to avoid being misled, given that the rule actually presents some regularities in certain places, such as on the left hand side.

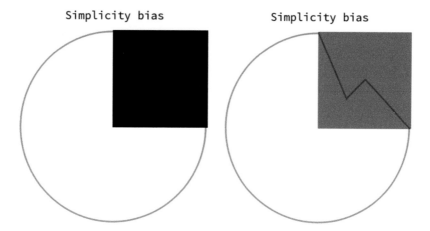

Figure 1.31 Simplicity bias: faced with the challenge of guessing what is behind the circle, the mind will tend to choose the shape that completes the picture in such a way as to minimise surprise, uncertainty, entropy, and algorithmic complexity. Most likely, you'd think that what is behind the black box (left figure) is the complete circle and you would react with surprise if what the black box conceals is a more complicated shape like the one on the right.

A simulation is, by definition, a series of causally connected events, a deterministic evolving system that we call a dynamical system. We will see in the following chapters how we can model and simulate evolving systems, and not only crack their generating codes from partial observations but also reconstruct these systems from disordered observations, and then even reprogram them to perform different computations and behave in different ways.

1.9   Inference and Complexity

In this section, we will briefly explain how causality and inferring functions and computer programs are related to complexity. Imagine you see an object with some part of it blocked by another object, in this case a black square (see Fig. 1.31). Typically, if you ask people what this object could be, they will complete the picture and suspect it to be a full circle. In some sense it seems that our minds are hardwired to complete a picture using the simplest possible shape. We seem deeply biased towards simple forms. If behind the black square there is something else, we would certainly be a little surprised.

Before getting into the technical details of entropy, as we will do later in the book, remember that we had said that entropy as defined by Shannon is traditionally taken as a measure of surprise. Now, something like classical information theory may be able to describe but not explain this bias towards simple forms by establishing that we tend to favour configurations that surprise us less, that is, that have low entropy. But why is this so? We have suggested that such hardwiring for simple things comes from living in a world that is not random, our minds having thus evolved with a high degree of algorithmic structure. We will discuss this in more detail, but this example is meant to show how inference is related to complexity, or rather to simplicity as opposed to randomness, and even to some form of subjectivity that may be cognitive or perhaps even more fundamental. And we will see that the type of randomness we are talking about is not statistical in nature but algorithmic.

Perhaps a useful way to explain how something is complex as opposed to simple is to evoke the way we classify certain human diseases, especially since in the last chapter we will be applying all these ideas, concepts, and the tools based on them to molecular biology and genetics, fields that are deeply relevant to human disease and syndromes. Most diseases are complex, with scientists dealing with challenges related to the observer, the quality and quantity of measurements, apparent noise, and highly interactive systems with multiple and intertwined causes.

Some diseases, such as multiple sclerosis, Alzheimer's, Parkinson's, and most cancers are very complex, in that they can be produced by multiple factors, not just one. They depend on many variables, both genetic and environmental, and are highly unpredictable. In contrast, simple diseases have single or easily identifiable and isolable causes. They may arise from punctual genetic mutations, as is the case with a certain type of breast cancer. The outcome of simple diseases is much easier to predict in that they have well-defined effects. Examples of simple diseases and conditions under this specific definition include cystic fibrosis, Down syndrome, and Huntington's disease. Table 9.1 illustrates what a complex disease looks like as opposed to a simple disease.

How may a classification of this type, contrasting simple and complex diseases, help us to, for example, treat these diseases and conditions? A basic implication of this classification is that one-for-all drugs can only work well for simple diseases, because having multiple causes and factors, complex diseases will appear for different reasons in different people. Hence the current pharmaceutical approach to and business model in medicine is inadequate, and a new paradigm is necessary. That paradigm is known as personalised medicine, and the idea is that one should be able to manufacture a specific drug for a specific person. So what scientists aim for, or should aim for, is an understanding of causes, so as to have a better chance of producing new drugs to direct the way in which a disease develops instead of merely controlling its effects.

This type of simple versus random behaviour can also be modelled mathematically with a view to studying ways of better understanding it, even if it will often be oversimplified. We will explore this issue in further detail later, but for the present we can again resort to cellular automata as an example. If you examine the response of the elementary cellular automaton with rule number 10 (see Fig. 1.32), you will find that no matter what the input is, the system separates the input signals into clear stripes that do not interact with each other, and always produces the same qualitative behaviour. So in a very simplified way it is modelling an aspect of a simple disease. In all four cases, the same rule is applied and the same behaviour obtained for very different initial conditions.

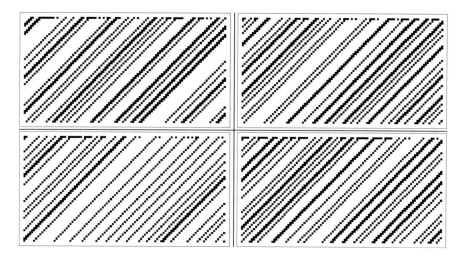

Figure 1.32 Behaviour of the ECA rule 10 for different initial conditions.

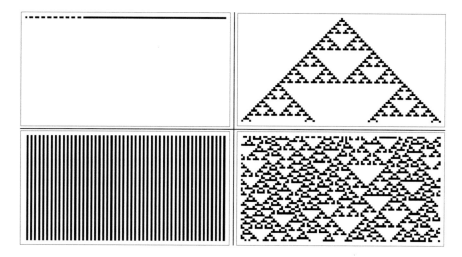

Figure 1.33 Behaviour of the ECA rule 22 for different initial conditions.

But if you take the behaviour of more complicated rules, such as rule 22 depicted in Fig. 1.33, you will find that even small changes in the initial conditions or small perturbations along the way will have a significant impact on the final outcome. This behaviour is more similar, for example, to the way in which cancer tumours evolve, often growing and spreading in unpredictable ways. In all four cases, the same rule, rule 22, was applied, but for different initial conditions. The result is highly unpredictable.

One of the topics that this book will cover is how to characterise these behavioural differences, especially when the rule is unknown but we want to understand the system from a hacker's perspective, that is, when we wish to know the underlying generating mechanism where we have no access to a source code – in artificial examples like this one, but also in real-world systems such as biological organisms like cells. The goal will be not only to attempt to crack the code behind natural phenomena but to try to manipulate and reprogram the way in which these systems evolve and behave, and to that end we will go back and forth between artificial and natural systems in applications to the biological and cognitive sciences.

1.10   The Future of Causality

In all fairness, dealing with causality is extremely difficult, and we used to do well with tools such as traditional statistics and classical probability theory. A major development in causality has been information theory, which is in some sense a clever extension of traditional statistics, one that has sometimes been dangerous because it may give the impression of dealing with causality in different ways than traditional statistics and probability theory, while in fact it does not. The concept of entropy was developed in the context of statistical mechanics in the early twentieth century, as an attempt to understand the average behaviour of small objects such as particles, atoms and molecules. In the late 1940s, Claude Shannon formulated a version of entropy similar to the one adopted in statistical mechanics, a branch of quantum physics, intended to help evaluate the capacity of communication channels. Shannon was working at the time at Bell Labs, a company concerned with communication lines and a precursor of today's AT&T. But it was Andrey Kolmogorov, standing on the shoulders of such giants as Richard von Mises and Alan Turing, who soon found out that probability theory was very limited in dealing with fundamental aspects of randomness and hence causality.

Yet, we have not made much use of what we know about mathematical randomness, as opposed to statistical randomness, either in data science or science in general, as we tackle the challenge of causality. To characterise randomness is key in the area of causation because it is exactly what one needs to rule out when using observations and models to come up with a candidate mechanism to explain the data at hand. One does not want to attribute chance where it does not obtain, and vice versa.

The founders of the theory of algorithmic complexity, namely Andrey Kolmogorov, Gregory Chaitin, Ray Solomonoff, and Leonid Levin, came up with a mathematical solution to the challenges of randomness, causality discovery, and inference.

Indeed, Solomonoff and Levin showed that the concept of algorithmic probability was the ultimate solution to the problem of causal inference, suggesting that intelligence was likely just computation, in a fundamental sense. The two concepts, algorithmic complexity and algorithmic probability, are faces of the same coin, deeply related to each other. However, while some applied scientists may have heard of algorithmic randomness, and a smaller number about algorithmic probability, very few actually use them in addressing the challenge of causation.

The advantages of algorithmic complexity and algorithmic probability in data science have been largely underused if not ignored, because unlike other, simpler, measures such as those derived from statistics, including Shannon entropy, measures of algorithmic complexity are much more difficult to estimate. One thing to bear in mind is that there will always be some statistics involved, because the only way to attain full certainty about causation is not only to witness a process of interest firsthand but also to isolate the system and have access to all the relevant processes at all scales, something that is difficult if not necessarily impossible in the real world.

In this book we will introduce what we think is an exciting – hopefully for you too – new approach to tackling the problem of causality based on algorithmic complexity by way of algorithmic probability. We will call this area Algorithmic Information Dynamics or Algorithmic Dynamics for short.

It is pertinent to return to the claim about algorithmic probability made by Marvin Minsky, a founding father of the field of artificial intelligence, just a few months before he passed away, and quoted above. Minsky recognised that the way forward was through AID and the work of Solomonoff and Chaitin.

Both Minsky and Solomonoff attended the Dartmouth College conference/workshop, the conference sometimes seen in the US as the inception of the new field of AI. Clearly, algorithmic probability is not just another pet topic in complexity science but a fundamental pillar of science, as we will see throughout this book.

1.11   Conclusion

In this textbook we offer an approach to causal discovery that is model driven and based on the theory of algorithmic probability in order to help discover and produce candidate generative models for observed data. We have made some progress in applying these new tools to data to deal with causation, and they themselves build on the progress made throughout history, from logic to statistics, probability, dynamical systems, and computation, including perturbation analysis, but they displace traditional statistics and the need for estimations of probability distributions from their central position.

We think that AID may be part of a shift away from the dominance that statistics and even equations have long exerted over science, and towards computation and machine learning, but a different kind of machine learning, rooted in algorithmic approaches rather than traditional statistics, as is the case with machine learning today.

We will see how AID can help predict and even reconstruct the phase space and space-time behaviour of complex non-linear dynamical systems with limited information. But first, in the next chapters we will cover many more topics needed to understand AID, now that we have offered a brief overview that highlights the significance and some of the many aspects and challenges of causality.

1.12   Practice Questions

1. The study of how things and events happen or come into being and are causally connected with each other as opposed to occurring by chance is called:
    (a) Non-linearity
    (b) Causality

     (c) Probability

     (d) Entropy

2. Scientists perform experiments mainly to:

     (a) Justify the costs of lab equipment

     (b) Discover causal knowledge

     (c) Find correlations

3. The interaction or evolution of non-trivial systems may often seem:

     (a) Trivial and easy

     (b) Complex and noisy

     (c) Quantum mechanical

4. Determinism implies that future events are easily predictable.

     (a) True

     (b) False

5. A mechanistic model means that a model:

     (a) Can be followed from cause to effect, like an algorithm, step by step

     (b) Cannot incorporate noise and is never complex

     (c) Is always the right explanation

     (d) Cannot run on a computer

6. A Rube Goldberg machine illustrates that:

     (a) Causality is always easy to identify no matter how laughable

     (b) Non-linearity is complex and interesting

     (c) Computation is universal and ubiquitous

     (d) Trivial outcomes can be produced by needlessly sophisticated but causally connected tasks

7. Both the geocentric and heliocentric models of the solar system are mechanistic models.

     (a) True

     (b) False

8. The geocentric model has a lot of predictive power even if considered wrong.

     (a) True

     (b) False

9. Occam's razor is also known as:

     (a) The law of big numbers

     (b) The law of logic

     (c) The law of parsimony

10. Occam's razor states that when presented with competing models one should favour:

     (a) The one that satisfactorily explains the observed phenomena with fewest assumptions

     (b) The one that satisfactorily explains the observed phenomena with more and more complex assumptions

11. Occam's razor is a criterion for:

     (a) Simplicity

     (b) Complexity

12. Occam's razor generally rules out Goldberg machine-like explanations for a given outcome.
    (a) True
    (b) False

13. If P = True and Q = False, P AND Q is:
    (a) True
    (b) False

14. If P = True and Q = False, P OR Q is:
    (a) True
    (b) False

15. The concept of algorithmic (Kolmogorov) complexity is colloquially related to Occam's razor.
    (a) True
    (b) False

16. In science, Occam's razor is used as:
    (a) The ultimate test for truth
    (b) A substitute for the experimental validation of theoretical models
    (c) A guiding principle for the production of scientific models

17. Causal inference draws conclusions based on:
    (a) Chains of effects
    (b) The detection of noise
    (c) Differential equations
    (d) The use of computers

18. Classical probability and traditional statistics usually mean the study of science:
    (a) Through the use of universal computation
    (b) Through assumptions of probability distributions
    (c) Through analysis of possible generating mechanistic models

19. A fundamental challenge in the study of causality is:
    (a) The difficulty of describing data probabilistically
    (b) The increasing entropy of the universe
    (c) Disentangling causes from effects and determining mechanistic models

20. A uniform prior distribution assumes that:
    (a) Some events are more likely than others
    (b) Rare events have probability equal to zero, no matter how many events are possible
    (c) All events are equally likely

21. Laplace's 'principle of insufficient reason' states that if there are n indistinguishable causes for an effect, then:
    (a) Each possible cause has probability equal to n
    (b) Each possible cause has probability equal to 1
    (c) Each possible cause has probability equal to 0
    (d) Each possible cause has probability equal to $1/n$

22. Shannon's entropy is a measure of:
    (a) Causality between observed variables

    (b) Lack of knowledge about the underlying probability distribution

    (c) Cost of building a universal computer

23. Correlation implies causation.

    (a) True

    (b) False

24. Which of the following is one of the latest practices in probabilistic causal discovery?

    (a) Applying the Wiener filter

    (b) Comparing time series to test correlation

    (c) Analysis of the response to perturbations and counterfactuals

    (d) Estimation of fractal dimensions

25. One way to test a causal relation between two time series is to:

    (a) Perform a perturbation on one and see if it has an effect on the other

    (b) Reorder the data

    (c) Merge the time series

    (d) Calculate the Spearman correlation coefficient

26. If we encounter a case where perturbations do not lead to a change in the effects, it is evidence of:

    (a) Strong causal relationship between the events

    (b) Lack of a causal relationship between the events

27. In a directed acyclic graph that represents causal relationships:

    (a) Nodes are events and events are linked with each other if there's a direct cause and effect between them

    (b) Nodes are causes and causes are linked with each other if there's an event between them

28. In a causal directed acyclic graph there may be no loops because:

    (a) It would become chaotic

    (b) It would make an effect redundant

    (c) It would make a cause into the cause of itself

29. Computer simulations in the context of causality are typically used to:

    (a) Find likely mechanisms for natural or artificial processes

    (b) Classify data

30. The Antikythera mechanism is an example of a _____ model used to predict celestial events.

    (a) Mechanistic

    (b) Statistical

31. The study of dynamical systems is an example of a:

    (a) Statistics-driven approach

    (b) Model-driven approach

32. Conventional machine learning (regression, support vector machines, etc.) are examples of a:

    (a) Statistics-driven approach

    (b) Model-driven approach

33. We often have to treat an object of study like a 'black box' because:
    (a) The real world is complicated and we rarely or never witness phenomena unfolding in real time at all scales
    (b) Concealing the underlying mechanism of a system allows us to use tools like statistics

34. A 'black box':
    (a) Exposes its underlying mechanisms
    (b) Conceals its underlying operating mechanism

35. 'The average of the results obtained from a large number of trials should be close to the expected value, and will tend to become closer as more trials are performed' is an enunciation of:
    (a) The law of parsimony
    (b) The law of large numbers
    (c) Bayes's theorem
    (d) The 2nd law of thermodynamics

36. When sampling, an increased amount of noise means that to provide a reliable estimate we need a _____ number of samples.
    (a) Small
    (b) Large

37. A cellular automaton is a _____ dynamical system.
    (a) Continuous
    (b) Discrete

38. Agent-based models and cellular automata are _____ approaches for constructing models of causal phenomena.
    (a) Top-down
    (b) Bottom-up

39. During the execution of a cellular automaton, each of the rows in its space-time diagram is updated:
    (a) Sequentially, from left to right
    (b) Sequentially, from right to left
    (c) In parallel

40. In a typical space-time evolution of an elementary cellular automaton, time flows:
    (a) From top to bottom
    (b) From bottom to top
    (c) From left to right
    (d) From right to left

41. The space-time diagram of an elementary cellular automaton shows a _____ view of the evolution of a _____ system.
    (a) One dimensional, two dimensional
    (b) Two dimensional, one dimensional
    (c) Three dimensional, two dimensional
    (d) Three dimensional, n-dimensional

42. Shannon's entropy is typically interpreted as a measure of:
    (a) Change over time
    (b) Meaning
    (c) Surprise
    (d) Complexity
43. What is one plausible explanation for why humans prefer simpler structures?
    (a) Human emotions block our ability to generate complex models of the world
    (b) We have adapted to environments that are not random
    (c) Neuron synapses fire less rapidly than thoughts
44. Diseases with well-defined causes and predictable effects are, in general, cases of
    (a) Simple diseases
    (b) More complex diseases
45. If a cellular automaton produces the same qualitative behaviour no matter what the input is, the cellular automaton's rule is:
    (a) Simple
    (b) Complex
46. If scientists want to develop personalised medicine, they should:
    (a) Focus on changing the 'source code' that causes diseases in the first place
    (b) Treat the conditions that these diseases cause by using generalised drugs
    (c) Study massive amounts of social data
47. The concept of entropy was first motivated by:
    (a) Developing mechanistic tools to understand the motion of celestial bodies
    (b) Computational methods to understand the behaviour of large social systems
    (c) The study and development of tools to understand the behaviour of small particles
48. Algorithmic probability is the optimal solution to:
    (a) Neural networks
    (b) Causal inference and characterising randomness
    (c) Statistics
49. Currently, data science does not usually use algorithmic complexity/probability because:
    (a) It is very difficult to calculate
    (b) Statistics already gives us an understanding of causality
    (c) Information theory has been unsuccessful
50. Algorithmic probability can be used practically because:
    (a) It is statistical in nature
    (b) It is possible to make approximations of what it is able to predict
    (c) It is widely used already

1.13   Discussion Questions

The purpose of the discussion points below is to facilitate your understanding of the material in the chapter by interrogating the concepts discussed in it. In particular, what are the limits of the concepts? Are there any hidden presuppositions in our reasoning when we think about causality, for example?

1. Bertrand Russell made the argument in a classic essay (1911) that causality was a relic of a bygone age: 'The law of causality, I believe, like much that passes muster among philosophers, is a relic of a bygone age, surviving, like the monarchy, only because it is erroneously supposed to do no harm.' What did he mean? Why did he make this statement? Do you think it is true or false?

2. In physics, at the most fundamental level, there are essentially fluctuating quantum fields. Here the idea or notion of causation appears to dissolve into nothingness. Do you agree with this description, and if not why not? Discuss its pros and cons. If this characterisation of the state of affairs has some validity, then why is it that we devote scientific effort to understanding causality in the specialised sciences? Might not our continuing to do so reflect a poor understanding of the world?

3. Gödel found closed, time-like solutions to Einstein's equations. Explain what he found. Discuss the implications such solutions, which allow time-travel into the past, have or do not have for the notion of causation and the scientific study of causality.

4. Discuss and enumerate sufficient and necessary assumptions about the world that would, in concert, imply that there is no causation in the world, and that the scientific study of causality is fundamentally flawed. Could this be a possibility? If not, why not?

5. Formulate some additional discussion questions in the spirit of the above that would, in effect, open new avenues for investigation beyond our current way of thinking and analysing causality.

2   Networks: From Structure to Dynamics

Chapter Summary

Here, we will review some key aspects of network theory and dynamical systems. This chapter serves as a bridge between the work of the previous chapter on the challenges of causality, and that of the following chapters on information theory, and algorithmic information theory in particular. The scientific analysis of causality can be translated into an inverse problem challenge: given certain observations, temporal or otherwise, how do we infer their driving causes and the effects of those causes – how they are related, and how they generate each other – from the data at hand. This in turn can be formulated as a network inference-analysis modelling problem. Here we consider all data points, or events, as discrete elements. We will also handle continuous data, either explicitly or as discrete elements as realised in a numerical approximation. Now, such discrete elements can be represented as nodes in a network, or equivalently, as vertices in a graph. The links in graphs, or edges in networks, represent the interactions between the discrete events, causes, or effects.

Obviously, there are several layers of complexity over and above this simplified picture, such as uncertainties in data, missing data, sparse sampling, and hidden variables, to name a few. Nevertheless, the challenge of causality can be formulated using the language of networks. We will therefore briefly survey key concepts and computational tools for describing and analysing networks. These include core definitions of graphs, originating in classical graph theory, and algorithms covered in discrete mathematics courses. For example, directed and undirected graphs, bipartite graphs, complete graphs, and the adjacency matrix formalism are introduced. Next, if we have a graph, we need to ask how to analyse such an object. Here, we review tools for finding topological substructures in graphs, graph spectra, and different network statistics, including random and scale-free graphs. Since these developments have been the subject of immense interest in the biological community, we briefly summarise their applications and the insights they have provided in the analysis of biological networks.

As causality only makes sense in the context of systems unfolding over time, we need to have a handle on dynamical systems. This is exactly what we introduce into algorithmic information theory (AIT) to transform it into algorithmic information dynamics (AID): a system changing over time as a result of perturbations and counterfactual simulations. This chapter therefore provides a brief introduction to the subject, using simple mathematical models to illustrate core concepts such as iterative maps and non-linear ordinary differential equations. The central tool of stability analysis is explained using several well-known model examples. Following our discussion of this classical continuous domain, we discuss corresponding work in the discrete dynamical domain, illustrated with Boolean networks in the context of gene regulatory networks. Formulating causality using the language of networks, including the dynamic component, makes it clear that to come to grips with causality we need powerful tools to analyse and understand complex dynamical networks. This chapter sets the stage to enable us to later address the extent to which AID can be useful in the study of causality.

2.1   Graphs

A system is defined as a regularly interacting or interdependent group of items forming a unified whole to achieve a common goal. Today, we live in a world of systems. There has to be some sort of organisation and order underlying the development of systems. Science studies systems in order to discover this ruling plan, or the causality in a system, which allows us to understand and predict its behaviour in response to perturbations. It is very unlikely that we will know everything about such complex systems as cells, tissues, whole organisms, or even ecosystems. Even if we did, it is not sufficient to identify and characterise the individual building blocks of a system. Therefore, we use models. Models describe our beliefs about how a system functions. When studying models, it is helpful to identify broad categories of models. Mathematical models can take many forms, including dynamical systems models, statistical models, differential equations, and game theoretic models. Modern complex network theory has proven to be a very effective approach to modelling these complex architectures. It can facilitate identification of emergent patterns and can provide hypotheses for relationships between structure and function in many systems and at different scales. Network theory is one of the most exciting and dynamic fields of science in the twenty-first century. Networks are everywhere, from the metabolites that fuel our bodies to the social networks that shape our lives. The mathematical structure used to model networks is called a graph. The graph is a fundamental concept in discrete mathematics. Informally, we may view a graph as a structure consisting of a set of points and a set of lines joining these points. Such points are called vertices or nodes, and the lines between them are called edges or links. In principle, a graph is a pair $G = (V, E)$ comprising a set V of vertices and a set E of edges, which are 2-element subsets of V. Graphs are very easy to represent and understand, and can be easily processed by computer programs. An edge between vertices u and v is given by the set (or unordered pair) $\{u, v\}$. Here, u and v are called endpoints of the edge. In the interests of simplicity, we may relabel an edge $\{u, v\}$ as uv by identifying its endpoints. An edge whose two endpoints coincide is

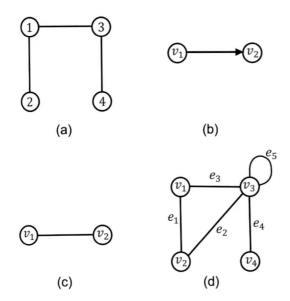

Figure 2.1 (a) Undirected Graph $G(V,E)$: circles represent nodes and lines represent edges. (b) A directed edge between node v_1 and node v_2. (c) An undirected edge between node v_1 and node v_2. (d) In the graph G_1, v_2 and v_3 are adjacent, e_2 is incident on v_2 and e_5 is a loop.

called a loop. Two vertices are referred to as adjacent or neighbouring if they are joined by an edge. A graph can be described by listing its vertices and edges, for example, by making $V = \{1,2,3,4\}$ and $E = \{\{1,2\},\{1,3\},\{3,4\}\}$ (see Fig. 2.1 (a)).

A directed edge is an ordered pair of vertices. A directed edge (v_1, v_2) between v_1 and v_2 is represented by an arrow from v_1 (the source vertex) to v_2 (the target vertex). Note that an (undirected) edge is just a pair of vertices with no specific orientation, as in Fig. 2.1 (a, c). We say an edge is incident on a vertex v if v is one of its endpoints (see Fig. 2.1 (d)). All graphs and networks we study can be divided into two broad classes based on their edge type: undirected and directed. An undirected graph consists of a non-empty (finite) set of vertices and a set of undirected edges. A directed graph, on the other hand, is formed by a set of vertices and a set of directed edges.

In biology, for example, transcriptional regulatory networks and metabolic networks would usually be modelled as directed graphs. In a transcriptional regulatory network, nodes and links would represent genes and the interactions between them, respectively. This would be best represented as a directed graph, because if gene A regulates gene B, then there is a natural direction associated with the edge between the corresponding nodes, starting at A and ending at B. Protein–protein interaction networks, however, are typically modelled as undirected graphs, in which nodes represent proteins and links represent interactions. Later in this chapter, and those following, we will explore these networks and their properties further.

There are many special graphs in graph theory, which are usually referred to by descriptive names. A null graph is a graph whose vertex set and edge set are empty. A cycle graph C_n is a graph consisting of a single cycle of n vertices joined with

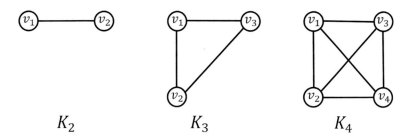

Figure 2.2 Complete graphs with two to four nodes.

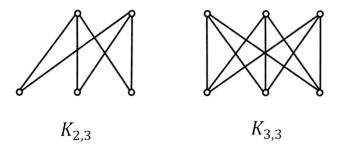

Figure 2.3 Two complete bipartite graphs.

n edges. Clearly, the number of vertices of a cycle graph should be greater than three. A path graph P_n is a graph with n vertices joined consecutively by a chain of $n-1$ edges. A complete graph with n vertices, as its name suggests, is a graph where all vertices are pairwise adjacent. This graph is usually denoted by K_n for $n > 1$ (see Fig. 2.2).

A multigraph is a graph that is permitted to have multiple (parallel) edges, that is, edges with the same end vertices. A simple graph or a strict graph contain no loops or multiple edges. A bipartite graph is a graph that has two parts, with edges joining nodes between the two parts. A complete bipartite graph $K_{m,n}$ has $m+n$ vertices in two parts, and the edges join all the mn pairs from the two parts (see Fig. 2.3).

The last two graph classes are Wheels and N-cubes. Wheels, denoted by W_n, are graphs formed by connecting a new single vertex to all vertices of a cycle, C_n (see Fig. 2.4). N-cubes, shown by Q_n, are graphs whose vertices are the 2^n binary strings of length n, with two vertices adjacent if and only if, the strings differ by exactly one coordinate (see Fig. 2.5).

The degree, d, of a vertex v in a graph G is the number of edges of G incident on V. A graph is d-regular if all its vertices have the same degree d. For directed graphs, we can rewrite our definition by dividing it into two: in-degree and out-degree graphs/edges. In-degree refers to the number of edges for which v is the terminal vertex, and out-degree to the number of edges for which v is the initial vertex (see Fig. 2.6). The number of vertices is denoted by $|V|$ and also referred to as the order of graph

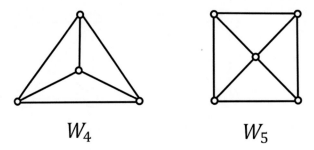

Figure 2.4 Two Wheels with 4 and 5 nodes.

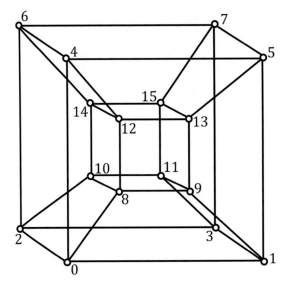

Figure 2.5 Examples of cube graphs.

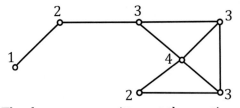

The degrees are written at the vertices.

Figure 2.6 Illustration of graph node degrees.

G, and the number of its edges, $|E|$, is referred to as the size of the graph. The degree sequence (the score) of a graph G is the collection of the degrees of the vertices of G.

In abstract graphs, vertices usually have no names or labels, and so we have to somehow specify their degree sequences. The particular order is not important. If the

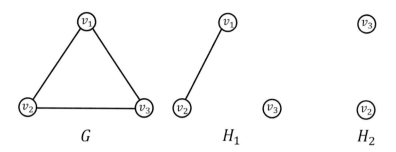

Figure 2.7 Graph G and two of its subgraphs, H_1 and H_2 (because all nodes and edges in H_1 and H_2 are included in G).

degree sequence of a graph is known, what can you say about the number of edges in the graph?

The handshaking theorem (lemma) is an answer to this question. It can be used to find a graph, given its sequence.

Let $G = (V, E)$ be a simple graph. Then

$$2|E| = \sum_{v \in V} \deg(v).$$

For example, the sequence 1, 2, 3, 4, 5 cannot be a degree sequence of a graph since its sum is not even.

2.2   Subgraphs and Graph Representations

Faced with the increasing size and complexity of real complex networks, it is essential to break them up into smaller entities called subgraphs in order to understand the functionality of these networks. Formally, a subgraph of a graph G is any graph H whose vertices and edges are a subset of the vertices and edges of G. We write $H \subseteq G$ for the set inclusion.

If H is a subgraph of G and v_1 and v_2 are vertices of H, then by the definition of a subgraph, v_1 and v_2 are also vertices of G. However, if v_1 and v_2 are adjacent in G (i.e., there is an edge of G joining them), the definition of a subgraph does not require that the edge joining them in G should also be an edge of H. If the subgraph H has the property that whenever two of its vertices are joined by an edge in G, this edge is also in H, then we say that H is an induced subgraph. In the example shown in Fig. 2.7, which one is the induced subgraph of G, H_1 or H_2?

As you may recall, graphs are mathematical structures that represent pairwise relationships between objects. A graph can be represented in many ways. The two most common ways of representing a graph are the adjacency matrix and the incidence matrix. When information about the vertices is more desirable than information about the edges, the adjacency matrix is more useful. However, if information about edges is more desirable, then the incidence matrix is more useful.

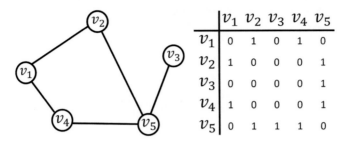

	v_1	v_2	v_3	v_4	v_5
v_1	0	1	0	1	0
v_2	1	0	0	0	1
v_3	0	0	0	0	1
v_4	1	0	0	0	1
v_5	0	1	1	1	0

Figure 2.8 A simple graph and its adjacency matrix.

$$m_{ij} = \begin{cases} 1 & \text{when edge } e_j \text{ is incident with } v_i \\ 0 & \text{otherwise} \end{cases}$$

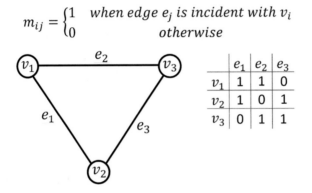

	e_1	e_2	e_3
v_1	1	1	0
v_2	1	0	1
v_3	0	1	1

Figure 2.9 A simple graph and its incidence matrix.

An adjacency matrix A is a $n \times n$ binary matrix. Element a_{ij} is 1 if there is an edge from vertex i to vertex j, otherwise a_{ij} is 0. For those of you who may be wondering what binary means in this context, a binary matrix is a matrix in which the cells can have only one of two possible values, either 0 or 1. A simple graph and its adjacency matrix are shown in Fig. 2.8.

Using adjacency matrices makes it easier to find subgraphs, and to reverse graphs if needed. For a simple graph with no loops, the adjacency matrix must have 0s on the diagonal. For an undirected graph, the adjacency matrix is symmetrical. Note that for multigraphs, a_{ij} is equal to the number of edges between vertices a_i and a_j. The concept of adjacency is based on the ordering of vertices. Hence, there are as many as $n!$ such matrices. When there are relatively few edges in the graph, the adjacency matrix is a sparse matrix.

The incidence matrix of a graph is the binary matrix that has a row for each vertex and a column for each edge. Element I_{ve} is 1 if, and only if, vertex v is incident upon edge e. It can be seen that the degree of each vertex is the sum of its corresponding row entries in either matrix (see Fig. 2.9).

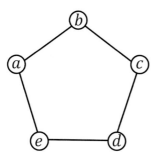 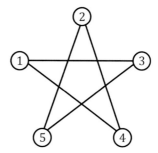

Figure 2.10 Two isomorphic graphs.

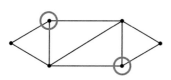 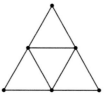

Figure 2.11 Two non-isomorphic graphs. The left graph has two vertices of degree 3, while the right graph has none, so they are not isomorphic.

If in a graph we assign a weight or cost to each edge, we have a weighted graph, which means that its adjacency matrix is not binary. For example, in a protein–protein interaction network whose nodes are proteins, the weight of an edge between two nodes can be the similarity score of these proteins. The two graphs in Fig. 2.10 look different but are they, from a graph theoretic point of view, structurally the 'same'?

If you look closely, you will see that they are indeed structurally the same. They are the same up to the renaming of the vertices. In fact, we can define a function that renames the vertices of the first graph so that they are identical to those of the second graph. These graphs are referred to as isomorphic.

Where two graphs are of the same order and size, it is not easy to determine whether they are isomorphic at first sight. One way to detect isomorphism between two graphs is to write their adjacency matrices. If we can write the adjacency matrices with vertices ordered in such a way that the matrices are identical, the corresponding graphs are isomorphic. Formally, $G_1 = (V_1, E_1)$ and $G_2 = (V_2, E_2)$ are isomorphic if there is a bijection f from V_1 to V_2, with the property that a and b are adjacent in G_1 if and only if $f(a)$ and $f(b)$ are adjacent in G_2, for all a and b in V_1. Function f is called an isomorphism. An isomorphism f 'preserves edges', thus preserving properties of graphs that depend on edges, which is to say that cycles and degrees are preserved.

It is often easy to show that two graphs are not isomorphic. For instance, if they have different numbers of vertices or edges, or if the degrees of the vertices do not match up (see Fig. 2.11).

But showing that they are isomorphic requires that an isomorphism can actually be produced, which is much harder.

2.3   Connectivity

Connectivity is one of the basic concepts of graph theory. The basic idea of connectivity is reachability among vertices of a graph by traversing its edges. This can be defined formally and categorised under five headings. We start by formally defining a path. A path is a sequence of edges that begins at a vertex of a graph and travels along edges of the graph, always connecting pairs of adjacent vertices (see Fig. 2.12).

Depending on the context, path length may either be the number of edges on the path or the sum of the weights of the edges on the path. Since a graph may have more than one path between two vertices, we may be interested in finding a path with a particular property. For example, a path with minimum length. The geodesic distance, or simply distance, from u to v in graph G is the length of the shortest path from u to v in G. The second way to traverse a graph is along a trail where no edge can be repeated, and the third is a walk where there is no restriction. If a walk or trail have the same endpoints, then we say they are closed. A circuit is a closed trail and a cycle is a closed path. We refer to a graph as cyclic if it contains a cycle; otherwise we call it an acyclic graph.

An undirected graph is connected if there is a path between every pair of its vertices. A directed graph is strongly connected if there is a path from v_1 to v_2 and from v_2 to v_1 whenever v_1 and v_2 are vertices of the graph.

A directed graph is weakly connected if there is a (undirected) path between every two vertices in the graph. A strongly connected graph is weakly connected, but the reverse is not true. For example, the graph in Fig. 2.12 (b) is an example of a graph that is not connected. If we add an edge between nodes 5 and 9, it will be connected. The minimum number of elements (nodes or edges) that need to be removed to disconnect the remaining nodes from each other in a graph is an important measure in studying the resilience of a network.

A maximal subgraph that is connected in a graph is called a connected component. Note that we cannot add vertices and edges from the original graph to the component and retain its connectedness. Therefore, it is clear that a connected graph has exactly one component. Going back to Fig. 2.12(b), it has two connected components (see Fig. 2.13).

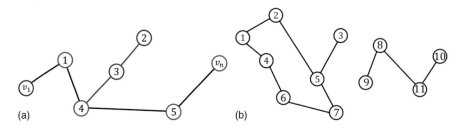

Figure 2.12 (a) A connected graph: If we want to go from v_1 to v_n we can start by traversing $\{v_1, 1\}, \{1, 4\}, \{4, 5\}, \{5, v_n\}$. Thus Path P from v_1 to v_n can be written as all edges in the order we traverse them inside a bracket $\{\{v_1, 1\}, \{1, 4\}, \{4, 5\}, \{5, v_n\}\}$ (b) A non-connected graph: We cannot always find a path between every two nodes of a graph. There is no path from nodes 2 to 9.

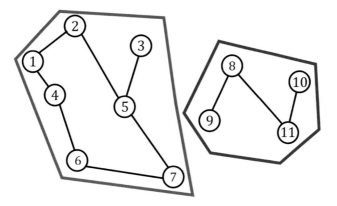

Figure 2.13 A graph with two connected components

Figure 2.14 The Königsberg bridge problem: finding a path over every one of seven bridges but without crossing any bridge twice. Courtesy of Historic Cities Research Project.

If a graph is cyclic, removal of an edge that is on a cycle does not affect its connectedness. Therefore, a connected subgraph with all vertices and a minimum number of edges has no cycles.

The practice of looking for cyclic subgraphs can be traced back to 1735, when the Swiss mathematician Euler solved the Königsberg bridge problem. The Königsberg bridge problem was an old puzzle concerning the possibility of finding a path over every one of seven bridges that spanned a forked river flowing past an island – but without crossing any bridge twice (see Fig. 2.14).

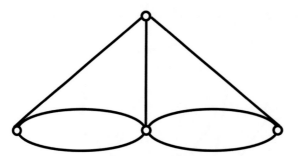

Figure 2.15 Graph representation of the Königsberg bridge problem.

What Euler realised was that most of the information on the maps had no impact on the answers to the answer. By thinking of each bank and island as a vertex and each bridge as an edge joining them, Euler was able to model the situation using the graph shown in Fig. 2.15. Euler argued that no such path existed. His proof involved only references to the physical arrangement of the bridges, but essentially he proved the first theorem in graph theory.

We say Graph G has an Euler Cycle if and only if every vertex of G has an even degree. It can be proved that a connected graph G has an Euler trail from node a to some other node b if and only if G is connected and $a \neq b$ are the only two nodes of odd degree.

2.4   Trees

Trees are an important class of graphs. The importance of trees is evident from their applications in various areas, especially theoretical computer science and molecular evolution.

A tree is a connected graph without any cycles, or equivalently, a tree is a connected acyclic graph. A disjoint union of trees is called a forest. A leaf is a vertex with a degree of 1. It follows immediately from its definition that a tree has to be a simple graph and has a unique simple path connecting each pair of its vertices. There are alternative approaches to defining a tree graph. It can be proved that if $G = (V, E)$ is a tree with $|V| = m$ and $|E| = n$, then $n = m - 1$.

The following theorem can be easily established and used as an alternative definition of a tree. Let $G = (V, E)$ be a graph with $n \geq 1$ vertices. The following statements are equivalent:

- G is connected and has no cycles.
- G is connected and has $n - 1$ edges.
- G has $n - 1$ edges and no cycles. For $u, v \in V$, G has exactly one path between u and v.

Sometimes you are given a graph, but only need one way to connect the nodes, for example, a network with redundant connections. When routing data, redundancy is

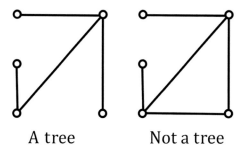

A tree   Not a tree

Figure 2.16 (Left) An example of a tree graph and (Right) an example of a graph that is not a tree.

beside the point; you simply need a way to send the data to its destination. So when making routing decisions, you would basically need to reduce a graph to a tree. The tree tells us exactly how we will send the data. Given a graph $G = (V, E)$, a subgraph of G that connects all of the vertices and is a tree is called a spanning tree. The spanning tree of a graph contains the minimum number of edges to keep the graph connected. A spanning subgraph of graph $G = (V, E)$ is a subgraph with vertex set V. A spanning graph does not need to be connected. An algorithm used to construct a spanning tree is non-deterministic in two ways:

- The choice of vertex u is arbitrary.
- The choice of the edge incident on u to add to the tree is arbitrary.

Therefore, a connected graph can have more than one spanning tree and all its spanning trees would have the same number of vertices and edges. A disconnected graph has no spanning tree. A tree has only one spanning tree that is itself. There are many algorithms to compute a spanning tree for a connected graph.

A vertex-centric algorithm is one of the common algorithms for finding a spanning tree for the connected graph G:

Step 1 Pick an arbitrary vertex and mark it as being on the tree.
Step 2 Repeat until all vertices are marked as on the tree:
- Pick an arbitrary vertex u on the tree, which is connected to vertex v not on the tree with edge uv. Add uv to the spanning tree and mark v as on the tree.
Step 3 Repeat Step 2 $n - 1$ times until there are $n - 1$ vertices to be added to the tree to obtain a spanning tree.

The minimum spanning tree is the one that has the minimum sum of edge weights. Kruskal's algorithm is often used to construct a minimum spanning tree. It is a greedy algorithm that runs in polynomial time. The input of this algorithm is a weighted connected graph and the idea is to enlarge a spanning subgraph H using edges with low weight, to form a spanning tree.

Kruskal's Algorithm for Finding a Minimum Spanning Tree
Consider the connected weighted graph $G = (V, E)$. Find the subgraph H of G using the following steps:

Step 1 Set $V(H) = V$ and $E(H) = \varnothing$.

Step 2 If the next cheapest edge joins two components of H, then include it. Otherwise discard it. Terminate when H is connected.

2.5   Properties of Graphs

Real-world graphs often span a large number of nodes and the interactions between them. Large graph comparisons are usually computationally difficult due to the NP-completeness of the underlying subgraph isomorphism problem. Thus, graph comparisons rely on easily computable heuristics, called 'graph properties'. Graph properties afford an overall view of a network, but may not be detailed enough to capture complex topological characteristics of large networks.

Here we review some of the most popular properties of graphs, and throughout the book you will learn alternative ways to compare and analyse graphs.

As you may recall, the degree of a node refers to the number of edges incident on the node. Now, if we average the degrees over all nodes in the graph, we will have a global measure for the whole system. However, it may not be representative, since the distribution of degrees may be skewed.

Let $P(k)$ be the percentage of nodes of degree k in the network. The degree distribution is the distribution of $P(k)$ over all k. Much of the recent research on the structure of biological networks and other real networks has focused on determining the form of their degree distributions. However, degree distributions are weak predictors of network structure. For instance, G_1 and G_2 in Fig. 2.17 are of the same size and same degree distribution, but they may have very different topologies or structures.

Clustering coefficients were introduced by Watts and Strogatz in 1998 as a way to measure how close a node (or vertex) and its neighbours are to being a complete subgraph [4].

Formally, the clustering coefficient C_v of a node v is defined as:

$$C_v = \frac{|E(N(v))|}{(\text{max possible number of edges in } N(v))},$$

where $N(v)$ is the open neighbourhood of v, i.e., the graph composed of the vertices adjacent to v and the edges connecting vertices adjacent to v. For vertex v with $|N(v)| < 2$, by definition $C_v = 0$.

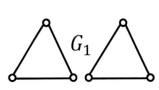
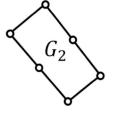

Figure 2.17 Two graphs with the same degree distribution and size.

The clustering coefficient for the entire system is the average of the clustering coefficients over all nodes in the network. C_v can be viewed as the probability that two neighbours of v are connected. Thus $0 \leq C_v \leq 1$. Average clustering coefficients are concerned with how densely clustered the edges in a network are. In a highly clustered network, the neighbours of a given node are very likely to themselves be linked by an edge. The clustering spectrum $C(k)$ is the distribution of the average of the clustering coefficients of all nodes of degree k in the network, over all k.

For example, consider the undirected graph with adjacency matrix

$$Q = \begin{bmatrix} 0 & 1 & 1 & 1 \\ 1 & 0 & 1 & 1 \\ 1 & 1 & 0 & 0 \\ 1 & 1 & 0 & 0 \end{bmatrix}.$$

The possible number of connections for nodes V_1 and V_2 in this matrix is 3 and the clustering coefficient for V_1 and V_2 is $2/3$. The possible number of connections for nodes V_3 and V_4 in this matrix is 1 and nodes V_3 and V_4 each have a clustering coefficient of $1/1$, which is 1. So the average of these four clustering coefficients is $5/6$.

Typically, the first step in studying the clustering and modular properties of a network is to calculate its average clustering coefficient and its distribution. It has been shown that the clustering coefficient of the metabolic network is at least an order of magnitude higher than that of the corresponding random network [5].

The distance between two nodes is the smallest number of links that have to be traversed to get from one node to the other. The path that covers that distance is called the shortest path. The average network diameter is the average of shortest path lengths over all pairs of nodes in a network.

Alternatively, we can represent a graph using the distance between its nodes, called the distance matrix, where each element d_{ij} is the distance between node i and node j.

In a sense, the average path length in a network is an indicator of how readily information can be transmitted through it. In biological networks, it has been observed that only a small number of intermediate steps are necessary for any one protein/gene/metabolite to influence the characteristics or behaviour of another [6].

We often have reason to believe that elements at the centre of every system are very important. The problem of identifying the most important nodes in a large complex network is of fundamental importance in a number of areas, including biology, communications, sociology, and management. To date, several measures have been devised for ranking the nodes in a complex network and quantifying their relative importance (Fig. 2.18). For instance, centrality measures have been used to predict the essentiality of a gene or protein based on network topology. A gene or protein is said to be essential for an organism if the organism cannot survive without it [7, 8]. In this section, we describe three standard centrality measures that capture a wide range of 'importance' in a network: degree, closeness, and betweenness.

The most intuitive notion of centrality focuses on degree: nodes with a large number of neighbours (i.e., edges) have high centrality. Therefore, we have degree centrality, $C_d(v) = \deg(v)$.

Nodes with a significantly higher degree are called hub nodes. The removal of these hub nodes has a far greater impact on the topology and connectedness of the network than the removal of nodes of low degree. Degree centrality, however, can be deceptive, because it is a purely local measure.

A second measure of centrality is closeness centrality [9]. A node is considered important if it is relatively close to all other nodes. Closeness centrality is defined in terms of the geodesic distance between nodes in a graph or network. Nodes with short paths to all other nodes in the network have high closeness centrality. The closeness of a node is formally defined as the inverse of the summation of distances from each node to every other node in the network, i.e.,

$$C_c(v) = \frac{1}{\sum\limits_{u \in V} \text{dist}(u, v)}.$$

Finally, yet importantly, we introduce betweenness centrality. The concept of betweenness centrality was introduced in 1978 by Freeman as a means of quantifying an individual's influence within a social network [10, 11]. The idea behind this centrality measure is that an important node will lie on a high proportion of paths between other nodes in the graph. Formally, betweenness centrality counts the number of shortest paths between i and k that node j resides on, i.e.,

$$C_b(v) = \sum\limits_{\substack{s \neq t, \\ s \neq v, \\ t \neq v}} \frac{\sigma st(v)}{\sigma st},$$

where σst is the number of shortest paths from s to t (they may or may not pass through node v) and $\sigma st(v)$ is the number of shortest paths from s to t that pass through v.

2.5.1   Graph Spectra

Eigenvalues are a special set of scalars associated with a matrix. The determination of the eigenvalues and eigenvectors of a system is extremely important in physics and engineering. Each eigenvalue is paired with a so-called eigenvector. To obtain the eigenvectors we should solve the equation for an eigenvalue, and v would be its eigenvector. If we solve this equation for the adjacency matrix of a graph, then we have the eigenvalue assigned to that graph. The set of all such scalars is called a graph spectrum.

However, in physics, the set of eigenvalues of a Laplacian matrix of a graph is called a graph spectrum. A graph Laplacian, denoted by L, is constructed by subtracting adjacency matrix from the degree, $L = D - A$, where D is the degree matrix and A the adjacency matrix. The smallest eigenvalue of L is 0, the corresponding eigenvector is the constant one, and L has n non-negative, real-valued eigenvalues. Two graphs are called cospectral if their adjacency matrices have equal multisets of eigenvalues. Cospectral graphs do not need to be isomorphic, but isomorphic graphs are always cospectral (see Fig. 2.19).

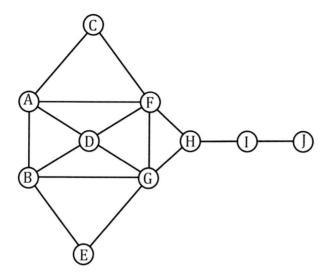

Figure 2.18 Different nodes assume importance depending on the centrality measure used. Here, D has the highest degree centrality, F and G have the highest closeness centrality, and H has the highest betweenness centrality.

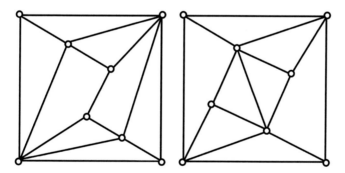

Figure 2.19 An example of two cospectral graphs (that are not isomorphic).

The final aspect of network structure that we will discuss here is concerned with small topological patterns. Milo et al. [12] showed that networks from diverse fields – biological and non-biological – contain several small topological patterns that are so frequent that they are unlikely to occur by chance. Moreover, different networks tend to have different sets of such frequent local structures. These patterns, referred to as 'network motifs', are recognised as 'the simple building blocks of complex networks'.

An algorithm for finding n-node network motifs is as follows:

1. First, find all n-node circuits in the real graph. For example, there are 13 3-node subgraphs, 199 4-node subgraphs and so on.

2. Then find all n-node subgraphs in a set of randomised graphs with the same distribution of incoming and outgoing arrows.
3. Finally, assign a p-value probability that any one of the n-node subgraphs will occur more randomly than in the real graph.

Thus a network motif is a small, over-represented partial subgraph of a real network. Here, 'over-represented' means that it is over-represented when compared to networks from a random graph model.

But what do we consider random? Specifically, which network 'null model' do we use to identify motifs? This will be discussed in the next section.

2.6   Network Model

Informally, a network model is a process (randomised or deterministic) for generating a graph. We can think in terms of models of static graphs or evolving graphs. Models of static graphs get a set of parameters Π and the size of the graph n as input, and return a graph as an output. Models of evolving graphs get a set of parameters Π and an initial graph G_0 and return a graph G_t for each time t as output. If the model is deterministic, then it defines a single graph for each value of n (or t). In contrast, a randomised model defines a probability space $< G_n, P >$ where G_n is the set of all graphs of size n, and P a probability distribution over the set G_n (similarly for t). We call this a family of random graphs R, and we will sometimes just call it a random graph R or random graph ER.

In the 1950s, Paul Erdős and Alfréd Rényi [13, 14] introduced their now classical notion of a random graph to model non-regular complex networks. The basic idea behind the Erdős–Rényi (ER) random graph model is the following: Assign the $G_{n,p}$ model the number of vertices n, and a parameter p, $0 \leq p \leq 1$, and for each pair (i,j), generate the edge (i,j) independently with probability p. In a related but not identical model, one can select m edges uniformly at random.

The degree distribution of an ER graph follows a binomial distribution, but for a large ER network it can be approximated by a Poisson distribution. The tails of degree distributions of ER graphs are typically narrow, meaning that the node degrees tend to be tightly clustered around the mean degree.

There will be no regions in the network that have a large density of edges. This is a beautiful and elegant theory that has been studied exhaustively since the 1960s, and many deep theorems about the properties of ER graphs have been formulated. Random graphs had been used as idealised network models, but unfortunately, they do not capture reality.

Let us look more closely at some real life networks and their properties. The degree distribution of many real life networks follow a power law. Such distributions are usually so skewed that it makes sense to plot the histogram in log-log form, where the characteristic distribution then becomes clear. Very low degrees are possible, and there are few nodes with large numbers of connections. Graphs having degree distributions that approximate power law distributions are referred to as scale free. In a scale–free

(SF) network, each node is connected to at least one other, most are connected to only one, while a few are connected to many.

The adjacency matrix of ER graphs is a uniform distribution of 1s, while in contrast an adjacency matrix of SF networks is 1s clustered in columns and rows for a few nodes.

The SF model focuses on the distance-reducing capacity of high-degree nodes, as 'hubs' create shortcuts that carry network flow. The most notable characteristic of a scale-free network is its hierarchy. The smaller ones closely follow the major hubs. In turn, other nodes with an even smaller degree follow these smaller hubs, and so on. This hierarchy allows for fault tolerant behaviour.

It seems that the SF network has what we are looking for, but how can we generate data with power law degree distributions? This question was first considered by Price in 1965 [15] when modelling citation networks. Each new paper is generated with m citations, and new papers cite previous papers with probability proportional to their in degree, each paper is considered to have a default citation, and the probability of citing a paper with degree k is proportional to $k + 1$. The degree distribution of a citation network is a power law with exponent $\alpha = 2 + \frac{1}{m}$. In a later paper [16], published in 1976, he proposed a mechanism to explain the occurrence of power laws in citation networks, which he called 'cumulative advantage' but which is today more commonly known under the name 'preferential attachment'.

A decade later, Barabasi and Albert (BA) suggested a model following the same idea [17]. The BA model gets an initial subgraph, G_0, and m number of edges per new node as input. Then nodes arrive one at a time, each node connects to m other nodes, selecting them with probability proportional to their degree. The result is a power law distribution with exponent $\alpha = 3$. So the BA model produces scale-free networks, which are time-invariant, and their properties (e.g., degree distribution) remain about the same as more nodes are added.

The removal of either assumption destroys the scale-free property, that is, without node addition over time, we end up with a fully connected network, and given enough time, without preferential attachment we get exponential connectivity.

If we compare the average length of shortest distances between any two vertices in the ER and BA models, we see that for the same number of connections and nodes, ER has a larger diameter than scale-free networks.

So far, we have focused on degree sequences, obtaining graphs with power law distributions. What of other properties? Real-life networks tend to be 'small worlds', a property found to be easy to produce.

Small world graphs are based on Milgram's (1967) famous work [18], the substantive point being that networks are structured such that even when most of our connections are local, any pair of people can be connected by a small number of relational steps. They work on two parameters: the clustering coefficient and the average distance (L) between all nodes in the network.

In a highly clustered, ordered network, a single random connection will create a shortcut that lowers L dramatically. Watts demonstrates that small world properties can occur in graphs with a surprisingly small number of shortcuts. According to Watts [4],

Table 2.1 Network models and their topological differences.

Topology	Average Path Length (L)	Clustering Coefficient (CC)	Degree Distribution $(P(k))$
ER Graph	$\log N / \log <K>$	$<k> / N$	Poisson distribution
Small World	$L_{sw} < L_{rand}$	78	Similar to random graph
Scale-Free network	$L_SF < L_{rand}$	778	Power law distribution

large natural networks ($n > 1$) have very sparse connectivity, no central node, but large clustering coefficients (larger than ER graphs of the same size), and they also have short average path length. To construct such a graph he suggested starting with a ring and $i = 1, \ldots, n$. Select a vertex j with probability proportional to R_{ij} and generate an edge (i,j). Repeat until z edges are added to each vertex, where R_{ij} is a measure inspired by the work of Solomonoff and Rapoport [19]. The logic behind it is that nodes that share neighbours should have a higher probability of being connected. The resulting graph has properties of both regular and random graphs. To summarise, in this section we have learned about three network models and their properties. Table 2.1 gives an overview of their topological differences.

2.7   Biological Networks

Many real world systems can be represented as networks, including biological networks. Networks are found in biological systems of varying scales, from the evolutionary tree of life, to ecological networks, to expression networks and regulatory networks. Can you name any other biological phenomenon that can be represented as a network? In a biological network, nodes and edges can represent different things. A node can be a protein, peptide, or non-protein biomolecule. Edges can be biological relationships, interactions, regulations, reactions, transformations, activation, inhibitions. One can construct bipartite or tripartite networks, for instance, between genes, proteins, and drugs. Protein or metabolic interaction in cells can be represented as a graph. In the study of a complex system, questions regarding some aspect of the structure or characteristics of a corresponding network have proven very useful. For example, finding protein complexes may be addressed as a graph-partitioning problem; questions involving the movement of information or spread of a disease can be modelled in terms of paths, and flows along these paths, of a network. The primary goal here is to connect the topological features of biological networks with biological function, the design principles of regulation mechanisms and the evolution of the systems. For example, studies suggest that cellular networks are scale free, and that small-world [20] or protein complexes as discrete units of the function of a system and can be found by community detection methods [21]. The graph shown in Fig. 2.20 is a protein-to-protein network (PPI) constructed using three different databases. This protein interaction network comprises 15,383 proteins and 337 413 interactions. Protein–protein

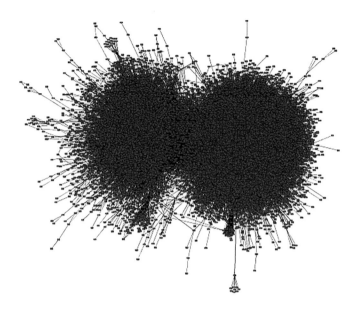

Figure 2.20 An example of a real-world biological network.

interactions, and transcriptional, regulatory, and metabolic networks have been studied, and their structural properties are similar. The observed properties are not in agreement with traditional random graph models for complex networks. Their degree distributions follow a power law: the rich get richer. They have a small average path length. They are very resilient and have strong resistance to failure on random attacks but are vulnerable to targeted attacks. They have hierarchical modularity and few central nodes. As we mentioned earlier, it has been observed that in biological networks, the more central a node, the more indispensable it is. PPI networks are typically modelled as undirected graphs, in which nodes represent proteins and edges represent interactions. The frequency of proteins having interactions with exactly k other proteins follows a power law. PPI networks exhibit small world phenomena: they can reach any node with a small number of hops, usually four or five hops. Although they are very sparse, they are very hard to study without breaking them up into to smaller entities.

However, we know that this network, like many other biological networks, is modular. Therefore, we can try to find highly clustered subgraphs of large biological networks, called communities, to get a high-level overview of the networks. This also allows us to predict unknown gene or protein functions based on communities (Fig. 2.21). Communities in a network are the groups of nodes that are more densely connected to each other than they are to the rest of the nodes in the network. There is no universal definition of community structure.

Diverse algorithms ranging from simple K-mean clustering to Markov chain-based models have been suggested in the literature as means to detect these modules in a complex network. Most of these techniques are based on the optimisation of objective functions. Modularity optimisation is one of the most widely used techniques among them. However, modularity optimisation remains an NP-hard problem.

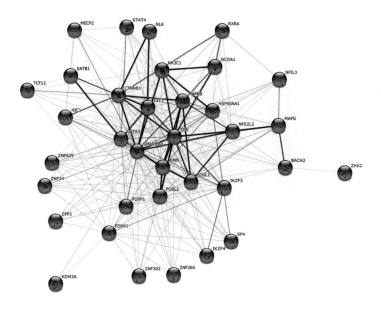

Figure 2.21 A module of the PPI network found by Markov clustering methods. Many of the genes here are known to work together in a particular biological pathway.

In spite of these studies, one of the main challenges in community detection is the number of communities. Most of these methods have parameters that need to be set before running them, and exact values of these parameters affect the number of modules detected, even though we may not have explicitly inquired after modules.

As we have seen, each of these biological networks has been reconstructed differently. A PPI network is a collection of diverse types of interaction, from experimentally validated physical interaction to computationally predicted interaction. The drug network is based on similarity measures. The exact definition of similarity can range from simple correlation to more sophisticated structural similarity.

Reconstruction of regulatory or signalling networks directly from experimental data using statistical or machine learning approaches has received significant attention in the last decade. A wide variety of different approaches are available that can be used to infer biological networks from experimental data. Approaches employed include Bayesian networks, auto-regressive models, correlation-based and mutual-information-based models, clustering techniques, and differential equation models. Most of these methods use perturbation to validate the reconstructed network. This can be performed by removing a node, gene, protein, or metabolite from a system or by preventing interaction among particular ones. Most graph-theoretic analyses of networks that we have discussed up to this point offer a static view of the systems of interest. Edges between nodes represent relations that occur at given times. For example in a cell, different gene expression activities occur at different time scales, and cellular functioning requires the precise coordination of a large number of events, but we ignore the timing when we reconstruct or analyse a network using graph theory. So for complete understanding and analysis of a complex system, its temporal aspects should be incorporated in the modelling framework. Understanding how a complex system's function emerges from the connectivity structure of its elements stands as one of the enduring challenges of systems theory. Many complex systems exhibit a very wide range of dynamic connectivity structures that enable the system to integrate and process information over time and in the face of perturbation. The temporal aspect of complex systems is a subject of study in another paradigm of mathematics, dynamical systems theory. The framework that has been used in dynamical systems theory is mostly differential equations and not networks, but after reviewing the basic concepts and introducing the terminology of dynamical systems theory we will return to combine both fields and devote a section to dynamic networks.

2.8 What Is a Dynamical System?

At the end of the seventeenth century, Leibniz (1646–1716) and Newton (1643–1727), independently of one another, invented a brilliant mathematical tool: infinitesimal calculus, or differential and integral calculus. This is an incredibly efficient crystal ball with which to predict the future, provided the system in question is governed by a differential equation. It enabled Poincaré's work on celestial mechanics, specifically a 270-page, prize-winning, albeit initially flawed paper [22] that inaugurated the qualitative theory of dynamical systems. The methods developed therein laid the foundation for the local and global analysis of non-linear differential equations, including the use of first-return (Poincaré) maps, stability theory for fixed points and periodic orbits, stable and unstable manifolds, and the Poincaré recurrence theorem. More strikingly, using the example of a periodically perturbed pendulum, Poincaré showed that mechanical systems with $n \geq 2$ degrees of freedom may not be integrable, due to the presence of homoclinic orbits. Within science and mathematics, dynamics is the study of how things change with respect to time, as opposed to the study of things in terms of their static properties. The patterns of change that we observe all around us afford an alternative way of describing phenomena. A dynamical system is a set of possible states, together with a rule that determines its present state in terms of past states. As examples of dynamical systems, think of any system that evolves in time. For example, a pendulum, the weather, the evolution of a population of bacteria, or any kind of seasonal or periodic phenomenon.

A dynamical system has two aspects, a state space and a function, and is described in terms of them. Let us see what they are.

As we mentioned earlier, dynamical systems are systems that change over time. Each change marks a state, and a state space is a model used to capture the gamut of states a system moves through over time.

Formally, a state space is the set of all possible states of a dynamical system. Each state of the system corresponds to a unique point in the state space. For example, the state of an idealised pendulum is uniquely defined by its angle and angular velocity, so its state space is the set of all possible pairs '(angle, velocity)', which forms a cylinder. In general, any abstract set could be a state space of some dynamical system.

A state space can be finite, consisting of just a few points, or else it could consist of an infinite number of points forming a smooth manifold, as is usually the case in ordinary differential equations and mappings. Such a state space is often called a phase space. A state space could be infinite-dimensional, as in partial differential equations and delay differential equations. And in symbolic dynamics it can be a Cantor set, which is zero-dimensional. The second aspect of a dynamical system, its function, tells us what the state of the system will be in the next instant of time, given its current state.

When investigating dynamical systems, it is necessary to specify some characteristics that would enable us to subdivide them into special classes. Specific methods are available for some of these classes, hence such a classification can help simplify analysis.

A dynamical system is deterministic if the present state can be determined uniquely from past states (no randomness is allowed). Stochastic models possess some inherent randomness. A chaotic model is a deterministic model with behaviour that cannot be entirely predicted. Chaotic models are predictable in the very short term, but appear random for longer periods.

Dynamical systems may be continuous or discrete. In a discrete system, the state variables change only at a countable number of points in time. These points in time are the ones at which an event/change in state occurs. In a continuous dynamical system, on the other hand, the state variables change in a continuous way, and not abruptly from one state to another (an infinite number of states). When real-value numbers are involved, the system is called a continuous dynamical system, and when there are only integers involved, the system is called a discrete dynamical system. Continuous systems are rendered by differential equations, whereas discrete dynamical systems (often called maps) are specified by difference equations.

Discrete dynamical systems can be solved using iteration calculations, called iterative maps. Iterative maps give us less information but are much simpler and better suited to dealing with composite entities, where feedback is important. A typical example is the annual activity in a bank account. If the initial deposit is 100 000 euros and the annual interest is 3%, then we can describe the system by

$$X(k) = 1.03^k \times 100000,$$

$$X(0) = 100000.$$

We denote time by k or n. In a continuous system, the time interval between our measurements is negligibly small, making it appear as one long continuum, and this is shown through the language of calculus, using a differential equation or a set of them. For example, a vertical throw is described by initial conditions $h(0)$, $v(0)$ and equations

$$h(t)' = v(t),$$

$$v(t)' = -g,$$

where h is height and v is the velocity of a body. Calculus and differential equations have formed a key part of the language of modern science since the days of Newton

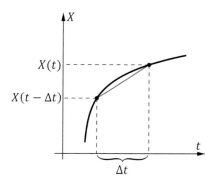

Figure 2.22 Illustration of the Euler method.

and Leibniz. Even though an analytical treatment of dynamical systems is usually very complicated, obtaining a numerical solution is (often) straightforward. There are a number of methods for solving differential equations numerically. Techniques for solving differential equations based on numerical approximations were developed before programmable computers existed. The easiest way involves Euler's first order method, which uses the idea of local linearity or linear approximation, where we use small tangent lines over a short distance to approximate the solution to an initial value problem (Fig. 2.22). If we minimise sufficiently, every curve looks like a straight line, and therefore the tangent is a great way to calculate what is happening over a period.

Today, we have many numerical methods for solving differential equations. And there is no 'best way' or 'best method', since the method to be chosen depends largely on the problem (stiff or non-stiff, equation or system, smooth or non-smooth, right-hand side, etc.). For a 'general' or 'standard', a 4th order Runge–Kutta is good; it's easy to add an error estimator to it at little or no additional cost. A predictor-corrector Adams method also does the job, and is quite popular.

Differential equations are a great representational system when only a few elements in the system are involved; they give us lots of information. But they also become very complicated very quickly. Whereas differential equations are central to modern science, iterative maps are central to the study of non-linear systems and their dynamics, as they allow us to take the output from the previous state of a system and feed it back into the next iteration, thus making them well equipped to capture the feedback characteristic of non-linear systems.

Another important classification of dynamical systems is based on linearity. A non-linear system is a set of (one or more) non-linear equations. Non-linear equations are equations where the unknown quantity we wish to solve for appears in a non-linear fashion. For example, if the quantity in question is a function $y(t)$, then terms such as y^2 or $\sin y$ would be non-linear. More precisely, a non-linear equation is one where a linear combination of solutions is not a new solution. In a linear system, a function describing the system's behaviour must satisfy two basic properties: additivity and homogeneity:

- Additivity

$$g(x + y) = g(x) + g(y),$$

- Homogeneity

$$\alpha g(x) = g(\alpha x).$$

A differential equation is linear if the coefficients are constant or functions of the independent variable alone.

And finally, an autonomous system is a system of ordinary differential equations that do not depend on the independent variable. If the independent variable is time, we call such a system a time-invariant system.

Condition: If the input signal $x(t)$ produces an output $y(t)$, then any time-shifted input, $x(t + \delta)$, results in a time-shifted output $y(t + \delta)$.

2.8.1 Modelling a Dynamical System

The first step in the analysis of a dynamical system is to derive its model. Models may assume different forms depending on the particular system and the circumstances in which we find it. A mathematical model of a dynamical system can often be expressed as a system of differential (difference, in the case of discrete-time systems) equations. The response of a dynamical system to an input may be obtained if these differential equations are solved. The differential equations can be obtained by utilising the physical laws governing a particular system, for example, Newtons laws for mechanical systems, Kirchhoff's laws for electrical systems, etc. In obtaining a model, we must make a compromise between the simplicity of the model and the accuracy of the results of analysis. As mentioned in the previous section, at any given time a dynamical system has a state given by a vector that can be represented as a point in a geometrical manifold. Future states that follow from the current state are described by the evolution rule of the dynamical system. As an example, we pick a real world dynamical system, a biological system containing two species – predators (foxes) and prey (rabbits). There are literally hundreds of examples of predator-prey relations in an ecosystem. Predation is a biological interaction where an organism hunts down and feeds on its prey. There is a continuous tussle between predators and their prey, and an inverse relationship between the number of predators and prey.

If we take the fox-rabbit pair as an example, then rabbits, left to themselves, would reproduce at a rate $dr/dt = 100r$. Foxes, without rabbits, would starve, and their population would decline with rate $dw/dt = -50f$. When brought into the same environment, foxes will catch and eat rabbits. Losses to the rabbit population will be proportional to the number of foxes f and number of rabbits r.

The predator-prey model was initially proposed by Lotka in the theory of autocatalytic chemical reactions in 1910 [23]. Volterra developed his model independently of Lotka and used it to explain d'Ancona's observation regarding the increase in Adriatic fauna during World War I [24]. The Lotka–Volterra model consists of a pair of

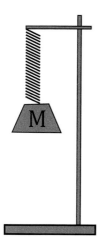

Figure 2.23 Motion of a one-dimensional particle subject to force, resulting in linear displacement Hooke's law: the force is proportional to the extension.

first-order non-linear differential equations, frequently used to describe the dynamics of biological systems in which two species interact, one as a predator and the other as prey. The populations change through time according to the pair of equations you see here:

$$dx/dt = x(\alpha - \beta y),$$

$$dy/dt = -y(\gamma - \delta x).$$

One of the kinds of motion you might encounter is periodic motion. For example, the motion of the planets around the sun is periodic. This type of periodic motion is very predictable and we can predict, say, eclipses, far out into the future, as well as precisely date eclipses that occurred in the distant past. In these systems, small disturbances are often rectified and do not increase sufficiently to alter the system's trajectory very much in the long run (Fig. 2.23). Changes in traffic lights, or the linear displacement of particles when subject to a force, are also examples of periodic motion.

A differential equation of motion, usually identified as some physical law and applying definitions of physical quantities, is used to set up an equation for the problem. Solving the differential equation will lead to a general solution with arbitrary constants, which results in a family of solutions. A particular solution can be obtained by setting the initial values, which fixes the values of the constants. As an example, we can study the motion of a two-dimensional particle subject to a force, where the displacement is linear:

$$x'' + (\omega_0)2x = 0,$$

$$y'' + (\omega_0)2y = 0.$$

Now, if we solve the equation for particle motion when subject to a force, then we can see that the solution is

$$x(t) = A \cos (\omega_0 t - \alpha),$$

$$y(t) = B \cos (\omega_0 t - \beta).$$

The amplitudes A, B, and the phases α, β are determined by initial conditions. The motion is simple harmonic motion in each of the two dimensions. Both oscillations have the same frequency, but (in general) different amplitudes.

Terminology

The set of all states of a system of ordinary differential equations is called a phase space. A phase space is a space in which all possible states of a system are represented, with each possible state of the system corresponding to one unique point in the phase space. The phase space of a two-dimensional system is called a phase plane. A phase curve is a plot of the solution of equations of motion in a phase space.

A phase portrait is a plot of a single phase curve or multiple phase curves corresponding to different initial conditions in the same phase plane. If we plot the solution of equations of motion in a phase space, then we have a phase curve. The phase curve is dependent on the initial state. If we plot single or multiple phase curves corresponding to different initial conditions in the same phase plane, then we have a phase portrait. To fully understand the behaviour of a dynamical system, we need to know how a system moves from one position to another, and this is described by trajectories through the state space. A trajectory or path is a set of positions in a state space through which a system might pass successively.

If we wish to plot the phase portrait for the two-dimensional motion system that we saw in the previous section we would need to find all of its paths. An equation for a path is obtained by eliminating t in the solution. And a trick for this system is defining a new parameter $\delta = \alpha - \beta$. Then we can see that the phase portrait for this system is a family of ellipses, each of which is a separate phase curve for different initial conditions. Studying such diagrams affords insight into the physics of particle motion.

2.8.2   Stability Analysis

We can solve equations for particular starting point(s), but this is often insufficient to enable us to understand a system. Therefore, we use complementary analysis focused on finding equilibrium states (or stationary/critical/fixed points) where the system remains unchanged over time. Analysis determines how the system behaves over time, in particular, by investigating the future behaviour of the system given any current state, i.e., the long-term behaviour of the system. Also, we try to classify these states/points as stable/unstable by investigating the behaviour of the system as it approaches them.

Let us start with a very simple example: flipping a coin. A coin has three equilibria points as it can be balanced on a table on its edge but can also rest on heads or tails. Small movements at its edge (perturbation) mean that the coin will end up in one of two different equilibria: heads or tails. Thus, leaving the coin on its edge is unstable.

However, resting on heads or tails is stable, as perturbations do not result in changes in state.

Similarly, think of a ball at rest in a dark landscape. Imagine that it's either on top of a hill or at the bottom of a valley. To find out which, push it (perturb it), and see if it comes back.

A set towards which a dynamical system evolves over time is called an attractor. It can be a point, a curve, or a more complicated structure. A perturbation is a small change in a physical system, most often in a physical system at equilibrium that is disturbed from the outside.

Each attractor has a basin of attraction, which contains all the initial conditions that will generate trajectories joining this attractor asymptotically. When studying a dynamical system, if we are only interested in its long-term behaviour, we will only study its attractors and determine their basin of attraction. One of the 'simplest' attractors is the point, a fixed point, i.e., the particular points of the phase space for which x doesn't change as time increases. For one-dimensional systems, $dx/dt = f(x, t)$ at fixed point $\hat{x}$, $dX/dt(\hat{x}) = 0$. The corresponding solution of the dynamical system does not depend on time. It is a stationary state.

For example, take the one-dimensional dynamical system described by $dx/dt = 6x(1 - x)$. If we solve $6x(1 - x) = 0$ we can find two fixed points: $x = 0$ or $x = 1$, check the stability, perturb them, and see what will happen.

If we use a difference equation, $x(t + h) = x(t) + h dx/dt$ from various different initial x's such as $x = 0 + 0.01$ and $x = 0 - 0.01$ for the first point and $x = 1 + 0.01$, $x = 1 - 0.01$ for the second point we can conclude that $x = 0$ is unstable and $x = 1$ stable.

There are general rules regarding the stability of a one-dimensional dynamical system. Given a fixed point $\hat{x}$:

- If $f'(\hat{x}) > 0$, $\hat{x}$ is unstable;
- If $f'(\hat{x}) < 0$, $\hat{x}$ is stable;
- If $f'(\hat{x}) = 0$, it is inconclusive,

where $f'(\hat{x})$ means the derivative of f, df/dx evaluated at $\hat{x}$. If $f'(\hat{x}) = 0$ one would need to use higher derivatives or other methods, and it can be semi-stable from above or below, or periodic.

So far we have learned that a fixed point is a special point of a dynamical system that does not change over time. It is also called an equilibrium, steady-state, or singular point of the system.

Terminology

If a system is defined by an equation $dx/dt = f(x)$, then the fixed point can be found by examining the condition $f(\hat{x}) = 0$. We do not need to have the analytic solution of $x(t)$.

A stable fixed point is defined as a point where for all starting values x_0 near it the system converges to $\hat{x}$ as $t \to \infty$.

A marginally stable fixed point is defined as a point where for all starting values x_0 near $\hat{x}$, the system stays near but does not converge to it.

And finally, an unstable fixed point is a point that for starting values x_0 very near $\hat{x}$, the system moves far away from it.

We conclude this section by looking at another biological model, a bacterial growth model.

Bacterial Growth Model

We place a nutritive solution and some bacteria in a petri dish.

Let b be the relative rate at which the bacteria reproduce, and p the relative rate at which they die. Then the population is growing at the rate $r = b - p$.

If there are x bacteria in the dish, then the rate at which the number of bacteria is increasing is $(b - p)x$, that is, $dx/dt = rx$. The solution of this equation for $x(0) = x_0$ is $x(t) = x_0 \exp rt$.

The model is not realistic because the bacterial population grows to infinity for $r > 0$. As the number of bacteria rises, they produce more toxic products. Instead of a constant relative perish rate p, we will assume a relative perish rate dependent on their number px.

Now the number of bacteria increases by bx and decreases by px^2. The new differential equation will be

$$dx/dt = bx - px^2.$$

To be able to find a fixed point, we have to set the right-hand side of the differential equation to zero.

There are two possible solutions, as we have two fixed points:

$$\hat{x} = 0, \quad \hat{x} = b/p.$$

At $\hat{x} = 0$, there are no bacteria, None can be born, none can die.

But after slight contamination (perturbation), which is smaller than b/p, the number of bacteria will increase by $dx/dt = bx - px^2 > 0$ and will never return to the zero state.

At the second point, $\hat{x} = b/p$, the population levels out. Bacteria are being born at a rate $b = b(b/p) = b^2/p$ and are dying at the same rate, so birth and death rates are exactly in balance. If the number of bacteria were to be slightly increased, then $dx/dt = bx - px^2 < 0$ and the system would return to equilibrium. If the number of bacteria were to be slightly decreased, then $dx/dt = bx - px^2 > 0$ and the system would return to equilibrium. Small perturbations away from $\hat{x} = b/p$ will self-correct back to b/p. Therefore, the second fixed point is stable.

2.9   2D Dynamical Systems

In this section, we review higher-dimensional systems. In higher-dimensional systems, movement trajectories can exhibit a wider range of dynamical behaviour. Fixed points still exist, but may be more interesting, depending on how trajectories approach or draw away from the equilibrium point. For example, a system could spiral in to a stable point. Other types of stability exist, such as saddle nodes, and importantly, cyclic/periodic

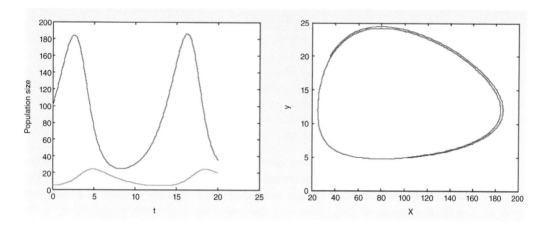

Figure 2.24 To examine behaviour at/near fixed points we can examine their change over time or in the phase plane. The view in phase-space (or phase-plane), where x is plotted against y rather than against time gives more information about the system. Cyclic behaviour is fixed but the system is not at a fixed point.

behaviour limit cycles. These are more interesting, but also more difficult to analyse. Suppose we have the following system:

$$dx/dt = f(x,t),$$

$$dy/dt = g(x,t).$$

First, we would need to learn how to find fixed points, then examine the stability of these fixed points, and finally, we would need to examine the phase plane and trajectories (Fig. 2.24).

To find fixed points, as before we would need to solve $dx/dt = 0$ and $dy/dt = 0$, which gives us fixed points (x_0, y_0).

Let us look at an example of a predator-prey system: Let $x(t)$ denote the population of the prey species, and $y(t)$ denote the population of the predator species. Then

$$x' = ax - \alpha xy,$$

$$y' = -cy + \gamma xy.$$

All parameters are positive constants. Note that in the absence of the predators (when $y = 0$), the prey population would grow exponentially. If the prey are absent (when $x = 0$), the predator population would decay exponentially to zero due to starvation. The system in general has two critical points. One is the origin, and the other is in $(c/\gamma, a/\alpha)$.

Before going forward, we need to define the Jacobian matrix. A Jacobian matrix is the matrix of all first-order partial derivatives of a vector- or scalar-valued function with respect to another vector. This matrix is frequently marked J, Df, or A. Jacobian matrices are used to classify fixed points of higher order linear dynamical systems,

$$\mathbb{J} = \begin{bmatrix} \dfrac{\partial f_1}{\partial X_1} & \dfrac{\partial f_1}{\partial X_2} & \cdots & \dfrac{\partial f_1}{\partial X_n} \\[2mm] \dfrac{\partial f_2}{\partial X_1} & \dfrac{\partial f_2}{\partial X_2} & \cdots & \dfrac{\partial f_2}{\partial X_n} \\[2mm] \vdots & \vdots & \ddots & \vdots \\[2mm] \dfrac{\partial f_n}{\partial X_1} & \dfrac{\partial f_n}{\partial X_2} & \cdots & \dfrac{\partial f_n}{\partial X_n} \end{bmatrix},$$

where $f_1, \ldots, f_n$ are the right-hand side of differential equations and $X_1, \ldots, X_n$ are the independent variables. The eigenvalues of the Jacobian matrix are used to classify the stability of the fixed point. Having found eigenvalues and eigenvectors of J evaluated at the fixed point:

1. If eigenvalues have negative real parts, x_0 is asymptotically stable
2. If at least one has a positive real part, x_0 is unstable
3. If eigenvalues are purely imaginary, x_0 can be stable or unstable

Different behaviours can be seen, depending on the eigenvalues and eigenvectors (Fig. 2.25). In general, points are attracted along negative eigenvalues and repelled along positive. The axes of attraction are eigenvectors.

If we go back to our predator-prey system, the Jacobian matrix is

$$J = \begin{bmatrix} a - \alpha y & -\alpha x \\ \gamma y & -c + \gamma x \end{bmatrix}.$$

At $(0,0)$, the eigenvalues are a and $-c$. Hence it is an unstable saddle point. At the second point, the eigenvalues of the Jacobian are $\pm\sqrt{ac}\, i$. It is a stable centre (Previously, we learned that the case of purely imaginary eigenvalues in a non-linear system is ambiguous, with several possible behaviours. But in this example it is a centre.)

Therefore, for a dynamical system to be stable the real parts of all eigenvalues must be negative.

All eigenvalues must lie in the left half of the complex plane.

A dynamical system is underdamped if there is a spiral fixed point (some complex eigenvalues), we say it is overdamped if it exhibits nodal behaviour (all eigenvalues are real), and we say it is critically damped at the boundary. For non-linear equations, behaviour near the fixed points will be 'almost like' the behaviour of a linear system, depending on how 'almost linear' it is. The less linear-like the behaviour becomes, the further away trajectories get from the fixed point.

2.9.1 The Lorenz System

In the early 1960s, Edward Lorenz, an MIT meteorologist, was working on developing a system to simplify the convection rolls in the upper atmosphere for long-range weather prediction (5+ days).

However, the weather is complicated! A theoretical simplification was necessary. In 1963, in seeking to model long-range predictions of the weather, he devised a three-dimensional system.

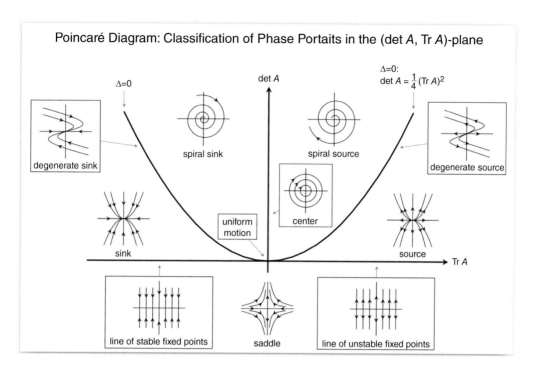

Figure 2.25 We can classify dynamical systems using the trace and determinant of the Jacobian. Source: https://texample.net//tikz/examples/poincare/.

Using this system he got into the problem of 'sensitivity to initial conditions', and sketched the outlines of one of the first recognised chaotic attractors. He came to the conclusion that either the model equations were inaccurate in their representation of some aspects of the weather, or else the model was accurate but some anomalous property of the equations made prediction difficult.

The Lorenz system describes the motion of a fluid between two layers at different temperatures. Specifically, the fluid is heated uniformly from below and cooled uniformly from above. Raising the temperature difference between the two surfaces, we initially observe a linear temperature gradient, and then the formation of Rayleigh–Bénard convection cells.

After convection, a turbulent regime is also observed. The Lorenz attractor defines a three-dimensional trajectory by the differential equations (Fig. 2.26): (the most frequently used parameter values are: $\sigma = 10$, $r = 28$ and $\beta = 8/3$)

$$\frac{dX}{dt} = \sigma(Y - X), \tag{2.1}$$

$$\frac{dY}{dt} = -XZ + rX - Y, \tag{2.2}$$

$$\frac{dZ}{dt} = XY - \beta Z. \tag{2.3}$$

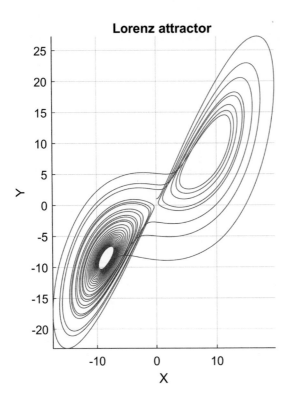

Figure 2.26 The Lorenz strange attractor of most frequently used values of parameters.

When calculating this model, Lorenz encountered a strange phenomenon. When entering slightly different input values for two successive attempts, he obtained completely different outputs (Fig. 2.27). This effect was later named the 'butterfly effect'. The 'butterfly effect' is the propensity of a system to be sensitive to initial conditions. Such systems become unpredictable over time, an idea captured in the notion of a butterfly flapping its wings in one area of the world and causing a tornado or some such weather event to occur in a different remote area of the world.

The Lorenz system has three fixed points that, for the recommended parameter values, are all unstable because they all include an eigenvalue with a positive real part.

A chaotic system is roughly defined by sensitivity to initial conditions: infinitesimal differences in the initial conditions of the system result in large differences in behaviour. Chaotic systems do not usually go out of control, but stay within bounded operating conditions. Chaos represents a balance between flexibility and stability, adaptiveness, and dependability. It lives on the edge between order and randomness.

As we have seen, dynamical systems can have different types of attractors. If the system evolves toward a single state and remains there, we call it a fixed point. Examples are a damped pendulum and a sphere at the bottom of a spherical bowl.

When a system evolves toward a limit cycle, it is said to have a periodic or quasi-periodic attractor. An example is an undamped pendulum, or a planet orbiting around

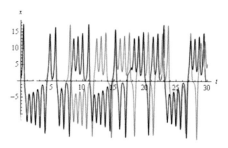 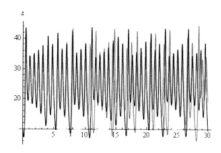

Figure 2.27 The graphs show time dependence of functions $x(t)$ and $z(t)$ in the Lorenz attractor for the recommended parameter values, while blue curves are related to initial conditions $x(0) = 1$; $y(0) = 1$; $z(0) = 10$ and red curves are related to initial conditions $x(0) = 1$; $y(0) = 1$; $z(0) = 10.01$. Very small changes in initial conditions result in large changes in the solution of the function.

the sun. If the system is very sensitive to initial conditions and we are not able to simply predict its behaviour, we say it has a *chaotic attractor*. An example is the Lorenz attractor.

Finally, a system has a strange attractor if it is also very sensitive to initial conditions and we are not able to simply predict its behaviour, and, furthermore, the system has the same properties as fractals. In other words, the strange attractor represents a fractal. An example is the Mandelbrot set.

A fractal is a non-regular geometric shape that has the same degree of non-regularity at all scales. Fractals can be thought of as never-ending patterns. The Hausdorff dimension of a fractal's border is higher than the topological dimension of the border. A fractal has the following features:

- It is self-similar, which means that when observing the shape at different magnifications we see the same characteristic shapes (or at least approximately the same shapes).
- It has a simple and recursive definition.
- It has a fine structure at arbitrarily small scales.
- It is too irregular to be easily described in traditional Euclidean geometric language.
- A Hausdorff dimension of its border is higher than the topological dimension of the border.
- If an object contains n copies of itself reduced to one kth of the original dimension, the Hausdorff dimension can be calculated as $\log(n)/\log(k)$.

An example of a fractal is a Cantor set, where the original line is divided into three parts while the middle part is erased, the same procedure being applied to newly created lines, etc. If we were to repeat this procedure to infinity, we would obtain an infinite number of points with topological dimension 0. The set contains $n = 2$ copies of itself reduced to $1/3$ of the original dimension ($k = 3$). The Hausdorff dimension is $\log(2)/\log(3) = 0.6309\ldots$, which is greater than 0. Generally speaking, we like to study qualitative features of the behaviour of dynamical systems and compare the behaviour of different dynamical systems. Studying bifurcations is one way of doing this. A bifurcation is defined as a qualitative change in a phase portrait in the area of attraction that can be achieved by changing the driving parameter when passing through the critical value. The critical points (equilibrium solutions) usually depend on the value of the model's parameter values. As the parameters steadily increase or decrease, it often happens that at certain values there is a bifurcation point, critical points come together or separate, and equilibrium solutions are either lost or gained.

We can distinguish two principal types of bifurcation:

1. Global bifurcation: its effects are not limited by the neighbourhood of a point or cycle in the phase space. It cannot be detected by analysis of fixed point stability.
2. Local bifurcation: its effects are limited by the neighbourhood of a point or cycle in the phase space. Fixed points may appear or disappear due to parameter change; they change their stability, or even break apart into periodic points. Such bifurcations can be analysed entirely through changes in the local stability properties of equilibria, periodic orbits, or other invariant sets.

We have learned that dynamical systems live in phase space and develop in time, following their dynamical law. We have seen examples of continuous dynamical systems and learned to analyse them. In the next section, we will do the same for discrete dynamical systems.

2.10   Dynamic Networks

Network dynamics are applicable to studies of changes in the network structure itself, and to studies of diffusion dynamics. The first is mainly aimed at understanding the building blocks of a system, whereas the second is focused on understanding processes that affect an elementary unit, nodes, in the system. Here, we outline different aspects of the second question. In general, two different approaches can be used to analyse this type of model: continuous dynamic methods, where the activity of the components is based on differential equations, or discrete dynamic strategies, in which each node can be characterised by only a few discrete states. In this section, we treat a dynamic network as a dynamical system where the state of the system evolves in discrete time steps, i.e., as a discrete dynamical system. A discrete dynamical system can be likened to taking snapshots of a system at a sequence of times. The snapshots could be taken once a year, once every month, or even irregularly, such as when someone coughs. When we take these snapshots, the idea is that we are recording whatever variables determine the state of the system: our chosen state variables that evolve through the state space. To complete the description of the dynamical system, we would need to specify a rule that, given an initial snapshot, determines what the resulting sequence of future snapshots must be. For example, the pendulum is a two-dimensional continuous dynamical system. A cellular automaton is a discrete dynamical system. Boolean network models, originally introduced by Kauffman [25], represent the simplest dis-

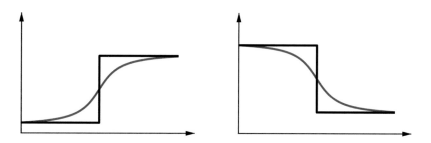

Figure 2.28 One can think about Boolean models as replacements of continuous functions (e.g., the Hill function) by step functions. Boolean models can be used as replacements of continuous models when we lack sufficient information for more detailed description or because of the increasing complexity of, and the computational effort required by, more specific models.

crete dynamic models (Fig. 2.28). Boolean models are discrete (in state and time) and deterministic. Each node can assume one of two states ON or OFF, corresponding to the logic values 1 (active) or 0 (not active, but not necessarily absent). That is why they are known as Boolean or logic models [26]. A Boolean network is a directed graph $G(V,E)$ characterised by a number of nodes N and number of inputs per node K. Boolean networks always have a finite number of possible states, 2^N, and therefore a finite number of state transitions.

As with any other dynamic system, a key aspect of the analysis of such systems is the determination of their steady-state behaviour. The dynamics of a Boolean network are described by a set of rules:

'if input value(s) at time t is(are)..., then the output value at $t + 1$ is...'

So we can define different rules for the same network. For example, Rule 0 says that independent of input, the output is always 0, and in rule 4 it is always 1, while in rule 2 it is negative and therefore the state of node B at time $t + 1$ equals the logical complement of the state of node A at time t. These transition rules are known as the Boolean functions. Depending on the output of the Boolean function, the state of a node can transit from one value to another as the simulation algorithm moves from one iteration to the next. Here, an iteration finishes when all the nodes in the network are updated according to their Boolean function. So Boolean functions consist of a set of rules specifying how the nodes' states change over time as a function of the current or past values of neighbouring nodes. The main operators of Boolean dynamics are AND, OR, and the negation NOT. Before going forward, we need to define some terminology: The state of a network is defined as the row listing the present value of all N elements (0 or 1). Since a Boolean network has a finite number of states 2^N, as the system passes along a sequence of states from an arbitrary initial state, it must eventually re-enter a state previously passed through, which makes for a cycle.

A cycle length is defined as the number of states on a re-entrant cycle of behaviour, and if a cycle contains merely one state, it is called a singleton attractor or equilibrium state; otherwise, it is an attractor cycle.

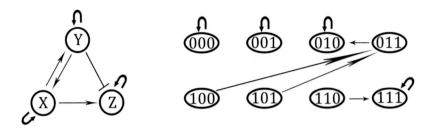

Figure 2.29 A state transition graph of the 3-node system (shown on the left-hand side).

The surrounding region in state space such that all trajectories starting in that region end up in the attractor is called the basin of attraction. A basin of attraction is a mathematical object that can be computed and shown as a graph for small networks. In biology, multiple attractors explain how the same genetic regulatory network can maintain different stable patterns of gene activation as well as the cell types in multicellular organisms. Perturbation can cause the dynamics to jump to alternative attractors. Transient (or run-in) length is defined as the number of states between the initial states and entrance into the cycle.

If we consider a 3-node system with the rules

- $X(t+1) = X(t)$ and $Y(t)$
- $Y(t+1) = X(t)$ or $Y(t)$
- $Z(t+1) = X(t)$ or (not $Y(t)$ and $Z(t)$),

then we can calculate all possible transient states for the given network. We can demonstrate all attractors and their basin of attraction using a graph called a state transition graph or STG (see Fig. 2.29).

So we can see that the number of accessible states is finite, 2^N, and that cyclic trajectories are possible. Not every state must be approachable from every other state. The successor state is unique; the predecessor state is not unique. But what if the rules for updating states are unknown?

2.10.1   Random Boolean Networks

Random Boolean networks (RBNs) are a class of generic discrete dynamic network models. In 1969, Stuart Kauffman proposed using random Boolean networks as abstract models of gene regulatory networks [25], where each vertex represents a gene and the 'on' state of a vertex indicates that the gene is expressed. An edge from one vertex to another implies the former gene regulates the latter, and '0' and '1' values on edges indicate the presence/absence of activating/repressing proteins. Boolean functions assigned to vertices represent rules of regulatory interactions between genes. The RBN model specifies both the topology and the rules for updating states of the network. The network topology is determined at random, subject to whether the in-degree for each node is constant or stochastically determined, given an average in-degree k. It is also possible to bias the network structure, for example, toward a scale-free degree distribution [27]. Each node in the network has a state value, which is updated in discrete time using a deterministic Boolean function of the current state values of the nodes from which it has incoming edges, its neighbours. This rule, by which each node computes its next state, is decided at random for each node when the network is initialised, subject to a probability p of producing 1 (on-state) output. Updating can be done synchronously or asynchronously. In classical RBNs the nodes all update their states synchronously. For example, $p = 0.5$ means that Boolean functions are assigned independently and uniformly at random from the set of 16 Boolean functions of 2 variables. With k connections, there are 2^{2^k} Boolean input functions. Once connections and rules are selected, they remain constant and the time evolution is deterministic. The number and length of their attractors depend on k and p. There is a critical average number of inputs defined as

$$k_c = \frac{1}{2p(1-p)},$$

which if:

- $k < k_c$ then the network will be in a frozen phase: many small attractors, not sensitive to changes in structure;
- $k > k_c$ then the network will be in a chaotic phase: few large attractors, very sensitive to changes in structure;
- $k = k_c$ then the network will be in phase transition behaviour: n attractors of length n, not sensitive to most changes, only to some rare ones.

If you think about the nodes of a network as bulbs, and their connections as wire, then

1. A chaotic network is one where the bulbs keep twinkling chaotically.
2. A frozen or periodic network is one where some flip on and off, but most soon stop.
3. A transition (complex) network is one where complex patterns appear, in which twinkling islands of stability develop, changing shape at their borders.

A network that is frozen or chaotic cannot transmit information and thus cannot adapt.

Gene regulatory networks of living cells are believed to exhibit phase transition behaviours on the border between the frozen and chaotic phases. Kauffman has shown that if $k = 2$ and $p = 0.5$, then the statistical features of RBNs match the characteristics of living cells, where the number of attractors can be interpreted as the number of cell types. The length of attractors can be interpreted as cell cycle time.

When we study a system, our motivation is usually to search for causal relations. Although in everyday life we frequently make causal statements such as 'I couldn't get up on time this morning because I was up late last night', in general we cannot 'see' causal relations but can only infer their existence. Our current systems theory, including all that is taken from physics or physical science, deals exclusively with simple systems or mechanisms. However, complex and simple systems are disjoint categories [28]. Von Neumann thought that a critical level of system size would trigger the onset of complexity, but complexity is more a function of system qualities than size. Complex systems require that all aspects be encoded in order to be more completely understood. This is not possible using only traditional parameter-dependent modelling. Biological networks are very complex and diverse, and we cannot analyse them by relying on statistically based methods alone. In the remainder of the book, you will learn about new ways of studying such complex networks.

2.11   Chapter Conclusions

At the beginning of the chapter we made the point that networks and their dynamics constitute a bridge between causality and algorithmic information theory. Let us elaborate on this, while summarising some key points from this chapter.

First, large amounts of data lend themselves readily to being represented as networks. Either there is naturally a relationship between the variables (nodes), such as being connected or not within a social network, or else, a relationship (an edge) can be directly computed from pairs of nodes (variables). Consider, for example, two measured variables such as calorie intake and weight over time. Computing the correlation in binned temporal intervals would most likely reveal a positive relationship. Note that both these examples are temporal networks, since the relationship in a social network or the calorie-weight relationship may change over time. In short, it is no surprise that networks are abundant in the world, or more precisely, in our attempts to understand the world. In the first half of the chapter we have therefore reviewed some of the major current tools with which to characterise and analyse networks from a statistical standpoint. These include notations (graph elements and different types of graphs; Sections 2.1–2.4) and statistical measures (clustering, diameter, centrality, spectra; Section 2.5). This part ended with an overview of generative network models (Section 2.6) and biological networks (Section 2.7).

Yet, at this juncture it is indeed a natural next step to ask which observations are causes and which observations are consequences (effects) of something else. This is a central question on two counts: for meeting the challenge of producing a minimal generative model explaining observations, and for determining how to control and engineer a networked system. Both these challenges require us to disentangle causes and effects from a web of complex interactions.

Being mindful of these challenges motivates us to distinguish what is really complex from what is not in data. Furthermore, what do we mean by complex, and how do we measure complexity? Before working through these questions with the help of information theory, computability theory, and algorithmic information theory in (Chapters 3 and 4), let us first summarise some relevant observations from the study of networks and dynamical systems.

If there is anything to take away from the theory of dynamical systems (Sections 2.8–2.10) it could possibly be summarised in the adage 'nothing is what it seems'. To unpack this we can reflect upon the transitions from simple to complex and from complex to simple, respectively. Dynamical systems, as in the form of differential equations, essentially capture the time evolution of state variables, which we can think of as nodes. On the right-hand side of the equations we find a description of the dependencies, i.e., edges (in a simplified manner) between the state variables. It was the genius of Newton to formulate the laws of motion and, intriguingly, only a small set of equations were sufficient to capture a vast set of empirical observations. Now, this clearly demonstrates the power of the language of differential equations, a success story that has been repeated in the development of physics. Yet, the discovery of an immense complexity, such as chaos or the generation of fractals, came as surprise. The bottom line was that seemingly very simple dynamical equations, or for that matter iterative maps, was sufficient to produce complex dynamics or patterns not readily predictable from the equations per se. Hence, systems that appeared to be 'simple' could produce complexity. Now, the reversal of this mathematical observation and associated theory is as remarkable. Here, large systems or smaller systems, like the Lorenz equations, could produce a dynamics such that the effective manifold in state-space is effectively a low-dimensional object. Hence, due to the non-linear coupling between the state-variables, simplicity results from the fact that the number of degrees of freedom is reduced. Note that this is also what makes classification and (machine) learning possible, since the objects at hand (e.g., images) effectively only occupy a subset of all possible states.

In conclusion, these lessons should make us cautious about having strong prior expectations about what to expect from a complex or simple looking dynamical system. In limiting regimes, we can find analytic results, as shown in the case of Boolean networks or in mean-field analysis of spin-glass systems in physics. Yet these results are most likely exceptions rather than the rule, at least as far as our current understanding goes. Finally, going beyond this level of analysis, we need to carefully consider what we really mean and should mean by the labels simple and complex. To this end we need to ground our analysis in algorithmic information theory and computability, the subject of the next chapter. Furthermore, once we have taken such a route, we need to reconnect an algorithmic information analysis to a dynamical and computable perspective, i.e., to theory, computations, and algorithms, for an algorithmic information dynamics – the subject of the book.

2.12 Discussion Questions

1. How can graph-theoretical measures be modified to incorporate the temporal aspects of a system?
2. Is there any fundamental difference in the number and basin of attractors between scale-free and ER random Boolean networks?
3. Do you think there is any classification for linear 3D systems using eigenvalues similar to those for 2D systems?
4. Discuss sufficient and necessary assumptions for a dynamical system to exhibit a linear increase followed by oscillatory behaviour for at least one state variable. What is the minimum state variable that this system should have?

5. How do you think algorithmic complexity can be used to analyse this type of system? What if you do not have equations describing the system and you only observe the output of the system close to its attractors?

6. What is the relationship – if any – between the average connectedness of genes and the ability of organisms to evolve?

7. Has evolution somehow selected only networks that are highly ordered circuits, which in themselves are sufficient to insure stability? How about flexibility and evolvability?

3 Information and Computability Theories

Chapter Summary

Thus far, we have arrived at the insight that to decipher causality in data or from observations we need powerful tools for analysing and understanding complex dynamical networks. It would seem that there is information present in complex systems and that its transmission over time could be modelled by dynamical networks. To mine this information in order to extract what's relevant to causality, we need to have an account of what computation is at a fundamental level. Computability is the quintessential theory of mechanistic behaviour, a theory that uses input-output mappings to define rule-based operations of a very general nature that can account for all forms of intuitively feasible computations transforming a discrete input into a discrete output according to a specified symbolic algorithm (even if chosen randomly). Classical information theory, as reviewed in this chapter, is a formalism of a combinatorial nature based on counting, which quantifies communicating bits across a channel. These two parallel streams of formalisms, bits and computation, can be examined to determine the extent to which they do or do not suggest directions our investigations might take as we attempt to develop a foundational formalism for analysing causality. This chapter therefore sets the stage for the subsequent chapter introducing algorithmic information theory, and in particular a dynamical formulation of that theory. Again, our aim is get at the roots of these input-output transformations, but using the language of interventions, programs, and code reconstruction. But before we can arrive at that summit of insight we need to survey Shannon information theory, entropy rate, Turing machines, and cellular automata, which we'll do in the current chapter. This chapter therefore defines the limits of what can be achieved using classical information theory and the limits of what is computable and what is not. We will see how entropy and Shannon entropy are measures of statistical patterns, how they are used to quantify typical or atypical statistical spatial configurations, and how they differ from those of an algorithmic nature based on automata theory. By the end of this chapter you should have a basic grasp of concepts and areas related to statistical mechanics, formal languages, symbolic computation, standard models of computation such as Turing machines and cellular automata, and enumeration methods.

3.1   Information Theory

The second law of thermodynamics, a principle derived from the work of scientists such as Carnot, Clausius, Maxwell, Boltzmann, and Szilard, is a fundamental concept in physics, and in science generally. It describes how energy and everything related to energy, such as heat and work, evolve over time and interact with the environment, how energy is transferred, and how it behaves in closed and open environments.

3.1.1   A Fundamental Probabilistic Principle

To help us understand this principle and its connections to the concept of information, let's perform what is called a thought experiment, that is, an experiment that one can run in one's own mind to reach some sort of logical conclusion. In this case, our thought experiment will involve an actual computer simulation. Let's imagine for an instant that we have a room filled with gas. Gas is made up of particles, so our room would be filled with particles, able to move around freely, colliding with each other, constrained only by the walls of the room. Figure 3.1 shows an illustration of 100 particles (shown in black) inside a square of size 100×100 pixels.

For ease of explanation (see Fig. 3.1) the room can be divided into 100×100 cells, each of the same size, amounting to ten thousand cells, each of which can contain one of the 100 particles. To describe the state of the room at any given time, we would have to provide the bit value of these ten thousand cells. Alternatively, we can simply specify the 100 coordinates identifying the location of the particles. If the room were empty, however, it would suffice to say simply that the room is empty; this would characterise the state of the room in a very simple description. But if the particles are distributed randomly, then one would need to provide the exact coordinate for each particle, and no simpler description would be possible if one wished to represent the state of the room.

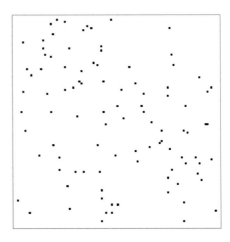

Figure 3.1 Particles represented by bit pixels in a 100×100 square.

Now, quantum physics affords us a means of partially circumventing the need to supply such detailed descriptions that are practically impossible, and completely impossible, even in principle, where very small particles are involved.

So, statistical mechanics makes qualitative assessments instead of providing exact locations. It would characterise the room statistically, in this case as typically 'disordered'. What the second law of thermodynamics tells us is that most configurations will look disordered, and that ordered configurations are unlikely to appear by chance and, if they did, would be unstable and quickly devolve into disordered configurations. The room would then be said to be in a high-entropy configuration. Such a qualitative description is statistical in nature. The idea is that if you plot the distribution of the particles across all the possible spaces they could occupy, the distribution would look the same for most random configurations, and thus anything with such a distribution could be taken to represent a state of disorder.

Figure 3.2 shows what the distribution of the particles in our room looks like when the room is divided into smaller squares of equal size. No particular square seems to have any distinguishing feature, meaning that the particles are, for the most part, distributed evenly. The same is true when the room is divided into 20, 10, or only 5 squares. Scanning the corners and the centre for signs of some distinct arrangement of particles, we come away with the sense that everything looks the same, in whatever direction we look.

Imagine that you were able to place the same number of particles into a smaller room. What would happen is that the same number of particles now have fewer

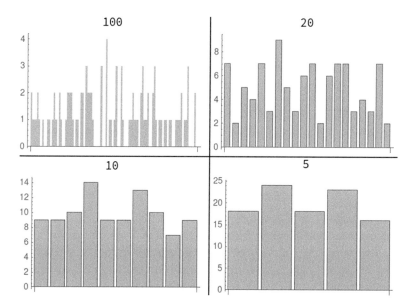

Figure 3.2 Distribution of particles per region for areas of 100, 20, 10, and 5 pixels/cells (i.e., different micro-configuration granularities) showing how evenly distributed the particles are and thus in a high-uncertainty or high-entropy configuration, no region being favoured other than by small random fluctuations.

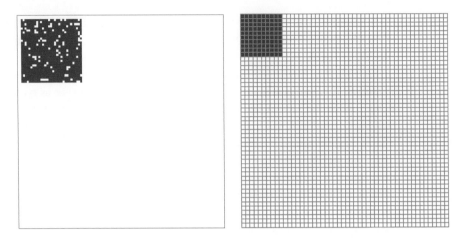

Figure 3.3 Entropy as a function of volume and statistical order. If one forces all particles into a corner, the entropy of the whole block is very low because it does not look disordered.

positions to occupy, fewer degrees of freedom, or fewer cells to occupy or move into. Because there is less room for the particles in the new room, it can be said that the room has a lower entropy compared to the larger one, for the same number of particles. If one reduced the number of particles, that would also lead to lower entropy, because one would need to describe fewer particles. Thus, entropy is a function of at least three elements: the number of particles, the size of the space, and the number of positions the particles can occupy.

Figure 3.3 shows the clustering of the particles when pushing all the particles into a smaller and smaller space in the corner of the room. Eventually the particles would have no place to go and would cluster together in an increasingly small and simple space that can be described by its boundaries in a succinct way. Then the room would be said to be in a low-entropy state because the configuration is not typical, in the sense that it would not be the state of the space if left alone, free of pressure. There are only a few cases among all possible ones in which the statistical distribution looks like the one confining particles to the corner of the room. The distribution plots (comparable to the ones we provided for the random configuration) are shown in Fig. 3.4.

The second law of thermodynamics tells us that if we looked at a room with some gas particles floating about in it, the chances of finding the room in a disarranged or disordered state, that is, in a high-entropy state, would be much greater, because there are many configurations that distribute particles in a random-looking fashion, as against a specific low-entropy arrangement like the one with all the particles confined to a corner. In other words, there are many more possible disordered arrangements than ordered ones, and low-entropy configurations will tend to produce higher entropy configurations by chance if no guided work is performed to position them in a specific (low probability), non-random–looking configuration. Thus, in order to reach a special configuration one has to invest work or apply energy, in this case to push all the particles into a corner.

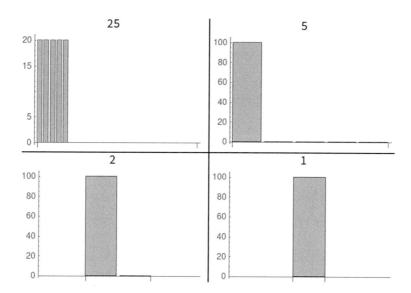

Figure 3.4 High-certainty/low-entropy configuration with most particles in specific areas of the grid, with different granularities emerging after particles are confined in a corner. Moving one particle outside that corner would make entropy increase.

Moreover, if the particles were moving at constant speed, when pushed into a corner they would collide with each other with higher frequency, producing heat that, when released, would naturally cool down the space, establishing a strong connection between energy and temperature.

This is why this law or guiding principle is so important, because it connects several fundamental order parameters. But, as we have seen, it is also all about information, entailing knowledge about the position of the particles and ways to steer them to one side or the other. Thus, we have connected all these measures – of information, heat, disorder, entropy, and temperature.

The second law of thermodynamics tells us that, in general, because particles will tend to maximal entropy, they will also dissipate heat over time instead of gaining it, and they will eventually reach thermodynamic equilibrium by matching the environment's temperature, just as happens with hot drinks, which eventually reach room temperature. Likewise, if you disconnect your fridge, it will reach room temperature soon because of the second law of thermodynamics, even though there is a small likelihood that it will remain cold or grow colder, though the likelihood of these outcomes is so incredibly small that they will almost never occur. It is right then to think of the second law of thermodynamics as a law based on probability and deeply connected to information about the specific and overall configuration of a system.

For a closed system such as our room, the second law of thermodynamics is formally rendered as follows: The increase of entropy represented by S over time t will remain constant or increase but not decrease.

$$\frac{dS}{dt} \geq 0,$$

where S is the entropy of the system and t is time.

3.1.2   Classical Information Theory

We have seen how entropy is defined to quantify the apparent disorder of a room with particles, and how it can provide statistical information on the average behaviour of these particles in the form of a description of their statistical properties. In information theory, the equivalent of our room with gas particles would be sequences or messages composed of symbols, and the question becomes how ordered or disordered a sequence of symbols is.

In the context of messages written in any given human language, the equivalent of a room would be the space of all possible well-formed sentences in that language.

As an example, the sequence of bits

$$s = 11111111111111111111111111111111111111111...$$

looks highly ordered because it is just a repetition of the same symbol, and there is very little surprise when the unfolding sequence continues to comprise 1s. However, the sequence

$$s = 00110100001100110000110000001101111101010100101000...$$

looks less ordered.

It can also be said that it looks more typical, because it is the kind of sequence one would expect to occur randomly. Yet every new bit comes as a surprise because no information from the first segment of the sequence gives any information about what is to come next. This is what one would expect, for example, when tossing a coin with 0 corresponding to heads and 1 to tails, with each toss independent of the others and not affording any clues as to future outcomes.

Notice that as in the case of the room, this kind of order and disorder is of a statistical nature, that is, having to do with how symbols seem to be distributed along the sequence of bits. Two sequences will have the same entropy if they have similar statistical properties and they will have similar statistical properties if the symbols they are composed of occur in the same proportions. In the case of a binary sequence, for example, if a binary sequence has about the same number of 1s as 0s then the sequence will be assigned high entropy.

Here is an example of a binary string with four 0s and six 1s:

$$1001110101$$

And here is a sample of 10 strings that are permutations of the bits of the same string. All these strings will have the same Shannon entropy in the set of all possible binary strings of the same length because all have exactly the same number of 0s and 1s as the original: 1101101010, 0011101110, 1100111001, 1010110101, 0111010101, 0101011011, 1101110010, 1011010101, 1101111000 and 0101010111.

The formula for calculating the Shannon entropy, denoted by the letter H, of a string or sequence s, is traditionally written as follows:

$$H(s) = -\sum_{i=1}^{n} P(s_i) \log_2 P(s_i),$$

where if $P(s_i) = 0$ for some i, then $P(s_i) \times \log_2(P(s_i)) = 0$.

In the case of arrays or matrices, s is a random variable in a set of arrays or matrices according to some probability distribution (usually the uniform distribution is assumed, given that Shannon entropy per se does not provide any means or methods for updating $P(s)$).

We can illustrate its application with the following example:

```
1  sequence="this is not a sentence";
2  StringLength[sequence]
```

```
1  Output: 22
```

Let's take as a sequence something written in English, such as 'this is not a sentence' and look at its distribution of letters.

```
1  Characters[sequence]
```

```
1  Output: {"t","h","i","s"," ","i","s"," ","n","o","t"," ",
2          "a"," ","s","e","n","t","e","n","c","e"}
```

We can then count how many times each symbol, in this case each letter, occurs, effectively producing a discrete distribution of letters from this message. We see that the letter 'a' occurs once, as do 'c', 'h', and 'o', followed by the letter 'i', which appears twice and then 'e', 'n', 's', and 't' , all of which appear thrice in the sentence, with the space appearing four times:

```
1  SortBy[Tally[Characters[sequence]],Last]
```

```
1  Output: {{"a",1},{"c",1},{"h",1},{"o",1},{"i",2},{"e",3},
2          {"n",3},{"s",3},{"t",3},{" ",4}}
```

The distribution looks like that shown in Fig. 3.5, with no particular order on the X axis. It does not look very special but it also does not look uniform, which means that the distribution may have giving up some information about the underlying process.

Then for every symbol s_i, in this case a letter, we can calculate its probability of occurrence in the sentence

```
1  Table[#[[i,2]]/Total[Last/@#],{i,Length[#]}]&@
2   Tally[Characters[sequence]]
```

```
1  Output: {3/22,1/22,1/11,3/22,2/11,3/22,1/22,1/22,3/22,1/22}
```

Finally, we multiply the values by the logarithm of the probabilities and take the sum over all the elements. Thus we have seen how to arrive at the entropy function, which we can write in the Wolfram Language as follows:

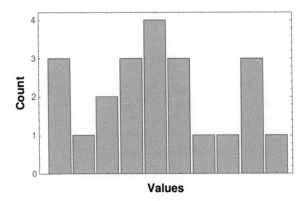

Figure 3.5 Distribution of letters in a short sentence.

```
1  MyEntropy[base_,s_]:=
2  N@Total[Table[-Log[base,#[[i,2]]/Total[Last/@ #]](#[[i,2]]/
     Total[Last/@#]),{i,Length[#]}]&@Tally[Characters@s]]
```

And we can directly obtain a value for the entropy of our sample sequence. In this case, our sequence of letters has a Shannon entropy of about 3.14

```
1  MyEntropy[2,sequence]
```

```
1  Output: 3.14036
```

So, if the distribution is not typical, in the sense of random-looking, which would mean that every letter occurs with the same frequency, what would this distribution and this entropy value tell us? Actually, the sentence in English tells us several things, despite being so short. It already suggests that 's' and 'e' are among the most frequently used letters in English, for instance. Thus we can see how a non-uniform distribution conveys some information about the underlying rules that made our sequence a sentence written in English. We will see more examples of this later.

The reason that a logarithm is used in the formula for entropy is because, besides having some useful mathematical properties such as being the inverse of exponentiation, it is also the result of integrating $1/x$, and because it can transform multiplication into addition, given that the logarithm of a product is the sum of the logarithms of the factors and the logarithm of the ratio or quotient of two numbers is the difference of the logarithms:

$$\log_a xy = \log_a x + \log_a y,$$

$$\log_a x/y = \log_a x - \log_a y.$$

The logarithm also helps in dealing with large numbers, as it is a sort of natural mathematical compression method that works by changing the alphabet base, thereby helping re-encode information in fewer symbols. Let us explain with the following example.

Imagine that you were asked to guess a number x and that the only information you had was that the number fell within the interval $1 \le x \le N$. In this situation, you could ask a series of questions: whether it is the number 1, or the number 2, the number 3, and so on up to N, and on average, if the number was picked at random, you would need about $(N + 1)/2$ questions to guess the correct number. However, the optimal algorithm for guessing a number is something called a binary search algorithm, which finds the number x in time $\log_2 N$ on average, also denoted by O, $O(\log_2 N)$, where O stands for 'order' as in the order of some quantity. Let's say the number is between 1 and 100. The $\log_2$ of 100 is 6.64, so rounding, we obtain 7. Which means that in about 7 'yes' or 'no' (hence binary) answers you may be able to guess such a number.

Suppose the number is 17. Then you could ask the following questions: Is the number...

1. between 1 and 50? – In this case, yes
2. between 1 and 25? – Yes
3. between 1 and 13? – No
4. between 14 and 20? – Yes
5. between 14 and 17? – Yes
6. between 14 and 16? – No

It is therefore 17 and it would have taken only 6 questions to guess. This strategy always works. Try it for yourself, picking some other number. Thus the $\log_2$ of an integer gives the number of binary questions that can encode any number in a decimal representation.

The Wolfram Language has its own entropy function that you can use by just typing:

```
N@Entropy[2,sequence]
```

```
Output: 3.14036
```

The first input parameter in the entropy function is the base, which will change the base of the logarithm in the formula. We will usually use base 2 as we are interested in bits in this book unless otherwise indicated. For sentences in human languages we can take the natural logarithm.

You can verify this function because it is the same as the function implemented in the Wolfram Language, but ours will allow you to modify the associated probability distribution:

```
MyEntropy[2,sequence]==Entropy[2,sequence]
```

```
Output: True
```

3.1.3   Shannon Entropy and Meaning

Shannon's original motivation was to study and introduce measures to quantify the channel capacity needed to send messages through electronic telephone lines. It is often said that Shannon's entropy does not quantify meaning, which is to some degree correct in the context of communication theory, because Shannon entropy seems to disregard the meaning of a message or its actual content. In other words, Shannon entropy is oblivious to the difference between a message such as 'I'll meet you for lunch on Tuesday 2pm', and a possibly random-looking number such as '84592646'. However, there is another angle to Shannon entropy that suggests exactly the contrary, and it is both what makes this measure interesting and also its limitation. Meaning in Shannon entropy is deeply encoded in the form of the underlying assumption, in other words, the context in which a question related to entropy is formulated.

For example, if the number '84592646' in our example is a telephone number, that would be significant because the underlying ensemble for the entropy measure is not the distribution of all possible numbers but the set of valid telephone numbers. Thus Shannon entropy can only distinguish between data if we start out knowing whether a message is a sentence written in English or a telephone number. This information makes meaning relevant to entropy, but at the same time it provides no tools to update the underlying assumptions or to seek to confirm them. We will see later that things are different with algorithmic complexity.

To put it differently, if you have any knowledge of the ensemble for your distribution, then Shannon entropy is all about meaning. For example, the sentence 'I'll meet you for lunch on Tuesday 2pm' only has meaning if you know that it is a sentence written in English, which means knowing that this sentence belongs to a subset of well-formed English sentences. Then the entropy of the sentence becomes significantly lower than if we were to assume that the string was in the space of all possible letters and words, in which case the entropy would be much larger.

The problem with entropy, so to speak, is not that it is unable to convey or capture meaning but that it is ambiguous and fragile, for exactly the same reason as probability distributions are – because Shannon entropy by itself does not provide any means to estimate the probability distributions and hence in practice relies on traditional statistics and the observer's beliefs, knowledge, or lack thereof. In general, one ends up using a general assumption, that is, the uniform distribution that makes entropy a trivial function ends up becoming a symbol-counting function. Indeed, if the uniform distribution is assumed, as it is in most cases, what Shannon entropy measures is the multiplicity of the different symbols used in a sequence, just as its counterpart in physics gauges the number of possible micro-states, such as particles or molecules, in a given volume of space.

Leaving aside these arguments related to meaning and the limitations of entropy, there are interesting properties of entropy worth mentioning and studying. For example, one of the general properties of Shannon entropy is that redundancy does not add new information as one would expect. Once the number of symbols or letters is fixed, the greater the redundancy the lower the entropy. For example, repeating the letter e at the end of some words does not provide any new information than the original sentence, and as a function of the sequence length the entropy drops, as can be seen (Fig. 3.6).

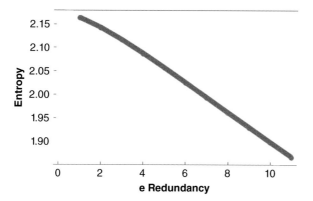

Figure 3.6 Redundancy: A growing sequence with an increasing number of trivial repetitions decreases the information content per letter.

```
1  sequence<>#&/@NestList["e"<>#&,"e",10]
```

```
1  Output: {"this is not a sentencee","this is not a sentenceee",
2         "this is not a sentenceeee","this is not a
3         sentenceeeee","this is not a sentenceeeeee",
4         "this is not a sentenceeeeeee","this is not a
5         sentenceeeeeeee","this is not a sentenceeeeeeeee",
6         "this is not a sentenceeeeeeeeee",
7         "this is not a sentenceeeeeeeeeee",
8         "this is not a sentenceeeeeeeeeeee"}
```

In practice, with no means to calculate or make an educated guess as to the underlying probability distributions, entropy is indeed blind to meaning and is a measure of combinatorial diversity. Hence, a sentence may have the same entropy as some other scrambled version of the same sentence, as long as it uses the same letters. For example, these two arrangements of letters have the same entropy when considering the ensemble of all possible sequences of the same length using the letters of the Latin alphabet:

```
1  N@Entropy["A quick brown fox jumps over the lazy dog"]
```

```
1  Output: 3.07114
```

```
1  rndmsentence=
2  StringJoin[RandomSample[StringSplit["A quick brown fox
      jumps over the lazy dog",""]]]
```

```
1  Output: "jpr mockiufz   yooqru svlbAeohwd xt nge a"
```

```
1  N@Entropy[rndmsentence]
```

```
1  Output: 3.07114
```

This is despite one of the letter arrangements being a sentence and having a meaning in English, because entropy by itself cannot know that this sequence may belong
to some sort of language. However, if we consider only the set of all valid English
sentences in building the underlying probability distribution on which entropy would
operate, then the sentence in English would have a much lower entropy because we
would know that in English it is very rare if not impossible to see the letter j next to
p and r, or to find a word starting with z, and other such things that we could bring to
bear from our knowledge of the distribution of letters in a language like English.

Now, a micro-state is a generalisation of the concept of a letter in a message. A
micro-state can be any unit, such as bits in a binary sequence. Assuming equal probability for all sequences of the same length, we have it that a pseudo-random sequence
of two thousand 0s and 1s produced by the function RandomInteger[ ] in the Wolfram
Language has almost the same Shannon entropy as the highly structured sequence of
01 repeated a thousand times:

```
1  s = RandomInteger[1,2000];
2  Style[StringJoin[ToString/@s],10]
```

```
1  Output:  11111001101011001000000010101011000110010001101111
2           01111100000001000100110100111100011100111110101010
3           10001001111010000010000001000011011000100111110000
4           00000010011111100011110100110010100101110111111101
5           00010000110011001011101001011111011001011000000000
6           11010100100000101001100101001111100111101101001100
7           00100010000100100010101101011110001110011101011011
8           ...
```

```
1  N@Entropy[2,s]
```

```
1  Output:  0.99974
```

```
1  Style[StringJoin[ToString/@Flatten[Table[{0,1},{1000}]]],10]
```

```
1  Output:  01010101010101010101010101010101010101010101010101
2           10101010101010101010101010101010101010101010101010
3           01010101010101010101010101010101010101010101010101
4           10101010101010101010101010101010101010101010101010
5           01010101010101010101010101010101010101010101010101
6           10101010101010101010101010101010101010101010101010
7           01010101010101010101010101010101010101010101010101
8           ...
```

```
1  N@Entropy[2,Flatten[Table[{0,1},{1000}]]]
```

```
1  Output:  1.
```

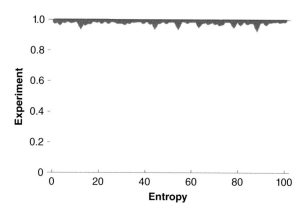

Figure 3.7 A hundred experiments producing pseudo-random sequences of 100 binary digits each, showing how all are close to 1, or maximum entropy.

This is because if taking single bits as micro-states or units for the application of entropy these sequences look equally diverse; they have about the same number of 1s and 0s. So we would need to take two bits as a micro-state in order to have entropy return a low value for the sequence of 01 repeated a thousand times, as would intuitively be expected, given that the second sequence looks ordered and not random.

This is because the intended behaviour of a pseudo-random generating function like RandomInteger[ ], as long as it is a good pseudo-random number generator, is to produce about the same number of 1s and 0s in a disarranged fashion (Fig. 3.7). So the Shannon entropy of these sequences will be log 2:

```
N@Log[2,2]
```

```
Output: 1.
```

with a very small standard deviation among several trials, that is, many different pseudo-random arrangements:

```
N@StandardDeviation[
  Entropy[2,#]&/@Table[RandomInteger[1,100],{100}]]
```

```
Output: 0.011018
```

Let's take another example. This last example on binary strings can help us understand another process. The entropy of a sequence resulting from tossing a coin is maximised when the coin is fair, producing about the same number of heads as tails, that is, the two possible outcomes have equal probability 1/2 or 50:50,

$$P(X = \text{head}) = \frac{1}{2}, \qquad P(X = \text{tail}) = \frac{1}{2}.$$

This is the situation of maximum uncertainty, as the outcome of tossing an unbiased coin is the most difficult to predict. The result of each toss of the coin is said to deliver one full bit of information because you have no idea what will come next. However, if

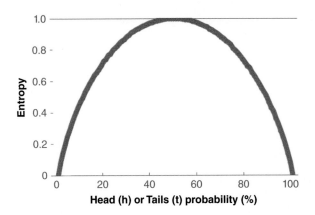

Figure 3.8 Shannon entropy of a binary random (Bernoulli) process as a function of outcome probability, called the binary entropy function. A function of its probability is minimum when its probability is 0 or 1 and maximum when its probability is 0.5 when it has an entropy of 1 bit. Intuitively, if a measurement is always false (or always true) there is full certainty, but if the outcome is as true as often false, the uncertainty is maximised.

the coin was biased and came up heads all the time, then each new toss would have an entropy of zero because there is no surprise; you know what you have a good chance of getting. Every time the coin is tossed, one side is more likely to come up than the other. This reduced uncertainty is quantified by entropy.

Maximum fairness in a random process, such as tossing a coin, is reached when the coin produces the same number of heads and tails, that is, when there is a 50% probability of being heads or tails when uncertainty is maximal, as illustrated in Figure 3.8. Note that the maximum value of the graph in the formula for entropy depends on the logarithmic base and the distribution. Here, the entropy is at most 1, as we are assuming a uniform distribution and base two, for which the result is said to be in bits.

From this plot, you can also see that entropy is a function of symbol density or symbol count. Here are some examples of binary and ternary sequences where both the number of symbols and the number of repetitions determine their Shannon entropy, both for natural and binary logarithms:

```
Entropy[2,#]&/@{"010101010101","000000111111","10","11000011
     ","012","0122"}
```

```
Output: {1,1,1,1,Log[3]/Log[2],3/2}
```

It is clear that, for this particular case, when taking every bit as the micro-state for entropy, the arrangement of the 0s and 1s is irrelevant as long as the number of 0s and 1s remains the same. Entropy values also vary significantly as a function of the number of potential symbols available. One way to tell apart cases such as a repetition of 0 and 1 in alternation or randomly arranged, having basically the same entropy as in this example, is by taking different micro-state lengths or coarse-graining the data, in this case a sequence, so that entropy can find the ordered nature of the sequence of alternating 0s and 1s. Taking units of two bits we ascertain that the entropy is indeed zero:

```
Flatten[Table[{0,1},100]]
```

```
Output: {0,1,0,1,0,1,0,1,0,1,0,1,0,1,0,1,0,1,0,1,0,1,0,
         1,0,1,0,1,0,1,0,1,0,1,0,1,0,1,0,1,0,1,0,1,0,1,
         0,1,0,1,0,1,0,1,0,1,0,1,0,1,0,1,0,1,0,1,0,1,0,
         1,0,1,0,1,0,1,0,1,0,1,0,1,0,1,0,1,0,1,0,1,0,1,
         0,1,0,1,0,1,0,1,0,1,0,1,0,1,0,1,0,1,0,1,0,1,0,
         1,0,1,0,1,0,1,0,1,0,1,0,1,0,1,0,1,0,1,0,1,0,1,
         0,1,0,1,0,1,0,1,0,1,0,1,0,1,0,1,0,1,0,1,0,1,0,
         1,0,1,0,1,0,1,0,1,0,1,0,1,0,1,0,1,0,1,0,1,0,1,
         0,1,0,1,0,1,0,1,0,1,0,1,0,1}
```

```
N@Entropy@Flatten[Table[{0,1},100]]
```

```
Output: 0.693147
```

```
randomintseq=RandomInteger[1,200]
```

```
Output: {0,1,0,1,1,1,0,0,0,0,1,1,0,1,0,1,1,1,0,1,0,1,0,
         0,0,1,0,0,0,1,1,0,0,0,1,1,0,1,0,0,0,0,1,0,1,0,
         1,1,0,1,1,1,0,1,1,0,0,0,0,0,1,0,1,1,0,1,1,0,1,
         0,1,0,0,1,1,0,1,1,1,1,0,1,0,0,0,0,1,0,0,0,1,0,
         1,1,1,1,0,0,0,0,1,0,1,1,1,1,0,1,0,0,1,0,0,1,0,
         1,0,1,0,1,1,0,0,0,1,1,1,0,1,1,0,0,1,1,0,1,1,0,
         0,1,0,0,1,0,1,0,1,1,0,1,1,1,0,1,1,1,0,1,0,1,0,
         1,0,1,1,1,0,1,0,1,0,0,0,0,0,1,0,0,1,1,1,0,1,1,
         0,1,0,1,1,1,1,1,0,1,1,0,0,1,0,0}
```

```
N@Entropy@randomintseq
```

```
Output: 0.692947
```

```
N@Entropy@Partition[Flatten[Table[{0,1},100]],2]
```

```
Output: 0.
```

But if we do the same with the random-looking case, we get a value that diverges from that for the ordered case:

```
N@Entropy@Partition[randomintseq,2]
```

```
Output: 1.3643
```

Which indicates that it is truly random – or is it? We will see that entropy, by itself, can only see one simple kind of randomness, statistical randomness, which is what we have been exhibiting by changing the coarse-graining. This type of randomness is periodic regularity.

So entropy is not only about the internal syntactic structure of a message but also about how much we know about the underlying ensembles and assumed distributions, and hence in a way is highly epistemological, although reduced to syntax in practice in the face of a complete lack of information.

3.1.4   Joint, Conditional and Mutual Information, and a Case Study

Joint Entropy

It is important to understand how entropy works in the face of multiple variables because one of the main applications of classical information theory is in attempting to find common or shared statistical properties across different objects. While we will use some basic concepts from entropy, we will also be replacing it in some applications with measures of an algorithmic nature – as opposed to measures of a statistical nature, which entropy is. Others have already done a great job describing the many versions and applications of entropy, so this is only a brief overview of the ways in which entropy and statistical information can be helpful to us when we move to algorithmic complexity.

In traditional information theory, one of the things that can be done is to ask what kind of similar statistical properties two or more objects may share. Joint entropy is a measure associated with a set of random variables. A random variable can be a binary sequence, with the events of the sequence its bits. Joint entropy is simply the entropy of two variables:

$$H(X,Y) = -\sum_x \sum_y P(x,y) \log_2 P(x,y).$$

And, in general, it can be written as follows for any number of variables:

$$H(X_1,\ldots,X_n) = -\sum_{x_1} \cdots \sum_{x_n} P(x_1,\ldots,x_n) \log_2 P(x_1,\ldots,x_n).$$

For example, we may want to know the entropy of two random variables related to weather. Let's say that we are interested in whether today will be a sunny or a rainy day, and whether it will be hot or cool, and that we have empirically calculated these probabilities based on the records of previous days and years:

$$P(\text{sunny},\text{hot}) = \frac{1}{2},$$

$$P(\text{sunny},\text{cool}) = \frac{1}{4},$$

$$P(\text{rainy}, \text{cool}) = \frac{1}{4},$$

$$P(\text{rainy}, \text{hot}) = 0.$$

Then the joint entropy of the variables X and Y according to the joint entropy formula is as follows:

$$H(X, Y) = -\left[\frac{1}{2}\log\frac{1}{2} + \frac{1}{4}\log\frac{1}{4} + \frac{1}{4}\log\frac{1}{4} + 0\log 0\right] = \frac{3}{2},$$

where $0\log 0 = 0$ by convention. The value indicates that these variables are not independent because the joint entropy is different from the entropies of the single variables alone. This is because there is a greater chance of it being a hot day if it is sunny than if it is rainy, and if it is rainy it is also more likely to be cooler, according to the empirical distributions that we assumed and calculated in this example. Something important to notice is that to apply entropy one needs to be able to properly calculate these joint distributions.

Conditional Entropy

Now, if you wished to know how much of the statistical properties of a certain sequence X you can guess from the statistical properties of another sequence Y, then we are in the realm of what is known as conditional entropy. Conditional entropy quantifies the amount of information needed to describe the outcome of a random variable Y given the value of another random variable X. For example, let's say we have two sequences, as follows:

$$X = \{1, 1, 1, 0, 1, 1, 0, 0\}, \qquad Y = \{1, 0, 0, 1, 0, 1, 0, 0\}.$$

Their conditional entropy should be very high because the sequences have very similar properties. You can write a function for conditional entropy in the Wolfram language as follows:

```
Mean[Entropy[2,Pick[Y,X,#]]&/@X]//N
```

```
Output: 0.951205
```

As extreme cases, the conditional entropy $H(Y|X) = 0$ if and only if the value of Y is completely determined by the value of X. Conversely, $H(Y|X) = H(Y)$ if and only if Y and X are independent random variables, which means that X does not provide any information about Y and thus seeking the entropy of Y given X is simply equal to asking for the entropy of Y alone, directly, with no access to X.

Mutual Information

The mutual information of two random variables or random sequences, as defined in terms of Shannon entropy and denoted by the letter I, is a measure of the variables' common statistical properties, that is, the amount of statistical information that one variable may convey about another. In terms of joint entropy, I can easily be defined

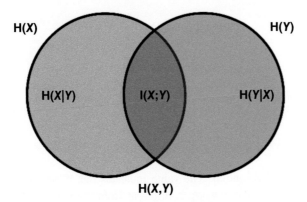

Figure 3.9 You will find this useful explanatory Venn diagram in most textbooks about information theory, to help explain each of these measures and their relationships.

by the joint entropy of X and Y minus the conditional entropy of X given Y minus the conditional entropy of Y given X. That is, all the information that can be known from knowing each variable (or sequence) from the other.

Mutual information is commutative, meaning that the order of the sequences does not matter. Whatever you can guess of X from Y you can do likewise for Y from X. This will be a problem that entropy carries with it. The fact that it is a symmetrical function means that it does not allow you to determine a causal direction, only correlation. There are ways to correct this by introducing time, but this will only partially fix the problem because entropy will miss anything that is not of a statistical nature, as we will see later, such as algorithmic connections between the variables.

Putting these Measures Together

Conditional entropy and mutual information can be thought of as complementary operations. $H(Y|X)$ is a measure of what X does not say about Y or the amount of information remaining about Y when X is known; and $I(X;Y)$ is a measure of what Y says about X or how much information X shares with Y (see Fig. 3.9).

Conditional Mutual Information and an Example

In this book and in science generally, we are interested in causal relationships, that is, in what gives rise to phenomena or events. What is the cause of lung cancer? Do people who smoke have a greater chance of developing lung cancer? Obviously, all this brings to mind mutual information and conditioning. For example, you can think about whether knowing that someone smokes two boxes of cigarettes a day will tell you anything about that person's chances of getting cancer, and you can calculate all the joint probabilities and apply information theory to arrive at an answer.

Imagine that you perform an experiment where you find out that people who are more stressed also have a greater chance of developing cancer. But what happens in reality is that people who smoke are more likely to smoke more when they are already stressed so, at best, stress is only indirectly a cause of cancer, not a direct cause, because

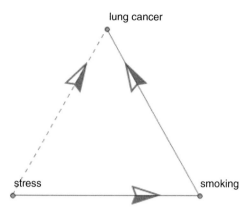

Figure 3.10 An oversimplified graphical example of a causal model diagram.

there may be stressed people who do not resort to smoking to relieve their stress. It is often the case that one cannot isolate causes in cases such as this due to lack of control over the variables. But by using conditional mutual information one may sometimes be able to tell direct from indirect causes. For example, to know what is statistically more informative, or to know whether being stressed or being a smoker increases the risks of developing lung cancer are matters we can explore with classical information theory and Shannon entropy. We basically need to unpack the causal scenario where we assign smoking the role of direct cause of cancer while at the same time wishing to test whether being stressed is equally informative or is only an indirect cause, the result of stressed people electing to smoke (Fig. 3.10).

The first thing to notice is that in this toy example there are only three random variables, which in this case may be binary for illustration purposes, but they can be given weights too. The variables are:

1. To be a smoker (yes and no)
2. To be stressed (yes and no)
3. To have lung cancer (yes and no)

If we want to test whether being stressed or being a smoker poses the greater risk of developing lung cancer, we can write both statements as follows:
Possible cases:

- $H(cancer, stressed) > H(cancer, smoker)$
- $H(cancer, stressed) < H(cancer, smoker)$
- $H(cancer, stressed) = H(cancer, smoker)$

The expectation may indicate that $H(cancer, stressed) > H(cancer, smoker)$. If the latter, then this specific test may fail to indicate the direct cause and we would need to come up with a more sophisticated one. One question to ask would be whether such an equality would tell us something about the entropy of being a smoker if one were already stressed versus if one was a smoker before one became stressed.

Clearly, if H(smoker, stressed) = H(stressed, smoker), then both would be equally informative, but would this imply that virtually every smoker is stressed and vice versa? Perhaps you can think about this.

What we may clearly derive from the evidence is that

$$H(cancer, smoker) > H(cancer, stressed),$$

which may mean that the association of cancer with smoking is greater than with being stressed. We know, however, that stressed people and smokers are not disjoint sets; many stressed people elect to smoke. So think about whether it makes sense to ask about the mutual information of cases such as I(cancer, smoker | stressed) versus only I(cancer, smoker), and what values of which would tell us something about the relationship among these three variables.

This approach can get more sophisticated with ideas from Judea Pearl and his do-calculus [29], but we can see how classical information theory and Shannon entropy provide a language for dealing with probability distributions and for posing this kind of question within a statistical framework. We can also see that how good an answer we obtain depends entirely on how much we know or how well we can approximate all the empirical distributions involved.

3.1.5   Entropy Rate, Languages, and Multidimensional Data

A first and popular application of Shannon entropy is in the study of the statistical properties of human languages. We owe to Shannon himself this very first application, contemporaneous with his introduction of his new ideas on communication theory. In linguistics, entropy is closely related to language redundancy and the frequencies of occurrence of certain combinations of letters and words, also called 'collocations' and technically known as 'grams'. n-grams can be letters, pairs of letters, triplets, words, or even sentences, and are the cornerstones of a kind of coarse-graining approach to studying languages. They are very useful in understanding entropy and classical information theory.

For example, the entropy of a language is an estimation of the probabilistic information content of each letter in that language, and hence is also a measure of its predictability and redundancy.

One can crack codes by simply plotting the distribution of words used in written or spoken language. One can think of a human language as a sort of encryption code used to describe the objects and events comprising human reality. If you wanted to crack such a code or language from scratch, one strategy would be to plot the distribution of words because there would be a strong tendency in any two languages to use words with about the same frequency when describing the same or a similar human reality. For example, if both languages have articles and prepositions, you might easily identify and match them even from a very short text because they are likely to be the most frequently used words, even if you may not know which are the articles and which the prepositions.

Figure 3.11 Word clouds illustrating word frequencies in two different languages. As it turns out, those words that have cognates across different languages will tend to occur with about the same frequency. Cognates can offer clues of what a word may mean in another language, even without any semantic information about it. This is at the core of classical cryptography and of information theory in linguistics. Just as words can be ranked, so can letters, syllables (called *n*-grams in computational linguistics), and even word collocations.

Here, for example, we plot two word clouds from the distribution of words in the United Nations Declaration of Humans Rights in English and Spanish. The size and colour of the word corresponds to the frequency of that word in its corresponding text. One can see that when running parallel texts, one finds a strong correspondence among word meaning and word size. And this would even work across different texts, not just a text and its translation. On average, the same words will have the highest frequencies. In this particular case, the top 20 most frequent words in these short texts reveal a lot about the structure of both languages. Through their respective word frequencies alone, each helps reveal the meaning of the other (see Fig. 3.11).

This frequency analysis is actually one of the most basic strategies used to crack codes, because any minimal information about an encrypted message will yield some of the secrets of its distributions. And this is also the basis of some theorems in classical information theory that, for example, guarantee perfect secrecy by replacing high frequency words with low frequency words when choosing a code to conceal the correspondence in distributions. But perfect secrecy is actually very difficult to achieve, and usually any human error leads to the eventual cracking of the code.

Coarse-grained and Multidimensional Entropy

As we said before, languages can be studied in terms of their distributions of letters or words, hence the choice of basic unit is very important. It is natural to ask what the entropy of a string would look like when focusing on blocks of different numbers of letters, from one letter to two letters, and so on. To illustrate this, let's take the periodic sequence of alternating 0s and 1s

01010101...01

As we have seen before, assuming a uniform distribution, this sequence would have maximal entropy, suggesting that it is random if we take single bits as units or micro states – because we know that there are as many 0s as 1s in this sequence. One bit is the finest granularity possible. However, this sequence should clearly not be random, and taking two bits at a time as blocks would capture its regularity.

Thus, it would seem that there are some granularities or blocks in a sequence the application of entropy to which would capture the statistical regularities of the sequence, and others where it will fail to do so. How then are we to deal with this high dependency on granularity or block size? One way around this problem is to always seek the regularity that minimises the entropy of the sequence. We will call this process of studying the entropy of an object as a function of increasing granularity the 'entropy rate', and to each of the values of entropy for different granularities or block lengths we will apply the term 'block entropy'. Notice, however, that variations of the same concept have different names, despite being identical or based on similar ideas.

If one looks at these plots, for example, one can see that for the string of alternating ones and zeros, there are coarse-graining lengths that capture the periodicity of the string, and others that occlude it, so that the minimums of entropy across all possible block sizes up to half the size of the sequence would capture a statistical regularity.

The best version of Shannon entropy can therefore be delivered by a function of variable block size where the minimum value best captures any (possible) periodicity of a string (and is actually one way to understand the basics of popular lossless compression algorithms). This can be illustrated with three strings of length 12 bits each as in Fig. 3.12, one regular, one periodic, and one random-looking. When there is some regularity, this is captured by a specific block size. Notice that one should never take minimums when the block size is equal to or larger than half the length of the string, because that results in only one block and thus automatically an entropy value of zero, but the entropy of a single block has no meaning and should not be taken into account.

One other useful application of entropy in our context is to multidimensional objects. For example, we will sometimes apply entropy to bi-dimensional objects where the granularity analysis will consist in breaking down a large array into smaller arrays or blocks. Here is an example of an array of 24 by 24 bits decomposed into smaller arrays or blocks of size 6 (Fig. 3.13).

```
1  ArrayPlot[block=RandomInteger[1,{24,24}]]
```

```
1  Grid[Map[ArrayPlot[#,ImageSize->Tiny]&,
2    Partition[block,{6,6}],{2}]]
```

We will do this because some of the methods that we will introduce may only deal with small blocks and we would wish to be able to compare Shannon entropy to these other new methods in order to approximate algorithmic complexity.

The application of entropy to objects other than strings and sequences will be exactly the same.

```
1  ArrayPlot[RandomInteger[{1,1},{10,10}],Mesh->All]
```

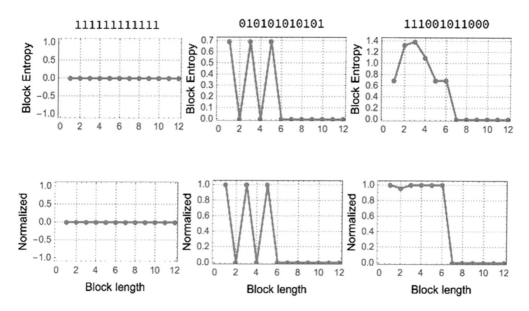

Figure 3.12 Three types of strings with different features that can or cannot be captured by block entropy, that is entropy over all microstate lengths. The maximum value is an indication of the statistical randomness of a string, while the minimum values up to half the length of the string (X-axis) are an indication of its lack of statistical randomness, i.e., the presence of some regularity or cyclicity – in the case of the leftmost string of period 1, the middle string with a cycle of 2 and the last one with no apparent cycle.

Figure 3.13 An example of decomposition by blocks from an original square block of length 24 to square blocks of length 6.

No matter how large a block like the one in Fig. 3.14, it will always have an entropy equal to zero:

```
Entropy[RandomInteger[{1,1},{100,100}]]
```

```
Output: 0
```

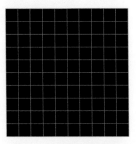

Figure 3.14 Applying entropy on a block of cells of a single colour like this one will have an entropy equal to zero.

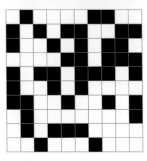

Figure 3.15 A random-looking square block of bits.

And this will, of course, be the case for any block granularity:

```
Entropy/@Partition[RandomInteger[{1,1},{100,100}],{10,10}]
```

```
Output: {0,0,0,0,0,0,0,0,0,0}
```

However, as soon as the object displays some diversity, Shannon entropy will diverge from zero, as in Fig. 3.15.

```
ArrayPlot[randomarray=RandomInteger[{0,1},{10,10}],Mesh->All]
```

```
N@Entropy[randomarray]
```

```
Output: 2.30259
```

3.1.6   Redundancy, Noise, and Biological Information

We have now learned how to measure entropy. One other contribution of the theory of information is to the understanding of the concepts of noise and redundancy and how noise and redundancy counteract and complement each other.

Imagine that you were given a message in a language in which every bit is fundamental if the message is to be understood. Such a language would be very fragile because the loss of a single bit would mean that you lose the entire message. Fortunately, most human languages are not of this type, instead being highly redundant. That's the reason that you can still understand a written or spoken message even when some letters or perhaps entire words are lost or deleted.

This may explain why some languages use grammatical gender, articles, and other linguistic features. They may be a way to add redundancy to the language, even though they may not appear fundamental in any intrinsic sense.

Let's take a piece of text from a popular source, such as an excerpt of 100 words from chapter one of Jane Austen's 'Pride and Prejudice', and let's replace every letter 'u' with an underscore:

```
Style[StringReplace[
  StringJoin[
    Riffle[Rest[
      Rest[StringSplit[ExampleData[{"Text",
      "PrideAndPrejudice"}], " "]]][[1 ;; 100]], " "]],
      (Characters[#] -> "_") & /@ {"u"}], Bold, 16]
```

```
Output: It is a tr_th _niversally acknowledged, that a
        single man in possession of a good fort_ne, m_st
        be in want of a wife. However little known the
        feelings or views of s_ch a man may be on his first
        entering a neighbo_rhood, this tr_th is so well
        fixed in the minds of the s_rro_nding families, that
        he is considered the rightf_l property of some one
        or other of their da_ghters. "My dear Mr. Bennet,"
        said his lady to him one day, "have yo_ heard that
        Netherfield Park is let at last?" Mr. Bennet replied
        that he had not. "B_t it
```

It is certain that you will still be able to recover all of the text with absolutely no loss of information or meaning, because you are able to mentally fill in the gaps. In this case we removed only one of the lowest frequency vowels, but what if in addition to the letter 'u' we removed the letter 'e', which is the most frequently occurring letter in English?

```
Style[StringReplace[
  StringJoin[
    Riffle[Rest[
      Rest[StringSplit[ExampleData[{"Text",
      "PrideAndPrejudice"}], " "]]][[1 ;; 100]],
      " "]], Characters["ue"] -> "_"], Bold, 16]
```

```
Output: It is a tr_th _niv_rsally acknowl_dg_d, that a
        singl_ man in poss_ssion of a good fort_n_, m_st
        b_ in want of a wif_. How_v_r littl_ known th_
        f__lings or vi_ws of s_ch a man may b_ on his
        first _nt_ring a n_ighbo_rhood, this tr_th is so
        w_ll fix_d in th_ minds of th_ s_rro_nding famili_s,
        that h_ is consid_r_d th_ rightf_l prop_rty of
```

```
8      som_ on_ or oth_r of th_ir da_ght_rs. "My d_ar
9      Mr. B_nn_t," said his lady to him on_ day, "hav_
10     yo_ h_ard that N_th_rfi_ld Park is l_t at last?"
11     Mr. B_nn_t r_pli_d that h_ had not. "B_t it
```

Now things start getting more difficult, but still, with a little extra effort one can fully recover the text. However, there is always a point of inflection when recovering information or meaning becomes impossible. The following text, for example, can still be recovered by hand, and even faster with the help of a computer looking for matches with real words in English to complete the missing letters, but the point of no recovery is not much farther away than this:

```
1  Style[StringReplace[
2    StringJoin[
3      Riffle[Rest[
4        Rest[StringSplit[ExampleData[{"Text",
5          "PrideAndPrejudice"}], " "]]][[1 ;; 100]], " "]],
6        Characters["aeiou"] -> "_"], Bold, 16]
```

```
1  Output: It _s _ tr_th _n_v_rs_lly _ckn_wl_dg_d, th_t _
2          s_ngl_ m_n _n p_ss_ss__n _f _ g__d f_rt_n_,
3          m_st b_ _n w_nt _f _ w_f_. H_w_v_r l_ttl_ kn_wn
4          th_ f__l_ngs _r v__ws _f s_ch _ m_n m_y b_ _n
5          h_s f_rst _nt_r_ng _ n__ghb__rh__d, th_s tr_th
6          _s s_ w_ll f_x_d _n th_ m_nds _f th_ s_rr__nd_ng
7          f_m_l__s, th_t h_ _s c_ns_d_r_d th_ r_ghtf_l
8          pr_p_rty _f s_m_ _n_ _r _th_r _f th_r d_ght_rs.
9          "My d__r Mr. B_nn_t," s__d h_s l_dy t_ h_m _n_
10         d_y, "h_v_ y__ h__rd th_t N_th_rf__ld P_rk
11         _s l_t _t l_st?" Mr. B_nn_t r_pl__d th_t h_
12         h_d n_t. "B_t _t
```

It turns out that the ultimate lower limit to this process of deletion, the point of no recovery, is given by Shannon's entropy. How removed a language is from that point of no recovery is referred to as its redundancy.

Redundancy in a message is related to the extent to which it is possible to compress the message. Our process of deleting letters is a form of compression, and as long as we are able to recover the original message, then that compression is said to be lossless.

What traditional lossless data compression does is to reduce the number of bits used to encode a message by identifying and eliminating statistical redundancy. When we compress data without losing any information, we remove redundancy. When we compress a message, what we do is to encode the same amount of information using fewer bits. So we end up having more Shannon information per symbol in the compressed message. A compressed message therefore looks less predictable because we have deleted all redundancies, such as repetitions and other statistical regularities.

transmission

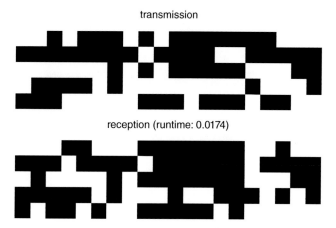

reception (runtime: 0.0174)

Figure 3.16 Here, we have two messages, one highly structured and one random looking. We can add noise and see the structured message begin to look random. To counteract this we can add redundancy, and thereby recover the original message.

On the one hand, the so-called Shannon source coding theorem states that a lossless data compression scheme cannot compress messages, on average, beyond the limit of one bit of Shannon's information per bit, that is, when the message looks random and all its elements are equally distributed. On the other hand, redundancy may often be a desirable feature in data, because data are always subject to errors in transmission.

These errors in transmission or additions to a transmission are usually referred to as 'noise', and Shannon's noisy-channel coding theorem establishes a fundamental trade-off between redundancy and noise. It tells us how much redundancy one needs in a message to accommodate a certain level of noise.

Figure 3.16 shows some code illustrating the theorem. The code introduces artificial noise, destroying the message, but one can increase the redundancy of the message in order to counteract the effect of the noise and thereby reconstruct the message in full, with no error on the receiving end. Therefore, increasing the noise in a message will inevitably make it lose information, but increasing the redundancy will increase its resiliency and robustness even in the face of additive noise.

Let's come back to human languages. One consequence of redundancy is that letters and sequences of letters will have different probabilities of appearance. If we assume we are dealing with 27 characters (that is, 26 letters plus the empty space), and that all of these characters are equally probable, then we will have an information entropy of about 4.8 bits per character. But we know that the characters are not equally probable. For instance, the letter 'e' has the highest frequency of occurrence, as we noted above, while the letter 'z' has the lowest. This is related to the concept of redundancy, which is none other than the number of constraints imposed on a text in the English language. For example, the letter 'q' is most of the time, if not always, followed by 'u' in English, and we also have rules such as 'i before e except after c', and so on.

The program created and linked to in Fig. 3.17 shows the frequency of sequences of *n* letters calculated from the text of the Universal Declaration of Human Rights in

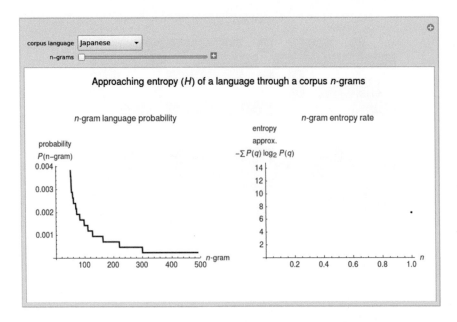

Figure 3.17 Prediction and entropy of languages based on *n*-gram distributions. Program by H. Zenil at https://demonstrations.wolfram.com/PredictionAndEntropyOfLanguages/.

20 languages, and illustrates the entropy rate calculated by taking different word block sizes. The entropy rate shows the entropy change as a function of the number of letters. The entropy of a language can be seen as an estimation of the probabilistic information content of each letter in that language.

By playing with this computer program to estimate the entropy of *n*-grams for 20 languages, you can see how different languages have different entropy rates, indicating different rates of redundancy.

Let's now turn our attention to biological information. How does biology store information? One main repository is of course the genome and the DNA. In some fundamental sense, DNA can be considered the source code of a living organism. An organism arises out of DNA and its interaction with the environment. So what about the redundancy in DNA? Would you expect to find redundancy in DNA? How much can one tamper with DNA before producing important changes in the phenotype? This is one of the main questions in molecular biology and genetics, and one of the answers is that different regions of the genome have different degrees of redundancy, according to whether or not they are under evolutionary selective pressure, that is, whether or not a DNA region plays an important role in the unfolding of the organism's biological development.

It turns out that, in general, the genome, and almost every repository of biological information, is highly redundant. Not only does the genome have many copies of the same genes, but also many proteins are encoded by several genes, and so on. Among the purposes of redundancy is that many genes can produce a stable number of proteins,

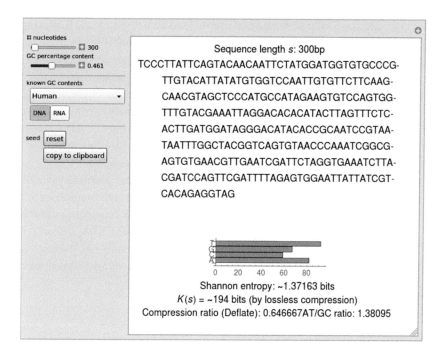

Figure 3.18 Generation of pseudo-random sequences of four letters, representing each nucleotide in the DNA by taking into consideration GC content, which is the number of Gs and Cs occurring in the sequence. Known GC content can also be chosen for popular organisms, including many mammals, but also a few bacteria, ranging in GC content from 20% to almost 52%. Even though the human genome GC content can vary from 35% to 60% from chromosome to chromosome, the average human genome GC content is 46.1%. Program by H. Zenil available at https://demonstrations.wolfram.com/ GeneratingRandomDNASequences/.

so that small variations to the source code have no significant effect. This is a highly desirable property in biology, conducive to biological robustness. Mistakes always happen in the real world, so the fact that biological organisms are nevertheless very resilient may therefore be explained by information theory.

The production of essential proteins in plants and animals is subject to different resilience rates and so different kinds of mutations that occur all the time as a consequence of copying errors or things like free radicals in the environment. Interestingly, this is reflected in the Shannon entropy of the genomes of different plants and animals. Indeed, each organism has a specific redundancy known as GC content, which is the number of times that two of the four nucleotides in the DNA are repeated in the genome. This repetition of GC content is closely related to entropy because it represents redundancy. And it turns out that each species has a specific GC content or redundancy, so that if two species have about the same rate of redundancy in GC content because they are evolutionarily related, then they will also have about the same genomic entropy.

With this computer program we can simulate that redundancy by creating artificial DNA sequences with exactly the same GC content and entropy as in real genomes, and we can do so for a wide range of species, and both for DNA and RNA (Fig. 3.18). And we can play with this GC content directly and see what it would mean to have a DNA sequence with GC content at 100% and how that would hurt the code of life by making it significantly less expressible because it would be unable to encode the same number of proteins with just two letters instead of four. On the other hand, just as for languages, having no redundancy at all is dangerous because errors in communication or storage may be fatal, and DNA segments key to the production of certain proteins would be unrecoverable. Hence, nature and organisms are always being optimised by evolution to find the sweet spot between redundancy and efficiency.

In this piece of software that we have written, and which is available online, we can generate pseudo-random sequences of four letters, representing each nucleotide in the DNA by taking into consideration GC content. GC content can also be chosen for popular organisms, including many mammals, but also a few bacteria, ranging from 20% to almost 52%. Even though the human genome's GC content can vary from 35% to 60% from chromosome to chromosome, the average GC content of the human genome is 46.1%.

3.2   Computability

3.2.1   Mechanistic Descriptions, Prediction, and Causation

In what way are Turing machines relevant to causality? Absolutely everything we do today, such as predicting and calculating solar or lunar eclipses with almost pinpoint precision, or estimating planetary motion so that engineers can achieve impressive accuracy in trajectory planning when sending probes to other planets, or the amazing achievements in weather forecasting, are based on continuous models that purportedly model continuous natural systems. Yet they are all, absolutely all, simulated in and numerically solved by digital computers that are fundamentally discrete, and in no essential way different from something as simple as a Turing machine.

The Turing machine model has been very successful because it has a very simple mechanical description. One can think of a Turing machine as some sort of actual device consisting of a head reading and rewriting the contents of a tape and performing different operations that we call states. The head can move to one side or the other according to certain basic rules.

3.2.2   Undecidability

The Turing machine idea emerged in response to a fundamental question in science asked by the famous mathematician David Hilbert at the Sorbonne. Hilbert was somehow certain that there had to be a way to automatise mathematics in such a way that every possible formula in a theory could be proven to be either true or false. In other words, he was proposing, although not explicitly, to build some sort of machine that could in effect produce all past and future theorems. In effect he asked: What is the power of mechanical reasoning, whether human or machine-like?

It took some time to get some answers, but eventually thy were forthcoming from mathematicians Kurt Gödel and Alan Turing, among others. Gödel's answer was in the negative, proving that there was no way to derive all possible truths from a theory without falling into contradiction.

Alan Turing arrived at the same answer by other means, showing that if one assumed that one could construct a machine to prove all mathematical theorems, then one fell into a contradiction, as we will see later.

But unlike Gödel's negative answer to Hilbert's question, Turing's negative answer bore a surprising secondary positive result: it led to the inauguration of a whole new field of science, what we know today as computer science.

3.2.3   The Turing Machine Model

To better understand and work with Turing machines, one has to be acquainted with certain formal details. Another way to describe a Turing machine is as a collection of five distinct elements, what is known as a 5-tuple, consisting of a finite set of states Q, including two special states that will denote an initial state or configuration of the machine when it starts, and one when it finishes or halts; a set Σ of symbols that are either written on the tape or that the head can write on the tape; and a transition function that we call δ which indicates how the machine will behave depending on the symbols it is reading off the tape and the current state of the Turing machine.

Let's see how it works. We denote by Σ^* the set of all possible binary strings, such as a sequence of 0s and 1s. We will define a formal language L as one that uses the symbols in an alphabet Σ, in this case binary. Then L is simply a subset of Σ^*, like English. On the one hand, you can have all possible words using all the letters in the English alphabet, but not all of them would be English words, only a subset would. So let's assume that the words that we want to recognise are those that start with 1 followed by 0 or more 0s and ending with a 1. So, for example, 0111 would not be a word in our language L because it is not a word that starts with 1 followed by zero or more 0s and ending with a 1 as required. Our goal is to build a Turing machine that recognises only words in our artificial language L.

To that end we will use a notation that resembles something like a flow diagram that treats all cases in a highly visual fashion. We can immediately recognise a few elements of a Turing machine. We said that there were two distinct states that could be identified as the initial one, in this case A, and the final one, in this case, the Accept state. Here, we also have another state that we can label Reject, which denotes the state of the machine when the input word is not a word in our language L.

Then there has to be a 1 on the tape for us to move to state B. Being in state B, if 0s appear on the tape we remain in the same state, but keep moving to the right until there is a 1, and if nothing else appears, i.e., if an empty cell follows on the tape, then we accept the word as part of the language L and halt the machine. In all other cases, if anything else other than a 0 or a 1 appears, then it goes to the Reject state. And this is how we have written and constructed a Turing machine that recognises the language L.

Notice that the machine represented by this state diagram is completely deterministic; there are no loose ends, such as the possibility that something unexpected could appear on the tape, or other cases where there could be ambiguity. Every case is properly covered, and this is why this particular Turing machine is said to be deterministic.

3.2.4   The Power of the Turing Machine

It is also interesting to note that this state diagram is very similar to the state diagrams of other types of machines, such as finite automata, but the key component in this diagram is that the rules mark the movement of a head as it is reading the contents of a tape. This memory is what makes the Turing machine model so powerful.

Indeed, it is the tape that makes all the difference from other models of computing that are less powerful than the Turing machine. If you limit the tape, or limit the head movement, or remove the tape altogether, it gives you machines of different power that build a hierarchy, better known as a 'Chomsky hierarchy'. Yes, the same Noam Chomsky who is a political analyst happens to also be one of the world's foremost linguists!

So, for example, even though a state diagram of a machine with no tape would look almost the same and can recognise our previous language L, only Turing machines can do certain things, such as recognise languages that are slightly more difficult than our language.

Let's consider another example of a Turing machine, but using slightly different tools. This time, the idea is to recognise all the words that start with a number of 0s followed by the same number of 1s and nothing else. Hence the word on the tape with four 0s and four 1s would be an example of a word in our new language L.

The strategy for accepting these kinds of words can be described in what in modern computer science is called a pseudo-code, that is, a code mostly in a natural language that can be translated into any specific computer language. The first pseudo-code was actually written by Turing himself, in his original Turing machine paper of 1936 [30].

The intuitive idea for a machine that accepts these words with the same number of 0s and 1s is to replace the first 0 with a marker, then move to the next 1 and mark it with another symbol as well, then repeat the process until there are no more 0s. If every new symbol for 0 matches a symbol for 1, then the word is accepted, and if not, it is rejected.

We can see this in a more visual format that is very useful in analysing the behaviour of Turing machines. It is called a space-time diagram. We start with the input word, in this case, the word with four 0s followed by four 1s. By convention, the machine always starts in State 1, or State A, as in the previous example. Then, as we said, we mark every 0 with an X and every 1 with a Y and we repeat the process, going back and forth until no more 0s remain. You can see what the history of the computation, which is the content of the Turing machine tape over time, looks like. The last row represents the output of the machine, and when they are all matched we accept the word. We can use colours to see things more clearly.

This is how a Turing machine as a function can be visualised for a specific input. Let's also denote the head location by an arrow. One can track down the behaviour of the Turing machine over time for this rule. For the rule says that if there is a blank and the machine is in state 1, then it leaves the blank and remains in the same state and moves to the left, with -1 denoting the left. So this machine enters into an infinite computation and keeps moving the head to the left forever, without changing any of the contents of the original tape.

We can hack the code of this Turing machine by looking at its behaviour if we assume that the number of positions of the head is the number of states. We can see that there are two states because we have the arrow pointing up and down. And when the head is in state 1 with the arrow pointing up and there is nothing on the tape, then it leaves a blank, moves to the right and changes to state 2. Then in state 2 and with a blank on the tape, it prints a blue square, for example, moves to the right, and keeps doing the same because it remains in state 2. So we now know what the mapping rule and source code of the Turing machine are.

The previous Turing machines were quite simple, but not all of them are the same. This is an example of a much more sophisticated machine, even though it has only three symbols or colours and two states. On your right you have the behaviour of this very simple Turing machine and on your left you have what is called a compressed version of the same space-time diagram, which only retains the rows where the head went further to the right. And it actually looks like another model of computation that you may or may not know that is called a 'cellular automaton', but that is a topic we will come to later.

Meanwhile, we encourage you to try to write your own Turing machine. For example, you can try to build a copy-machine that, given any input, reads it and writes it again after the input, that is, it repeats the input twice on the tape or makes a copy of the input before halting. The best way to understand a subject such as this is to try to write a program by yourself.

3.2.5   The Undecidability of the Halting Problem

Notice that a Turing machine is just a computer program like any other. Now, one may ask whether all problems that can be written as computer programs can be computed or solved by a computer like a Turing machine.

For example, have you ever wondered if there was any way to get rid of certain kinds of messages on your computer? How about this one: the scary blue screen of some versions of Microsoft Windows telling you that your computer has just crashed?

It turns out that they cannot be completely eliminated; it turns out that these kinds of problems are not solvable or computable if you are considering the behaviour of Turing machines or computer programs in general. Alan Turing himself showed this, precisely by way of showing that David Hilbert's vision of automatising mathematics was unrealisable. Imagine that we were able to know when a computer program will never halt, as is the case when you get a message from your computer telling you that there is a problem running a program. In other words, we could build a Turing machine

U that accepts as input the description of another machine U and then U halts and tells you that U does not halt if U does not halt, or U does not halt if U halts. But this may lead to a contradiction, if we replace U with U itself, so that basically U gets a description of itself! If U halts then U does not halt and U does not halt if U halts! This means that our assumption was wrong and indeed it is not possible to know in general whether or not a Turing machine U will ever halt. This explains why companies cannot write completely flawless software. What we've just laid out has the fancy technical name of the 'Undecidability of the Halting Problem'. And this a simplified answer to Hilbert's question related to mathematics, because mathematical problems are like computer programs.

3.2.6   Turing Universality

The positive outcome we referenced above has tremendous consequences. Before the era of modern computers it was thought that different tasks required different machines: a typewriter if you wanted to type a novel, a radio device if you wanted to listen to music, a phone if you wanted to reach someone. But these days all these tasks and many more are performed by a single machine. Isn't this amazing? In other words, there is one machine for everything, and this property of machines is called Turing universality.

In the exercise we undertook to prove the undecidability of the Halting problem, we encoded the description of the Turing machine M as an input to the machine U. This was not an obvious trick. One would need to prove that indeed one could do such a thing, and that by doing so U would be made to follow the instructions of M and behave like M. This is precisely what Alan Turing did. He proved, by writing such a universal Turing machine, that given any Turing machine, one could encode it as the input on a tape for a special kind of machine that could emulate all other machines. These days this is rather obvious because it is what we do all the time. We reprogram our computers to behave in different ways when we open an application, and this is what Turing understood and made possible. He showed us that there were no fundamental differences, other than operational ones, between programs and data, because one could be written as the other.

So how difficult is it to build or actually write a universal Turing machine? Here is the code of the current shortest (285 characters long) universal Turing machine implementation in C/C++ (by Alex Stangl and John Tromp [31]):

```
1  #include<iostream>
2  #define Z atoi(b[a++])
3  int main(int a,char**b){
4      int c=a=1,d=0,e=2;
5      char*f,*g=0,*h,*i=0;
6      while(c){
7          if(i<=g|i>=g+e)
8              f=g,g=(char*)calloc(e,2),i=f
```

```
9        ? (memcpy(g+e/2,f,e),i-f+e/2+g):
10          g+e,h=f?h-f+e/2+g:i,e*=2;
11      a=Z-c|Z-*i\%2?a+3:(c=Z,*i=Z|48,i+=Z,h-=i<h,++d,1);
12      }
13      printf("\%s\n\%d",h,d);
14 }
```

You can see how small it is. Do not hope to understand it because the code is quite obscure, as the goal of the contest was to elicit the smallest possible Turing machine in every programming language.

Another question worth asking: How might one find or construct universal Turing machines in general? Given a Turing machine, how does one prove universality or non-universality? If one can decide whether or not a machine will always halt, then that machine is said to be decidable. In some cases this is very easy to do. Here, the rule δ sends any input to the halting state 0, so we know that the machine halts for any input and therefore is decidable. If a machine is decidable then it cannot be universal.

However, proofs of universality are much more difficult, and a common way to prove that a model is universal is by showing that one can construct a universal machine following that model the emulates a universal Turing machine. By chaining to universal models, one proves universality.

Turing's original universal machine used at least 18 states, but we know today that universal Turing machines can be so small as to have only 5 states and 2 symbols. And under special conditions, one may build a universal machine with only 3 symbols or colours and 2 states. The precise boundary is not yet completely known, but the known results are on this plot, with every dot showing a machine that is universal, the coloured ones being universal in a slightly different way. We know, for example, that no universal Turing machine exists with only two states and two symbols and that Claude Shannon himself proved that one can always exchange symbols for states [32], so there is some sort of nonlinear symmetry among the machines on either side of these axes.

3.2.7 Busy Beavers and the Behaviour of Turing Machines

Let us start with the first question on the behaviour of Turing machines. And this will actually involve a sort of game that some people, even researchers, study. A game because it was believed to have no theoretical or practical value, or any application. What drives the game is the quest to find the Turing machine that, given a number of symbols and states, takes the longest computing time to halt, among all the machines of the same size that halt. A similar quest involves a machine that prints more 1s on its tape, than all other machines of the same size. Such a Turing machine is called a Busy Beaver, because it looks very busy, more so than any other machine of the same size. Fixing the number of states and symbols means that the number of Turing machines is finite, so for small numbers of states and symbols one can almost proceed one-by-one in trying to identify the Busy Beaver machines. Some such machines are known.

The Busy Beaver is an example of an uncomputable problem because of the halting problem. If you were able to find all the Busy Beavers you would know whether or not a Turing machine would halt simply by looking at its number of states and symbols, but we know this is not possible because of the undecidability of the halting problem.

We have a conjecture about Busy Beaver Turing machines, wherein we claim that they are natural candidates for Turing universality because they are non-trivial Turing machines [33, 34]; it is quite difficult to avoid not constructing universal Turing machines [34].

It is worth noting that the Busy Beaver problem is defined for Turing machines with initial empty tapes, and Turing machines studied in this document are all provided with initially empty tapes too. Turing universality tells us, however, that for every Turing machine with an arbitrary input, there is a Turing machine with empty input computing the same function. Hence, Turing machines with empty tapes cover all possible cases (the translation may only result in some extra states).

Figure 3.19 is a nice visualisation that we created that shows the behaviour of a typical finite set of Turing machines for a given number of states and symbols. It shows a slide of the computational universe, and does so in a clever way. Each dot is arranged according to a space-filling curve called a Peano curve that has the advantage of preserving, as much as is possible in two dimensions, the linear distance between Turing machines that are close to each other in an enumeration.

3.2.8   Pervasiveness of Turing Universality and Reprogrammability

The actual extent and pervasiveness of universality among all possible computer programs remains an open question. For example, if you generate a valid computer program at random, will it turn out to be Turing universal? We recently shed light on this by proving that with a probability of almost 1, this is indeed the case [34], which means that almost every computer program can emulate any other computer program – which has all sorts of interesting consequences. We did this precisely using Turing's seminal concept of emulation, finding encodings, or 'compilers' as they are known in computer science, with which to translate computer programs into other computer programs.

In fact this was done using another model, slightly different from that of Turing, called a cellular automaton, though it turns out that the Turing machine model and the cellular automaton are equivalent. For each and every Turing machine there is a cellular automaton computing the same function and for every cellular automaton there is a Turing machine computing the same function. So in fact one can completely replace the model of the Turing machine by some other equivalent, and nothing, or very little adaptation would be required by the theory of computation.

3.2.9   Cellular Automata

A cellular automaton is a computer program comprising a collection of cells on a grid of specified shape that evolves through a number of discrete time steps according to a

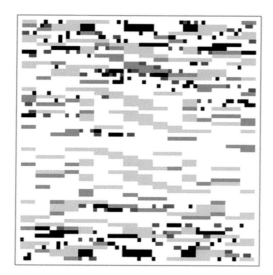

Figure 3.19 Visualising the computational universe: Runtime space in a Peano curve. It represents a small slice of the computational universe of all possible computer programs. Each dot represents a small Turing machine with 2 states and 2 symbols, of which there are ten thousand. The darker the dot, the longer the machine takes to halt. White dots are machines that never halt and red dots are the Busy Beavers. It is evident that most Turing machines either never halt or halt quickly, and only a few do a lot of work before halting. This should give you an idea of the typical behaviour of a Turing machine for an empty tape. From H. Zenil, From Computer Runtimes to the Length of Proofs: With an Al-gorithmic Probabilistic Application to Waiting Times in Automatic Theorem Proving In M.J. Dinneen, B. Khousainov, and A. Nies (Eds.), *Computation, Physics and Beyond International Workshop on Theoretical Computer Science*, WTCS 2012, LNCS 7160, pp. 223–240, Springer, 2012, with permission.

set of rules based on the states of neighbouring cells. The rules are applied iteratively for as many time steps as desired. The number of states (or colours) (k) of a cellular automaton is a non-negative integer. In addition to the grid on which a cellular automaton evolves and the colours its cells may assume, the neighbourhood over which cells affect one another must also be specified and can vary in many ways. Cellular automata can also be defined in n different dimensions, evolving in space-time diagrams that can be visualised in $n + 1$ dimensions.

3.2.10 Elementary Cellular Automata

The simplest type of cellular automaton (CA) is the one-dimensional automaton, $k = 2$ (binary), nearest-neighbour, (called *Elementary* by S. Wolfram). There are 256 such automata, each of which can be indexed by a unique binary number whose decimal representation is known as the rule for the particular automaton.

A one-dimensional CA can be represented by an array of *cells* x_i where $i \in \mathbb{Z}$ (integer set) and each x takes a value from a finite alphabet Σ. Thus, a sequence of cells $\{x_i\}$ of finite length n describes a string or *global configuration* c on Σ. This way, the set of finite configurations will be expressed as Σ^n. An evolution comprises a sequence of configurations $\{c_i\}$ produced by the mapping $\Phi \colon \Sigma^n \rightarrow \Sigma^n$; thus the global relation is symbolised as:

$$\Phi(c^t) \rightarrow c^{t+1}, \tag{3.1}$$

where t represents time and every global state of c is defined by a sequence of cell states. The global relation is determined over the cell states in configuration c^t updated simultaneously at the next configuration c^{t+1} by a local function φ as follows:

$$\varphi(x_{i-r}^t, \dots, x_i^t, \dots, x_{i+r}^t) \rightarrow x_i^{t+1}. \tag{3.2}$$

Wolfram [35] represents one-dimensional CA with two parameters (k, r), where $k = |\Sigma|$ is the number of states, and r is the neighbourhood radius. Hence this type of CA is defined by the parameters $(2, 1)$. There are Σ^n different neighbourhoods (where $n = 2r + 1$) and k^{k^n} distinct evolution rules. The evolutions of these cellular automata usually have periodic boundary conditions. Wolfram calls this type of CA an elementary cellular automaton (denoted simply by ECA), and there are exactly $k^{k^n} = 256$ rules of this type. They are considered the most simple cellular automata (and among the simplest computing programs) but are capable of great behavioural richness.

3.2.11   Computable Numbers

One other question of relevance here concerns the kinds of things that cannot be computed, even by reprogramming machines with the power of Turing universality. There are several ways to illustrate the number of uncomputable problems unsolvable by mechanistic or algorithmic means implemented by a Turing machine. To illustrate, let us introduce the concept of a computable number. You may remember the rational numbers: rational numbers are those that can be expressed as a fraction of integers, integers being numbers like 500, -2, or 1 million. So, for example, 1 over 2, or 1 over 3 are examples of rational numbers. In the case of $1/3$ we have a fraction whose decimal expansion is infinite even if periodic, but the number is said to be computable because we know how to calculate and generate every one of its digits, no matter how many there are. Unlike rational numbers, irrational numbers cannot be described as a fraction of integers. Examples are the square root of 2 or the mathematical constant π, that characteristically have no periodic decimal expansion, though some of them, such as π, can still be calculated digit by digit by a computer using one of the many short formulas that generate the digits of π. However, most numbers are not of this type. For most irrational numbers there is no formula to generate the digits, not because we do not know the formula but because the formula does not exist. The way to grasp this is to think of enumerating all possible programs. We know, for example, that all computer programs are stored in a computer in binary. So technically, an executable program is just a finite string of 0s and 1s. You also may know that you can convert any binary string into a unique integer, e.g., 101 in binary is 3 in decimal notation. So each possible computer program has an integer associated with it, an index, and there is an infinite number of these programs, but they are enumerable, meaning you can count them using their decimal notation. But as you may also know, the number of irrational numbers is not enumerable. There is no way to count them because there are always more and more numbers between any 2 irrational numbers, hence there are not enough computer programs or formulas for every possible irrational number. This means that most numbers are non-computable, and actually this result can be generalised, and it turns out that most problems that can be written as an input for a Turing machine or computer program are not computable. We have just given you an example with numbers.

3.2.12   Enumeration and Exploration of a System's Behaviour

Consider a Turing machine with the binary alphabet $\Sigma = \{0, 1\}$ and n states $\{1, 2, \ldots n\}$ and an additional halt state denoted by 0 (as defined by Rado in his original Busy Beaver paper [36]).

The machine runs on a two-way unbounded tape. At each step:

1. the machine's current 'state' (instruction); and
2. the tape symbol the machine's head is scanning

define each of the following:

1. a unique symbol to write (the machine can overwrite a 1 on a 0, a 0 on a 1, a 1 on a 1, and a 0 on a 0);
2. a direction to move in: -1 (left), 1 (right), or 0 (none, when halting); and
3. a state to transition into (which may be the same as the one it was in).

The machine halts if and when it reaches the special halt state 0. There are $(4n+2)^{2n}$ Turing machines with n states and two symbols, according to the formalism described above. The output string is taken from the number of contiguous cells on the tape the head of the halting n-state machine has gone through. A Turing machine is considered to produce an output string only if it halts. The output is what the machine has written on the tape.

Let $T(0), T(1), \ldots, T(n), \ldots$ be a natural recursive enumeration of all 2-symbol deterministic Turing machines. One can establish an enumeration by number of states from smaller to greater number, starting with all 2-state Turing machines (the set of 1-state Turing machines is trivial), then 3-state, and so on. Let n, t, and k be three integers. Let $s(T(n), t)$ be the part of the contiguous tape cells that the head visited after t steps. Let's consider all the k-tuples, i.e., all the substrings of length k from $s(T(n), t) = \{s_1, s_2, \ldots, s_u\}$, i.e., the following $u - k + 1$ k-tuples: $\{(s_1, \ldots, s_k), (s_2, \ldots, s_{k+1}), \ldots, (s_{u-k+1}, \ldots, s_u)\}$.

Because the enumeration T for all states eventually traverses all the space of all computer programs, it is said to be universal meaning that every computable model will eventually appear. Following this enumeration, we will define some methods in the next sections that will help establish the foundations to build and explore the space of computable models key in what we will propose as a method for causal discovery and causal analysis.

3.3   Practice and Discussion Questions

1. What does $dS/dt \geq 0$ mean?
   (a) That the universe starts chaotic and then self-organises
   (b) That entropy of an isolated system can never decrease over time
   (c) Isolated systems spontaneously evolve towards order
2. When particles are all distributed in a corner of a room, moving a single particle in a disordered fashion outside the corner will tend to:
   (a) Increase entropy
   (b) Keep entropy the same
   (c) Decrease entropy
3. What does the second law of thermodynamics say?
   (a) That any open and closed system will remain chaotic if it starts from complete randomness
   (b) That a closed system such as the universe started disordered and will remain so out of equilibrium
   (c) That a closed isolated system will tend to increase its entropy until it reaches equilibrium with its environment
4. What is the entropy of the sequence 11111111...?
   (a) zero no matter the granularity
   (b) maximum
   (c) log in base two of two plus 1
5. What is Shannon entropy when no particular probability distribution is provided and the uniform distribution is assumed?
   (a) A counting symbol function that may be thought of as a measure of combinatorial complexity and diversity
   (b) A tool designed to help find empirical probability distributions and measure complexity
   (c) An objective measure that does not depend on probability distributions and can detect patterns that are not only statistical in nature
6. In the formula for entropy, apart from its desirable mathematical properties, the logarithm is useful because:
   (a) It is a function that makes all values look flatter
   (b) It is some sort of natural mathematical compression method
   (c) It is a sophisticated random-looking measure
7. Shannon's original purpose for his definition of entropy was to:
   (a) Compress files and data for computers to work faster
   (b) Quantify randomness and perform interesting analysis
   (c) Measure the channel capacity needed to communicate messages
8. In what way is meaning is represented in Shannon's entropy?
   (a) In the application of the measure to different parts of a variable
   (b) In the assumptions about the underlying ensemble and probability distribution
   (c) In the number of variables involved and the associated joint distributions

9. Joint entropy can quantify:
   (a) How rainy a day may be and how cool the next day will be
   (b) What kind of similar statistical properties two or more objects may share
10. Can Conditional entropy tell you if one event happened before another?
    (a) Yes, because correlation is causation
    (b) No, because whatever you can guess of X from Y you can do so for Y from X with conditional entropy
    (c) No, because causation is correlation
11. If $H(Y|X) = H(Y)$ then:
    (a) X and Y are independent variables
    (b) $H(Y|X) = 1$
    (c) $H(Y|X) = \log Y$
12. Can Shannon entropy help to crack codes?
    (a) No, only cryptography can do that
    (b) Yes, by looking at probability distributions like in the frequency of words and letters
    (c) Yes, but only to compress a message without understanding it
13. The frequency of words in languages close to each other will be:
    (a) Roughly similar because they share a reality and possibly some history
    (b) Very dissimilar because of cultural differences
    (c) Inversely proportional
14. Can entropy be applied to other objects than sequences?
    (a) Yes, such as arrays
    (b) No, only to sequences
    (c) Yes, to measure complex numbers
15. What is the best version of Shannon entropy?
    (a) One that takes into consideration all possible partitions
    (b) Joint entropy
    (c) Mutual information
16. Does redundancy counteract noise?
    (a) The question does not make sense
    (b) Yes
    (c) No, it adds noise
17. If a language has redundancy zero, if you loose some words or letters you would not be able to:
    (a) Compress it to zero length
    (b) Fully reconstruct the message and meaning with full certainty
    (c) Transmit instantaneously to the other end
18. What does the Shannon's noisy-channel coding theorem establish?
    (a) A fundamental tradeoff between redundancy and noise
    (b) A lossless compression cutoff value
    (c) The amount of noise that people can tolerate
19. Biological information is highly redundant because:
    (a) Biology is faulty and species require diversity

     (b) Organisms need a resilient and robust code that can deal with noise

     (c) DNA is full of errors and junk

20. The so-called GC content provides an estimation of:

     (a) Genetic redundancy

     (b) Fitness

     (c) Reproducibility

21. Elements of a Turing machine include:

     (a) A halting state and at least three other states

     (b) An alphabet, a tape, a reading/writing head

     (c) A motor

22. Where is one main source of power in the Turing machine model?

     (a) The binary code

     (b) The halting state

     (c) The memory from the tape

23. Does adding more tapes and heads lead to a more computationally powerful machine than Turing machines?

     (a) Yes, if more symbols are added

     (b) No

     (c) Only if more states are added

24. Can we always determine if a Turing machine will halt?

     (a) No

     (b) Yes, if it does not halt

     (c) No, only if we add more states

25. Why can it not be guaranteed that computer programs such as MacOS or MS Windows will never crash?

     (a) Because Bill Gates never got things right and kept making the same mistakes

     (b) Because solving it is equivalent to the halting problem, which we know is unsolvable

     (c) Because software programmers are unable to write perfect code

26. Is there a Turing machine that can emulate all other Turing machines?

     (a) No, that would contradict the halting problem

     (b) Yes, a Rube Goldberg machine

     (c) Yes, a universal one

27. What is a Busy Beaver?

     (a) A wild animal

     (b) A Turing machine that performs a lot of work and halts

     (c) A solvable problem

28. How can Turing universality be proven?

     (a) By computing the Busy Beaver avoiding halting

     (b) By emulating another Turing universal machine

     (c) By proving that the Turing machine always halts

29. An enumeration can help:

     (a) Count the number of objects even if there are many repetitions

     (b) Explore a space of objects and study their behaviour

    (c) Avoid missing some objects but not all

30. Are countable sets enumerable?

    (a) Yes

    (b) No, they may be infinite

    (c) No, they are always infinite

31. Are infinite sets not enumerable?

    (a) Yes, because they are too large

    (b) No, only infinite sets that are not countable like the set of irrational numbers

    (c) Yes, because of the halting problem

32. Are cellular automata enumerable?

    (a) No, because they are too many

    (b) Yes

    (c) Yes, but only those that are two-dimensional

33. What is an example of non-computable numbers:

    (a) Natural numbers

    (b) Real numbers

    (c) Integer numbers

34. Are rational numbers enumerable?

    (a) No

    (b) Sometimes

    (c) Yes

35. Is the mathematical constant pi computable?

    (a) Yes, in principle, because its digits can be produced one by one by a computer

    (b) No, because it is infinite and cannot be printed in finite time

    (c) No, because it does not fit in any finite number of pages

36. What is an example of a non-computable number?

    (a) Chaitin's Ω

    (b) The mathematical constant π

    (c) The golden ratio

Part II

Theory and Methods

4 Algorithmic Information Theory

Chapter Summary

In order to assess the content of an object that can be identified with a generating algorithm and causal mechanism, we need a precise criterion for deciding whether or not an object is random. Algorithmic complexity enables us to make this distinction and offers a mathematical framework within which to understand the limits of classical probability theory. In exploring this, we will use cellular automata, introduced in prior chapters, as a (Turing-universal) model of computation able to generate statistical but also algorithmic patterns. Our analysis demonstrates differences between randomness, pseudo-randomness, and statistical patterns. Intuitively, a non-random object can be compressed. As in science, we can find a shorter description than the phenomenon itself, or science would always fail. A first approximation to estimate the complexity of objects leads to a discussion of compression algorithms. However, we will go further, and show why the use of statistical lossless compression algorithms such as Lempel–Ziv–Welch, to mention one example, is unsatisfactory (especially under certain regimes, such as short strings, e.g., shorter than 1K bits), thus necessitating the development of new approaches, such as the use of what is referred to as the coding theorem method (or CTM) for estimating the lower bounds of algorithmic randomness via algorithmic probability, and a block decomposition method (or BDM), which extends the power of CTM to cover an object/system using classical information theory. We will see how an algorithmic approach to randomness is fundamentally sound and mathematically unbiased. Here, we arrive at the insight captured in the equivalence of three formal concepts, incompressibility, unpredictability, and typicality, linking various previous topics of general formal and informal discussion. Typicality refers to the lack of properties in an object that can be partly or fully explained by mechanistic causes. In this sense, such a property serves us in practice as a proxy for traces, or the lack thereof, of causation in objects. Having covered applications to finite strings, we will turn our attention to infinite strings. Working with infinite objects, we touch base with deep issues such as the halting and decidability problems. What we learn by way of what may at first seem an esoteric digression is that these deep issues explain how and why algorithmic probability and complexity are fundamentally linked to epistemological problems at the core of the challenge of causation.

> In summary, this chapter presents the literature needed to understand what follows, and it begins to introduce the novel techniques associated with algorithmic information dynamics, so that we can now start putting things together and appreciate how signals produced by a complex non-linear dynamical networked system can be inspected and dissected. As we will further demonstrate, the inner workings of causal systems can be characterised using the machinery presented in this chapter, thus linking back to our seminal opening topic, causality. Concepts to understand in this chapter include the so-called invariance theorem, Chaitin's Ω, the coding theorem, algorithmic probability, the universal distribution, CTM, and BDM.

4.1   Intuitive Properties of Randomness

In this chapter we will explore the concept of randomness, both intuitively and mathematically. We will see that algorithmic randomness provides the accepted mathematical definition of randomness – a very robust definition.

Among the most common properties associated with randomness are the following:

A random object or random event traditionally occurs for no particular reason, almost by accident; it has no causal explanation.

A random object or random event traditionally has no patterns, otherwise it would be hard to consider it random.

A random object or random event is not something predictable.

How connected are these concepts to one another? Is there a mathematical definition characterising each or all of these notions? For example, can something be predictable but not causal, or be unpredictable but display patterns? Are any of these notions more fundamental than the others? Can one or a combination of these notions yield another? We will see how a theory of algorithmic information can answer these questions.

A useful strategy for studying randomness is to focus on random and non-random sequences. For example, the result of tossing a coin can be encoded as a binary sequence, with each bit encoding an outcome. Assuming tails is 0 and heads is 1, a typical random run may look like this:

$$\{0,1,1,1,0,1,0,0,1,1,1,1,0,1,0,1,1,1,0,0\}$$

Likewise, throwing a dice can be encoded with an alphabet of six symbols, in this case numbers:

$$\{2,5,6,6,3,4,1,2,1,2,2,5,6,6,2,4,3,2,6,6\}$$

And even the behaviour of more sophisticated systems, such as weather, can be encoded with N symbols, each corresponding to a possible forecast with, for example, rain encoded by 1, cold by 2, warm by 3, sunny by 4, and so on:

$$\{10,6,4,1,6,6,9,3,8,7,6,5,7,1,5,10,4,7,8,6\}$$

4.1.1   Limitations of Classical Probability

As we have seen, its inability to make an objective distinction among the elements of a distribution is a limitation of classical probability theory. For example, under a uniform distribution, the sequence of four 0s:

$$0000$$

has exactly the same probability of occurring as any other sequence of length four. This is because, according to probability theory, each has a probability of $1/n$, with n being the number of elements, in this case 16. Hence 1 over 16: $1/16$.

However, we do feel that a string such as a sequence of only 0s should be less random than another sequence of the same length, for example:

$$0111010011$$

when produced by a random process, because the sequence of 0s somehow looks atypical, predictable, and patterned. This is because one can describe it in only a few words, while the latter seems to require more effort and a longer description.

It can also be said that the latter sequence, the more random-looking one, seems to contain more information, as if it were encoding something meaningful, but it may also just happen to be random, and we will see how algorithmic complexity can help us decide.

As we have seen and will continue to see, classical information theory inherits some limitations from probability theory. For example, classical information theory can only recover some features associated with randomness, just like any other computable measure that will be ultimately limited.

Computable, as we saw in the last chapter and elsewhere, means that one can compute something with a computer program such as a Turing machine. We will see that to characterise randomness we would need to introduce more powerful measures, which will also turn out not to be entirely computable, meaning that in general it would not be possible to encode and run them as computer programs to get definite answers.

Let us demonstrate how we will introduce the concept of computation in relation to randomness and sequences. Let us start with a simple statistical example, a type of property that can be captured by a computer program. For example, a finite sequence of bits with a repeating pattern can be generated by a short computer program implementing a loop routine. In plain English, one may say that the sequence of alternating 0s and 1s, 01010101..., and so on, can be described using the phrase 'the repetition of zero and one n times'. This description would require a computer program implementing a loop that prints 0 and 1 n times. In contrast, a more random-looking sequence would require a longer description, perhaps requiring that every bit be identified individually. This is something that Shannon entropy in classical information theory can also capture with the right parameters, in the case of this sequence of alternating 0s and 1s. By, for example, taking units of 2-bits at a time, Shannon entropy can tell us that this sequence is indeed very simple. But we will come across examples in which Shannon entropy will fail.

This is because Shannon entropy can only deal with a rather small subset of properties associated with randomness. That subset is the subset of statistical properties. For example, sequences with period 1, sequences with period 2, or sequences overlapping by n bits, and so on. But not all possible properties are of this type. Thus, for an atypical sequence to pass a randomness test based on Shannon entropy, it would suffice that it be devoid of any statistical regularity.

4.2   Sources of Randomness

Let us talk about the different sources that seem able to generate randomness. Figure 4.1 shows some examples, in each case a square with 40 thousand bits from different sources, one from atmospheric noise from static generated by a radio receiver with no noise filter, with a fraction of it coming from a chaotic feedback loop and another fraction possibly from the cosmic noise that is a result of the Big Bang. Indeed, the Big Bang left some background noise that can get into certain channels. On another square we have bits from a quantum source detected by a Geiger counter timing radioactive decay. Both atmospheric and quantum bits were obtained from a website and service called RANDOM.ORG [37]. Finally, another window with the digits of the mathematical constant π in binary.

The set of atmospheric bits comes from a chaotic but deterministic system, one that obeys the laws of classical mechanics, and thus is of a causal nature. It arises not by accident but as a result of a long sequence of cause and effect, probably going all the way back to the big bang. The bits in this window are deterministic, in the sense that there is a state prior to the one on the screen that can fully explain it, no matter how complicated. Indeed, the only reason this window of random bits looks random is because it is extremely difficult to reproduce the initial condition for these atmospheric bits in precisely the same way, while reversing the process is also extremely difficult. Thus the causal origin, even if well defined, is not reachable. In contrast, in the third

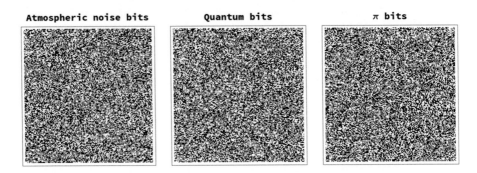

Figure 4.1 Bits arranged in 2D from different sources extracted from a radio receiver with no filter, hence with atmospheric noise; another from a Geiger counter, and a third from the digits of the mathematical constant π. They are visually indistinguishable but their origin is, in principle, very different in nature.

window, we have binary digits from the mathematical constant π, which are deterministic bits as we can reproduce them any number of times from the same source without any loss of information, in this case from a formula for π reproducing it digit by digit. In fact, we can reproduce the same bits from any formula for π, and not only is there nothing unpredictable or even chaotic about the digits of π, but we have access to their source and cause. And yet the arrangement of bits appears random. The causal origins of these digits can be explained in various ways, including as the relationship between the ratio of the length of a circle and its diameter. Finally, in the second window, we have bits from a quantum source whose digits may be explained by completely different mechanics than classical mechanics, namely quantum mechanics, and may have been produced not only in an unpredictable fashion but also, according to some theories, by a non-deterministic source. However, one can see that the images look very similar despite being very different in nature. Statistical tests are sometimes useful but clearly fail to capture the nature of the sources. In the application of Shannon entropy to these windows, in the absence of knowledge of the sources entropy values would not tell these cases apart, and we will see what algorithmic randomness can tell us about them.

One can emulate some types of randomness using computer programs such as cellular automata. Traditionally, it was thought that simply feeding a computer program with a trivial input would always produce a trivial output, as is the case with Wolfram's elementary cellular automaton (ECA) with rule 18 starting from a single black cell, which is the simplest possible initial condition (see Fig. 4.2). It may be claimed that only by feeding random initial conditions (Fig. 4.3, top row) may one get a random-looking output, such as from rule 22 starting from a random initial condition.

However, this is not always the case. Not all computer programs, no matter how simple, reproduce only random-looking output from random-looking input. A small computer program, an elementary cellular automaton with rule 30, was found by Stephen Wolfram to be able to intrinsically generate random-looking behaviour from even the simplest input. Depicted in Fig. 4.4 is one side of the evolution of the cellular automaton that shows no apparent regularities. This random-looking behaviour is also called pseudo-randomness, because it is deterministic randomness, hence simulated.

We will see how algorithmic complexity provides a framework to distinguish between these sources and cases that traditional approaches cannot easily tell apart. It won't be easy for algorithmic complexity either and we will face some challenges, but we will see how it offers directions for improvement, whereas Shannon entropy is for the most part a dead end.

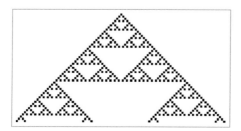

Figure 4.2 Elementary cellular automaton (ECA) with rule 18 displaying an ordered space-time evolution pattern for an ordered simple input.

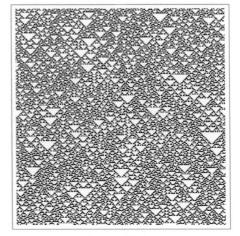

Figure 4.3 ECA rule 22 shows a bi-stable behaviour: for some symmetric initial conditions as inputs it behaves like rule 18, but for most other initial conditions, such as the one shown, it displays random-looking behaviour with no apparent order.

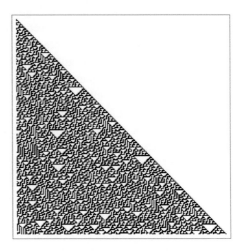

Figure 4.4 Right side of ECA rule 30 that shows no patterns and appears random-looking even for the simplest initial condition (a single black cell). Despite the discrete nature of the space and time in which it runs, this incredibly simple initial condition and computer program is capable of apparently unbounded statistical disorder.

Let us consider some of the statistical properties of one of the objects used in our previous example, the mathematical constant π. But this time in decimals, as it is most typically shown. The constant π is actually believed to be Borel normal, a concept introduced by the mathematician Émile Borel, one that it will be quite important to understand because we will later use it to explain some instances where the application of entropy can be proven to be deceptive. Borel normality means that each digit appears exactly the same number of times, so the number 0 appears as many times as the number 1 and the number 2 and 3 and so on up to 9 with equal likelihood, and not only each digit but each pair of digits and then each triplet and so on. Thus if π is Borel normal, this means that it has no statistical regularity at the limit; no sub-sequence is over-represented except temporarily, for as the sequence gets larger local regularities vanish. The value of π is actually believed to be Borel normal in all bases, including in binary, and this property is called absolute Borel normality. While it is not known for sure whether π is absolutely Borel normal, all the statistical evidence suggests it is.

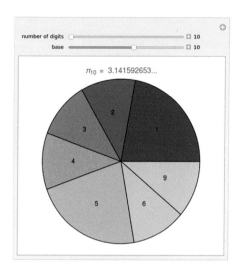

Figure 4.5 The mathematical constant π distributes its digits in equal parts. Although not proven, it is believed to be absolutely Borel normal, meaning that it does distribute any subsequence of digits in any base of equal length in equal parts of this π pie chart.

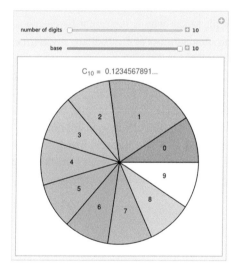

Figure 4.6 A proven Borel normal number discovered and named after Champernowne. It consists of the concatenation of all the natural numbers as the expansion of a number in the interval $(0, 1)$ that leads to an irrational number that turns our to be Borel normal and also computable and with a short description, as it is generated recursively from the successor function $x = 1; f(x): x + 1$.

In Fig 4.5, for example, we can go through a long segment of the digits of π and see that all the pieces of the pie chart (no pun intended) representing each of the digits of π are about the same size the greater the number of digits that are produced, and only at the very beginning are there small noticeable fluctuations – and this happens in all bases.

However, normality does not capture the concept of randomness. For example, the Champernowne constant is produced by putting all positive integers together in a sequence, as if it were the expansion of a real number. The result is something like 0.123456789101112. . . . When the number of digits grows, the pie chart (see Fig. 4.6) is cut into parts that become more nearly equal, indicating the constant's statistical normality in base 10 for single digits. No proof of normality is known for this number in other bases, but it is normal in base 10 by design. Here, we see the statistical

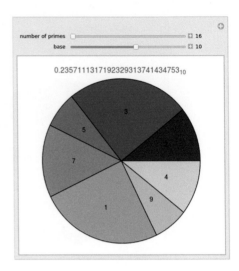

Figure 4.7 Pie chart for the digits of the so-called Copeland and Erdős constant.

evidence, and it is not too difficult to see why it is normal by design – because we will asymptotically use about the same number of digits at any given time when going through all positive integers.

Another constant, called the Copeland–Erdős constant, is obtained by concatenating the digits of the primes and looks like 0.23571113171923... (see Fig. 4.7). Copeland and Erdős proved that this constant was Borel normal in base 10, meaning that at the limit the frequency of each digit was 1/10. But clearly it is not random.

So how to characterise randomness if sound measures such as Borel normality, Shannon entropy, and even the whole body of traditional statistics cannot properly characterise it in the way in which we would intuitively characterise it?

4.3   Pseudo-randomness

In the last section we saw how some processes that are not random may appear random to an observer. Some authors have taken advantage of this phenomenon to produce fake randomness, or what is officially known as pseudo-randomness. Pseudo-random number generators have applications in many areas, from simulation to game-playing, cryptography, statistical sampling, optimisation algorithms, and machine learning. The first documented pseudo-random generator or PRNG for short, was that of John von Neumann, a mathematician who made contributions to many areas of science, including quantum mechanics and cellular automata. The piece of code shown in Fig. 4.8 reproduces his famous PRNG.

John von Neumann suggested the code in 1946 for the purpose of devising, together with Stan Ulam, what would later become known as the Monte Carlo method for the simulation of physical processes, motivated by their work on the bomb at Los Alamos National laboratory. Iterating von Neumann's procedure produces a series of numbers generated by a deterministic process intended merely to imitate a random sequence. The procedure is very simple:

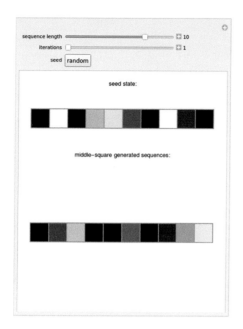

Figure 4.8 Von Neumann's PRNG, one of the very first (and most simple) algorithmic pseudo-random generators. We made the computer program available for you to play with at http://demonstrations.wolfram.com/ JohnVonNeumannsFirstPseudorandom-NumberGenerator/.

1. Take any n-digit number
2. Square it
3. Take the middle n digits of the resulting number as the 'random number'
4. Use that number as the seed for the next iteration

Eventually the whole sequence repeats in the same order and a number comes up that was squared before. Another flaw manifests itself when the sequence reaches $000\ldots$, from which point all squares are 0 and the resulting sequences (after padding) are also 0. In other words, the method eventually reaches a fixed point. One way to avoid this is to pad with 1s rather than 0s, as in this computer program.

Each iteration starts from a random seed and produces a sequence of sequences of numbers generating the grid shown. The irregularity of the grid shows that the procedure succeeds in producing a random-looking output in spite of its simplicity and deterministic nature.

There are many different methods for generating random bits and testing their quality. Clearly, when generating randomness one does not wish only to enforce Borel normality, one wishes also to allow for cases such as the Champernowne or the Copeland–Erdös constants. These methods may vary as to how unpredictable or statistically random they look and how quickly they work.

Nowadays, the kind of arithmetical PRNGs, such as the one suggested by von Neumann, also called a congruential PRNG, fail under basic statistical tests, and better PRNGs have been developed.

In version 4 and older versions of Mathematica, for example, the central column of ECA rule 30 was used as a pseudo-random number generator, but beginning with Mathematica 6 a new default RNG based on another cellular automaton has been used. Figure 4.9 shows a pie chart similar to the ones above, showing how the central column

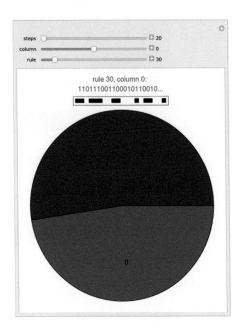

Figure 4.9 Distribution of bits in the central column of EVA rule 30. The number of 0s is about the same as the number of 1s.

of ECA rule 30 produces about the same number of 0s as 1s, which would be a first basic property of a good pseudo-random number generator, in this case, a pseudo-random bit generator.

Modern versions of the Wolfram Language allow the user to choose among six different PRNGs, called the 'congruential', 'ExtendedCA', 'Legacy', 'MersenneTwister', 'MKL' and 'Rule30CA' PRNGs, as well as from eight other RNG methods for 'MKL'. The full tutorial is available online at: http://reference.wolfram.com/mathematica/tutorial/RandomNumberGeneration.html. Here are some examples:

- ECA Rule 30 based (the default RNG before version 6):
  {0, 1, 0, 1, 1, 1, 1, 1, 0, 0, 1, 0, 0, 1, 0, 0, 0, 1, 1, 0, 0, 0, 0, 1, 0, 0, 1, 0, 0, 0, 1, 1, 1, 1, 1, 1, 0, 0, 1, 1, 0, 1, 1, 0, 0, 0, 0, 1, 1, 0, 1, 1, 0,1, 1, 0, 1, 0, 0, 1, 1, 1, 1, 0, 0, 1, 1, 0, 1, 0, 0, 1, 1, 1, 1, 0, 1, 1, 0, 0, 1, 1, 0, 1, 0, 0, 1, 0, 0, 0, 1, 1, 0, 0, 1, 1, 0, 0, 0, 1}
- Cellular automaton PRNG (ExtendedCA), from version 6 on:
  {0, 1, 1, 1, 0, 1, 0, 0, 1, 0, 1, 0, 1, 0, 1, 0, 0, 0, 0, 0, 0, 1, 0, 0, 1, 1, 0, 0, 1, 1, 1, 1, 0, 0, 1, 1, 0, 0, 0, 0, 0, 1, 1, 1, 0, 0, 1, 1, 0, 0, 0, 1, 1, 1, 1, 1, 0, 1, 0, 0, 1, 0, 1, 0, 0, 1, 0, 1, 1, 0, 0, 1, 1, 0, 0, 1, 0, 1, 0, 0, 1, 1, 0, 0, 0, 1, 0, 0, 1, 1, 0, 0, 0, 1, 0, 0, 0, 0, 0}
- MersenneTwister based (introduced in version 6):
  {0, 1, 0, 0, 0, 1, 1, 1, 0, 1, 1, 1, 0, 0, 1, 1, 0, 0, 1, 1, 1, 0, 0, 0, 1, 1, 0, 1, 1, 0, 0, 1, 1, 1, 1, 1, 1, 1, 0, 1, 1, 1, 1, 1, 1, 0, 0, 0, 1, 0, 0, 0, 0, 1, 1, 1, 0, 1, 1, 1, 0, 1, 0, 0, 0, 1, 1, 0, 0, 1, 0, 1, 1, 0, 0, 0, 1, 1, 1, 0, 0, 1, 0, 1, 0, 0, 1, 0, 1, 0, 0, 1, 0, 1, 1, 1, 1, 1, 1}

You can see that all these sequences look random.

Figure 4.10 shows a small computer program that allows you to visualise and even test some of these PRNGs using some simple statistical tests where you can see that the so-called congruential PRNGs, such as the one proposed by von Neumann, fail, and

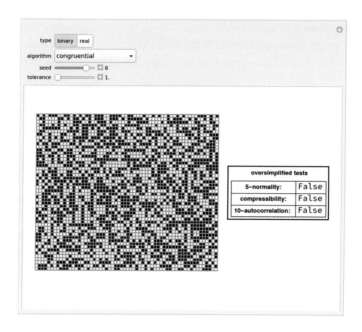

Figure 4.10 Different PRNGs of varying sophistication, some of which are amongst the most frequently used in all manner of applications. This computer program written by Zenil illustrates them, including some based on cellular automata. We made the computer program available for you to play with at https://demonstrations.wolfram.com/RandomNumberGeneration/.

how some more modern methods pass. In general it is a cat and mouse game, where better PRNGs also suggest better statistical tests and vice versa.

Among the most popular statistical tests are: normality, compressibility, linear complexity, autocorrelation, Fourier coefficients, run and gap tests, partitioning the set of m-bit strings and counting hits in subsets, serial tests, rank of a binary matrix, longest run of 1s, Hamming weights, random walk tests, and close pairs.

All of them can also be used to test PRNGs and thus make better PRNGs. Popular statistical batteries can run up to 40 tests or more. A piece of data fails one of these tests when it is found to have an atypical characteristic defined by what the particular test evaluates, such as an over-representation of certain symbols, to give one example. A good pseudo-random sequence can pass any finite set of tests, but a truly random sequence would pass all known and future tests. In the next section we will see what true mathematical randomness is, as opposed to pseudo-randomness.

4.4   Algorithmic Complexity and Compressibility

As we saw in the previous chapter, traditional probability theory and classical information theory attempt to quantify some properties related to randomness by asking how many bits are needed to encode or communicate a sequence. Algorithmic information theory or the theory of algorithmic complexity is, however, of a very different nature. What drives algorithmic complexity is the wish to ascertain what it means for an object to be truly random, to possess the following properties, in the broadest and most general terms.

UNPREDICTABILITY: the impossibility of predicting an outcome or an object.

INCOMPRESSIBILITY: that one cannot describe something random in a simple or brief fashion.

TYPICALITY: that a random event does not have anything special about it.

A sequence of coin tosses, for example, would be called typical if it looks unbiased, that is, random. But if it comes up tails every time then it would look atypical, compressible, and highly predictable. In contrast to traditional statistics and classical information theory, algorithmic information theory avoids probability distributions and provides a very different and, in a fundamental sense, general and open approach to randomness by focusing on the properties of data that can be modelled by computer programs. Thus, something random will be something that no computer program shorter than the data can model or reproduce.

4.4.1   Program-Size Characterisation

We have arrived at one of the main concepts in this work, that is, the concept of algorithmic complexity. Defined by Solomonoff, Kolmogorov, Chaitin, and Levin, program-size complexity, also known as algorithmic complexity or Kolmogorov complexity, is a measure that quantifies algorithmic randomness, a type of randomness that is strictly stronger than statistical randomness.

Formally, the algorithmic complexity, which we will denote by K, of a string s is the length of the shortest computer program p running on a universal Turing machine U that generates the string as output and halts [38, 39]:

$$K(s) = \min\{|p| : U(p) = s\}.$$

The largest or smallest difference between the original length and the compressed length, that is the length of the Turing machine, determines the complexity of the string. The length of the computer program and the size of the string are not very different, with both measured in bits. A string s is said to be random if $K(s)$ (in bits) $\sim |s|$.

A technical inconvenience of K as a function taking s to be the length of the shortest program that produces s, is the fact that K is upper semi-computable. In other words, there is no effective algorithm that takes a string s as input and produces the integer $K(s)$ as output. K is uncomputable because of the halting problem, given that one cannot always find the shortest programs in finite time without having to run every computer program, which means having to wait forever in case they never halt.

This is usually considered a major problem, but the theory of algorithmic randomness [40] ascribes uncomputability to any universal measure of complexity, that is, a measure that is at least capable of characterising mathematical randomness [41]. However, because it is upper semi-computable, $K(s)$ can be approximated from above, or in other words, upper bounds can be found, for example, by finding and displaying a small computer program (measured in bits) relative to the length of a bit string.

In practice, one follows pragmatic approaches such as the wide use of lossless compression algorithms to approximate K. Lossless means that there is no loss of information after compressing, with decompression retrieving the original object unchanged. The compressed version of an object is an upper bound of its algorithmic complexity, which means that the actual algorithmic complexity of the string cannot be greater than its compressed length. Hence, finding a short representation of an object by compressing it is a sufficient test of non-randomness.

For finite strings it is sufficient to calculate the length in bits of the compressed version and compare it to the original string length. This function roughly encodes a string in binary. For example, a non-random binary string will be highly compressible, but a pseudo-random sequence is harder to compress. For an algorithmically random string, its compressed length in binary is actually not shorter than its uncompressed length.

The problem, of course, is cases in which no compressed version is found. Because K is not computable, we cannot guarantee that the lack of a compressed version of an object means that no such object exists, and so not much can be said about it. Given this limitation, it is in general more important and significantly more informative to discover that something is compressible than that it is not.

As a generalisation about ever-growing sequences, a sequence s can be said to be algorithmically random if all its initial segments are not compressible by more than a constant. Eventually, the growing sequence will be mostly incompressible from a certain length on.

This means, for example, that unlike in the case of random strings, for non-random strings such as 0s and 1s alternating two thousand times, one can find a constant that makes all initial segments assume low algorithmic complexity from a certain point on. For a random sequence no such constant can be found.

The Kolmogorov complexity characterisation captures the intuition that a random sequence is not compressible. So you can see how algorithmic information theory draws heavily on the theory of computation – as formulated by Alan Turing – by looking at the computer programs behind the data that produce said data. Algorithmic complexity can help to formulate and tackle questions that traditional tools from information theory are poorly equipped to deal with. For example, the string

$$0110100110010110100101100110100 1\ldots$$

may look random when viewed through the lens of Shannon entropy, but the sequence is simply generated by starting with 0 and then successively appending the 'Boolean complement', that is, 1 where there is a 0 and 0 where there is a 1. Another look at the way this string is generated may reveal the process. You start from 0, then the Boolean complement of 0 is 1, you end up with the sequence 01, then you take the Boolean complement of 01, which is 10, and append it to the sequence, then you have 0110, and so on. The resulting infinite sequence is called the Thue–Morse sequence, after its discoverers.

Despite the potentially infinite length of this sequence, the generating mechanism is a very short computer program of fixed length that can generate every bit, analogous to the short description we just gave in plain English. Such a computer program, however, would never be taken into consideration by Shannon entropy, which would assign it a large value because the number of different sub-strings keeps growing, in spite of the fact that the generating program remains of the same (small) size.

Shannon entropy and algorithmic complexity use and work with the same units: bits. But information is interpreted in very different ways. For example, algorithmic complexity considers individual objects independent of any probability distribution.

Thus, in another example, if you didn't know the possible deterministic source of a sequence and assumed a uniform distribution, different initial segments of a deterministic sequence, such as the Fibonacci sequence, would have different entropy values:

$$1, 1, 2, 3, 5, 8, 13, 21, 34, 55, 89, 144, 233, 377, 610, 987, 1597, 2584, 4181, 6765, \ldots$$

This is because the larger the numbers, the more the bits required to encode them in binary. However, for algorithmic complexity, the size of the generating program remains the same:

$$F(n) = F(n-1) + F(n-2) \quad \text{for } n \geq 2, \quad F(0) = 1, \ F(1) = 1.$$

This means that the Fibonacci sequence has a low and constant algorithmic complexity, and so any segment of the Fibonacci sequence would have the same algorithmic complexity.

This example reveals a great difference: two messages can have different entropy values, yet they can have the same algorithmic complexity.

4.4.2   Most Relevant Properties of Algorithmic Complexity

What are the most salient properties of algorithmic complexity? The central idea behind algorithmic complexity is that a sequence of bits is random if there is no computer program whose length in bits is shorter than the sequence itself in bits – this basically means that random sequences are those that cannot be compressed. By a simple combinatorial argument, one can see that almost all strings are random. That is, the shortest program producing the output rarely has a much shorter representation than the length of the string itself.

For instance, the number of strings of size 20 bits is 1048576, and strings that can be encoded in less than 20 bits total 1048575, i.e., at least one string cannot be compressed at all. But programs of less than 19 bits are significantly fewer, totalling 524287. It turns out that half (524288) of the original strings are compressible by just one bit, half of the half (262144) of the original strings are compressible only by two bits, and so on. So most strings are close to the maximal complexity because they are not compressible except for a few bits. This means that one cannot pair off all n-length binary strings with binary programs of much shorter length because there simply aren't enough short programs to encode all strings in shorter strings, even under optimal circumstances.

Another important property of algorithmic complexity is also commonly seen as its greatest drawback. That is its uncomputable nature. A function is uncomputable if there is no Turing machine that is guaranteed to produce an output for all inputs, or in other words, if, for a number of inputs, the machine computing the function doesn't halt. For Kolmogorov complexity this means that the function $s \mapsto K(s)$ has no effective procedure (or Turing machine). That is, there is no general function that, given a specific string, can generate the shortest program that produces that string.

This uncomputability of the function s to $K(s)$ is, however, also the source of its greatest strength. Contrary to the common belief that the greatest drawback of K is its uncomputability, it is this very uncomputability that provides K with its greatest power. Algorithmic information theory proves that no computable measure will be up to the task of finding all possible regularities among all possible infinite sequences, or even among all finite strings. This is because there is a countable but infinite number of possible (computable) regularities in the set of all possible finite strings (the set is infinite) each of which can be matched to one in a countable but also infinite number of possible Turing machines (or computer programs). Hence no finite set of computer programs is capable of dealing with every possible (computable) regularity. Any computable measure can be fully implemented by a set of finite computer programs (none of which needs to be Turing-universal).

It is more precise and appropriate to refer to the uncomputability of the function s to $K(s)$ as semi-computability, because one can actually approximate $K(s)$ from above, i.e., one can calculate upper bounds of K. One traditional way to calculate upper bounds on K is, as we have said before, with the use of lossless compression algorithms. A trivial upper bound on K for any string s is simply the program print(s). If a string s does not allow any other shorter program than print(s), then s can be said to be incompressible or algorithmically random. A proof of the uncomputability of K is sometimes given in terms of Berry's paradox.

4.4.3 Lossless Compression as an Estimation of K

Popular lossless compression algorithms such as Lempel–Ziv and Compress, which are behind formats such as zip, are regularly used to estimate algorithmic complexity. As noted, the usefulness of lossless compression algorithms as a method for approximating algorithmic complexity derives from the fact that compression is a sufficient test for non-randomness. For example, the difference in length between the compressed and uncompressed forms of the output of a cellular automaton is an approximation of its algorithmic complexity. In most cases, the length of the compressed form levels off, indicating that the cellular automaton's output is repetitive and can be easily described. However, in cases like rules 30, 45, 73, or 110, the length of the compressed form grows rapidly, corresponding to the apparent randomness and lack of structure in the display (Fig. 4.11).

According to algorithmic complexity, the shortest description of an unfolding object is the length of the shortest generating rule from which it evolves. Hence, the algorithmic complexity of all the 256 ECAs is no longer than about 8 bits plus the size of

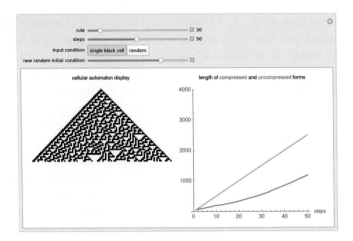

Figure 4.11 Attempting to compress random-looking outputs asymptotically approximates the length of the uncompressed versions, while those that are compressible diverge as a function of time. We have written a computer program for you to play with that is available at http://demonstrations.wolfram.com/CellularAutomatonCompressibility/.

the `CellularAutomaton[]` function. This does not mean that all these rules are equally algorithmically random or complex, because it can easily be seen how rule 0, which evolves exclusively into blank cells, can actually be encoded in less than 8 bits plus the size of the function `CellularAutomaton[]`, and is thus very likely to be less algorithmically complex than, say, rule 30 (Fig. 4.12).

Figures 4.13–4.16 show tables of ECA rules sorted from highest compressibility and thus lowest estimated algorithmic complexity by lossless compression, to lowest compressibility or highest estimated algorithmic complexity.

4.5   The Invariance Theorem

We saw in previous sections how we could quantify algorithmic randomness by using computer programs running on some reference universal Turing machine. One might suppose that it is always possible to find a language in which a particular object has a short encoding, no matter how random. For example, Alice and Bob could agree to compress the entire contents of the Encyclopaedia Britannica into a few symbols, so that when Alice presents Bob with these symbols, Bob would know Alice meant the Encyclopaedia Britannica. Algorithmic information theory (AIT) would mean little if the complexity of something could be determined so arbitrarily, simply by renaming things with an arbitrary number of symbols. AIT requires the program to reconstruct the original object from scratch, so that there is no cheating. Bob would need a computer program to reconstruct the Encyclopaedia Britannica without any external help from Alice or anyone else. It would also appear that one would need to specify the programming language and the particular universal Turing machine on which these computer programs would run for algorithmic complexity to work or make sense.

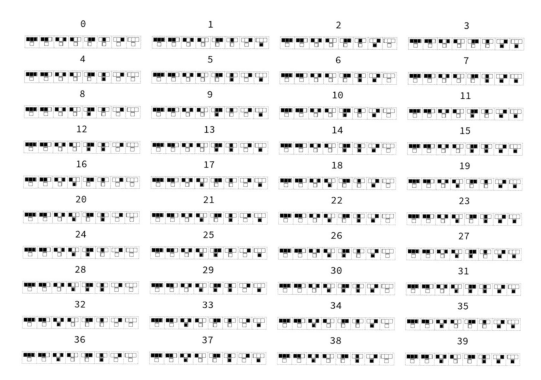

Figure 4.12 Iconic representation of ECA generating rules. They can all be represented in a few bits (once the top row is fixed only the lower 8 bits change). Still, some rules can be further compressed, like rule 0 or rule 1. Those with an equal distribution of zero and non-zero elements will tend to have longer descriptions and order parameters. Langton's λ, for example, takes advantage of this simple phenomenon.

Otherwise, one would be able to 'rename' things and make anything look random or simple by changing the underlying language and Turing machine.

Let us consider the following sequence based on three symbols (0, 1, and 2): 1, 0, 2, 1, 0, 2, 1, 0, 2, 1, 0, 2, 1, 0, 2, 1, 0, 2, 1, 0, 2, 1, 0, 2, 1, 0, 2, 1, 0, 2, 1, 0, 2, 1, 0, 2, 1, 0, 2, 1, 0, 2, 1, 0, 2, 1, 0, 2.

We can show that this sequence can be generated by a small computer program like this one:

```
NestList[Mod[#+2,3]&,1,50]
```

which takes no more than 56 bytes, no matter how long the sequence, to produce the same pattern.

```
ByteCount["NestList[Mod[#+2,3]&,1,50]"]
```

However, this is in the Wolfram Language running on Mathematica. What if we had used Lisp, Java, or Visual Basic instead? It seems that the result would depend on the computer language chosen. And even in a single computer language there may be different computer programs of different sizes that can produce the same object.

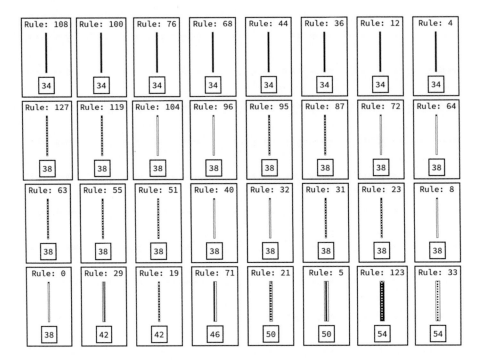

Figure 4.13 Highly compressible evolutions of ECAs from the simplest initial condition.

For example, the following two computer programs can generate any number of digits of the Thue–Morse sequence and they differ in length, even though both are small and generate the same sequence. So how do we deal with different computer programs in possibly different computer languages with different lengths?

```
Mod[Table[1/2(-1)^n+(-3)^n Sqrt[Pi] Hypergeometric2F1[3/2,-n
    ,3/2-n,-1/3]/(4n! Gamma[3/2-n]),{n,0,2^7-1}],2]
```

```
Nest[Join[#, 1 - #] &, {1}, 7]
```

Fortunately, the invariance theorem, as proven by Kolmogorov, Solomonoff, and Chaitin, tells us that the difference between any two computer programs producing the same object is at most of constant length.

More formally, the so-called invariance theorem establishes that there exist optimal languages or reference machines on which the difference in lengths of the minimal programs in any two computer languages L_1 and L_2 is always bounded by a constant that depends on L_1 and L_2 but not on s:

$$K_{L_1}(s) - K_{L_2}(s) \leq c_{L_1,L_2}.$$

The way to grasp this is by thinking of a translator program between two languages L_1 and L_2. Because what the invariance theorem says is that one can always write a 3rd computer program of fixed length capable of translating between any two languages L_1 and L_2. This is actually very clear with the computer language Java, because one

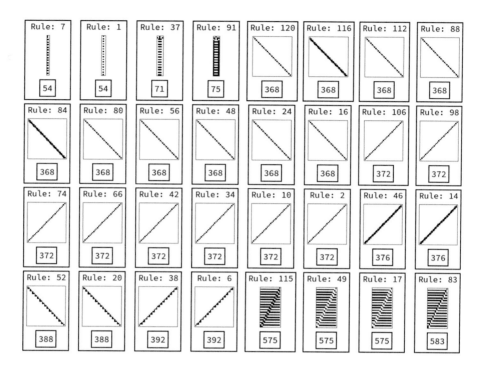

Figure 4.14 Second highly compressible evolutions of ECAs from the simplest initial condition.

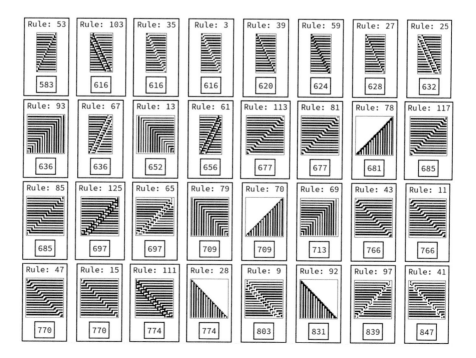

Figure 4.15 Less compressible evolutions of ECAs from the simplest initial condition.

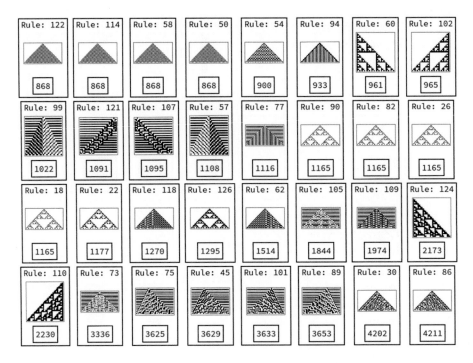

Figure 4.16 Most incompressible evolutions of ECAs from the simplest initial condition.

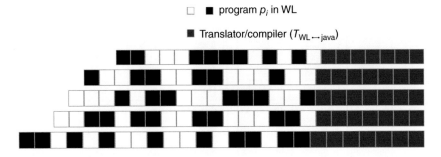

Figure 4.17 Visual sketch of the translation concept behind the invariance theorem.

of the innovations introduced by Java was a Java Virtual Machine or JVM that made Java multi-platform, in the sense that you could run your Java program on any platform without changing the language code (Fig. 4.17). This is because the many JVMs were acting as translators – called compilers – between the Java programming language and the machine codes for different platforms, such as Unix and Windows. So if the translator is shown in blue, one has only to add the translator to every computer program written in a given language to convert it into another program in another language, with both languages producing the same sequence of interest. The length of the compiler or the length of the blue part is the same for all input strings, and so it is constant. Thus, if L_1 is the Wolfram Language and L_2 is Java, then the invariance theorem tells us that the shortest computer programs to measure the algorithmic complexity of a string are about the same length up to a constant.

The longer the string, the less important c is (i.e., the choice of programming language or UTM). However, in practice c can be arbitrarily large, thus having a great impact, particularly on short strings, and never revealing at which point one starts approaching a stable K or when one is diverging before finally monotonically converging.

The invariance theorem tells us that it is impossible to guarantee convergence, but it does not imply that one cannot study the behaviour of such a constant for different reference universal Turing machines or that K cannot be approximated from above.

4.6   Convergence Rate and Invariance of Description Length for Deterministic Systems

In a more practical sense, what the c in the invariance theorem characterises in a biological context, for example, is that the DNA and a set of chemical reading and writing machines called ribosomes form a couple comprising language and compiler, able to transcribe and translate DNA into proteins effectively from one language into another. So ribosomes can be thought of as compilers between the DNA, RNA, and the protein space, with the ribosomes always having the same complexity, independent of what they are reading or writing.

Now, remember that we said that the sequence:

$$1, 0, 2, 1, 0, 2, 1, 0, 2, 1, 0, 2, 1, 0, 2, 1, 0, 2, 1, 0, 2, 1, 0, 2, 1, 0,$$
$$2, 1, 0, 2, 1, 0, 2, 1, 0, 2, 1, 0, 2, 1, 0, 2, 1, 0, 2, 1, 0, 2 \ldots$$

could be reproduced with a code no longer than 56 bytes, no matter how long a sequence with this pattern we require.

Well, that is not entirely true. Actually the code has to grow by a little bit if we want to produce longer sequences with the same pattern. It does grow a little bit because we have to determine the end of the algorithm, that is, the point at which we wish the sequence to come to an end. And that number does make the code grow a bit. In fact, classical information theory tells us that we can encode an integer n in about $\log n$ bits because, if you remember, we can always guess a number in about $\log n$ yes and no answers.

You can see how when running this code with very large numbers, the byte count does actually change:

```
Table[ByteCount["NestList[Mod[#+2,3]&,1,"<>ToString[i]<>"]"],
    {i,1,10000000000000000/9,10000000000000}]
```

```
Output: {56, 64, 64, 64, 64, 64, 64, 64, 64, 64, 64, 64,
        64, 64, 64, 64, 64, 64, 64, 64, 64, 64, 64, 64,
        64, 64, 64, 64, 64, 64, 64, 64, 64, 64, 64, 64,
```

```
 4          64, 64, 64, 64, 64, 64, 64, 64, 64, 64, 64, 64,
 5          64, 64, 64, 64, 64, 64, 64, 64, 64, 64, 64, 64,
 6          64, 64, 64, 64, 64, 64, 64, 64, 64, 64, 64, 64,
 7          64, 64, 64, 64, 64, 64, 64, 64, 64, 64, 64, 64,
 8          64, 64, 64, 64, 64, 64, 64, 64, 64, 64, 64, 64,
 9          64, 64, 64, 64, 72, 72, 72, 72, 72, 72, 72, 72,
10          72, 72, 72, 72}
```

Thus, despite how trivial this extremely little piece of information may seem, and how slight the difference it makes, it is not only mathematically precise but it will turn out to be the most important little piece of information in the field of algorithmic information dynamics, and you will soon see why. Not to keep you in suspense, the main idea, without getting into details, is that unlike a deterministic system that has a well-defined generating mechanism, such as this sequence and its generating formula, in the case of systems that may be subject to some source of randomness or are interacting with other systems, this logarithmic change will no longer be observed, and the more removed from the logarithmic term, the more valuable its insights into the system's behaviour and its underlying causes. To understand this better, in the chapter on dynamical systems you will learn what a dynamical system is, before we finally get into the subject of algorithmic information dynamics.

4.7   Convergence of Definitions

One may ask how fundamental these measures are. Perhaps there are better measures for randomness? One of the most surprising results in algorithmic complexity is what is known in mathematics as the 'convergence in definitions'. This is a phenomenon similar to other types of convergences in definition, such as in the notion of an algorithm in the 1930s, when people such as Kurt Gödel, Alonzo Church, Alan Turing, Emil Post, Stephen Kleene, and Rózsa Péter, among many others tried to characterise the notion of an algorithm using different, independent approaches that turned out to be equivalent in computational power, giving the sense that the concept of the algorithm had been mathematically captured by all these characterisations, and leading to what is known today as the Church–Turing thesis, that is, the strong belief that any practical definition of an algorithm will collapse into a definition equivalent to the one that Turing or Church provided.

Something similar has happened with algorithmic randomness, leading to what Jean-Paul Delahaye calls the Martin-Löf–Chaitin thesis, which is itself similar to the Church–Turing thesis. The Martin-Löf–Chaitin thesis is the thesis that all definitions of randomness will be equivalent to one of the previous characterisations. And indeed, when people such as Per Martin-Löf, Greg Chaitin, Andrei Kolmogorov, Leonid Levin, Claus-Peter Schnorr, among many others, independently proposed characterisations of randomness, they did find that all these definitions were essentially the same, as they were equivalent to one another when the Church–Turing thesis was assumed – which almost everybody does in the field.

So not only is each of these definitions able to characterise intuitive notions of randomness such as compression, predictability, and typicality as discussed above, but they also do so in a very general and comprehensive way.

Notice, however, that statistical randomness is not on the list of equivalent definitions of randomness because, surprisingly, even though its use, misuse, and abuse are pervasive in science, statistical randomness and measures such as Shannon entropy are approaches that do not provide the accepted mathematical characterisation of randomness.

We have seen how, according to algorithmic complexity, if an object is random then it is impossible to compress it. We have also seen how compressibility is a sufficient test for non-randomness. That is, if you find a short computer program for some data, then you know that said data are not algorithmically random.

On the other hand, we also briefly mentioned the concept of lack of special properties, which we referred to as 'typicality'. We don't call something random if it is atypical, because it can be described in terms of its lack of typicality. It turns out that this intuitive concept is thus also related to other intuitive properties of randomness. In particular, it is clear how being atypical can be used to compress an object. The basic idea is that if something is not typical, then the non-typical feature gives you some sort of handle with which to distinguish the object among more typical objects, which contradicts the intuitive idea that it is random and relates it to the concept of compression.

One can also devise statistical tests for these kinds of properties, but it is first necessary to formalise the kinds of properties allowed, such as the so-called recursive properties, that is, properties that can be characterised by computer programs, which is a generalisation of the properties that can be characterised by traditional statistics, given that computer programs can easily capture any statistical regularity with even low computational power such as those found in regular languages, though statistics cannot characterise recursive properties. It was Per Martin-Löf, the Swedish mathematician and student of Kolmogorov himself, who devised a universal test to assess a sequence for any recursive or computable property [41], thereby technically achieving another formal characterisation of randomness.

An example of a recursive property can be whether a sequence has an even number of 1s, or whether the digits of a sequence are the digits of a mathematical constant, such as the mathematical constant π that comes from a short computer program implementing one of the many formulas that can generate the digits of π. So random sequences can then be characterised by their failure to meet any property that a computer program can encode.

Finally, another item on the list of intuitive properties of randomness that we started out with was the unpredictability of a random sequence, which is also a characterisation of Shannon entropy. What Claus Peter Schnorr and others have proved mathematically is that it is impossible to make money by guessing the next digits in a truly random sequence by using a recursive or computable betting strategy [42]. This is to be expected, but if you were using Shannon entropy, you would fail in practice because you could produce a random sequence with no statistical patterns but generated by a pseudo-random generator, and you could predict every digit, yet Shannon entropy would suggest that it was random. However, it is when there are no predictable patterns, statistical or computable, that a sequence can truly be deemed random.

This convergence in definitions of mathematical randomness as opposed to Shannon entropy means that each definition assigns exactly the same randomness as the other. In other words, the extension of each definition is the same; the set that each definition characterises contains exactly the same objects, thereby strongly suggesting that each definition has proven itself to be fundamental in a mathematical way. We can write this elegant result in a compact manner as a causal chain:

$$\text{incompressibility} \longleftrightarrow \text{unpredictability} \longleftrightarrow \text{typicality}$$

A series of universal results, both in the sense of being general and in the sense of Turing-universality, leads to the conclusion that the definition of algorithmic randomness is mathematically objective.

In summary:

- It has been proven that there is a universal statistical test that can test for all computable properties of an object but is uncomputable or semi-computable. This definition of randomness is therefore general enough to encompass all effective tests for randomness.
- It has been shown that a predictability approach based on betting strategies leads to another characterisation of randomness, which in turn is equivalent to Martin-Löf randomness.
- It is known that an algorithmically incompressible sequence is also Martin-Löf random, hence showing all the equivalences.

4.8   Algorithmic Probability

The classical probability of production of a bit string s among all possible 2^n bit strings of length n is given by $P(s) = 1/2^n$. The concept of algorithmic probability (also known as Levin's semi-measure) replaces the random production of outputs with the random production of programs that produce an output. The algorithmic probability of a string s is thus a measure that estimates the probability of a random program p producing a string s when run on a universal (prefix-free) Turing machine U.

The algorithmic probability $m(s)$ of a binary string s is the sum over all the (prefix-free) programs p for which a universal Turing machine U running p outputs s and halts [39, 43, 44]. It replaces n (the length of s) with $|p|$, the length of the program p that produces s:

$$m(s) = \sum_{p\,:\,U(p)=s} 1/2^{|p|}. \tag{4.1}$$

Here, $m(s)$ can be considered an approximation of $K(s)$ because the greatest contributor to $m(s)$ is the shortest program p that generates s using U. So if s is of low algorithmic complexity, then $|p| < n$, and will be considered random if $|p| \sim n$.

4.9   The (Algorithmic) Coding Theorem

The coding theorem [43, 44] further establishes the connection between $m(s)$ and $K(s)$,

$$\left| - \log_2 m(s) - K(s) \right| < c, \tag{4.2}$$

where c is a fixed constant, independent of s.

The coding theorem implies [45, 46] that the output frequency distribution of random computer programs to approximate $m(s)$ can be converted into estimations of $K(s)$ using the following re-written version of Eq. (4.2) (under assumptions of optimality):

$$K(s) = - \log_2 m(s) + O(1). \tag{4.3}$$

Among the properties of algorithmic probability and $m(s)$ that make it optimal is the fact that the data do not need to be stationary or ergodic, and that it is universal (stronger than ergodic), in the sense that it will work for any string and can deal with missing and multidimensional data [47, 48, 49, 50]. There is no underfitting or overfitting because the method is parameter-free, and the data need not be divided into training and test sets.

4.10   Chaitin's Omega Number

So far, we have mostly talked about finite sequences or strings, but let us now momentarily move to the sub-field of algorithmic complexity that studies the algorithmic randomness of infinite objects. One such object is the so-called halting probability number [51], also known as Chaitin's Ω (omega) number, also often wrongly called a constant, defined as follows:

$$\Omega_U = \sum_p \frac{1}{2^{|p|}},$$

where p is a computer program running on a universal Turing machine U and the sum is a probability between 0 and 1 that computer programs running on U will halt or not. Clearly, calling Ω a constant is misleading because it depends on the choice of universal Turing machine U, and for different choices there will be a different Ω number. However, the Turing machine must be somehow special, a type of universal Turing machine that is often called a prefix-free Turing machine. Let us explain what a prefix-free machine is.

4.10.1   Prefix-free Codes and Turing Machines

There is a special type of computer called a prefix-free computer that has to be used for objects such as Chaitin's Ω and is also sometimes used to define algorithmic complexity. The need for such a computer stems from the fact that it wouldn't do to trivially count the same program more than once. The problem is that it is very easy to generate an infinite number of programs as an extension of a given program. Take the program that prints s and then prints an extra 1 at the end, only to delete it before halting. The new program prints s, but it is only a spurious variation of a more compact program. The number of such spurious programs that produce the same output is infinite, and for a measure to be called a probability the sum of the probabilities has to be 1. It would not have any mathematical meaning to say that something has more than a 100% likelihood of occurring.

To circumvent this problem, Leonid Levin in 1974 [44], and then Gregory Chaitin in 1975 [51], devised a way to consider only significant programs, determined by a certain rule. The rule is that new programs should never be initial sub-programs of any other programs. These types of sets are called 'prefix-free domains'. A classic example is the set of telephone numbers. We can reach someone by calling their telephone number because no telephone number is part of any other telephone number. Imagine that Alice's number is 12345 and Bob's telephone number is 123. No one would be able to reach Alice because they would be connected to Bob as soon as they finished dialling 123.

There are different ways of avoiding this. For example, telephone numbers are all of the same length, so that any other number of a different length is not a valid telephone number. Another way is to choose a special character to indicate the termination of a number. For example, some online banking systems ask customers to use the # sign to indicate that they have finished keying in their bank account numbers. Prefix codes are guaranteed to exist for a countable set and the so-called Kraft, or Kraft–Chaitin inequality guarantees that taking the sum of all the probabilities of the series will converge to 1, which is the necessary condition for a probability measure. For algorithmic complexity or Kolmogorov–Chaitin complexity, there are versions that can be defined both on regular universal Turing machines and also on prefix-free Turing machines, but they do not differ much from each other and do not dramatically affect the definition of algorithmic complexity. However, this does change this probability measure more fundamentally.

4.10.2   The Infinite Wisdom Number

Coming back to the definition of Chaitin's Ω, every time that a computer program p halts, it contributes to the value of Ω by determining a binary sequence that can be seen as the binary expansion of a real number between 0 and 1 – because remember, this is a probability. For example, in 2007, Cristian Calude computed the first 43 bits of a Chaitin Ω for a certain universal Turing machine. The first digits of this Ω are as follows:

$$0.0001000000010000101001110111000011111110101$$

The longer the program length $|p|$, the smaller its contribution to Ω, so the sequence is not only very difficult to calculate for increasing program lengths but it is also uncomputable, because we know that given the halting problem we can never really know which computer programs will halt. This is why this Ω number is also called the halting probability. From the formula it can be seen that short programs have the greatest weight in the fraction, because the smaller the denominator gets, the larger the values will be, and so the shortest computer programs contribute the most significant values of Ω. But like algorithmic complexity, Ω is also semi-computable, meaning that one can estimate it by, for example, fixing a programming language framework and running random programs, just as Calude, and later we ourselves did in 2007 [52].

4.11   Epistemological Aspects of Infinite Wisdom

There is a question, of course, about whether there is any value in attempting to estimate uncomputable objects such as Ω numbers or uncomputable functions such as the Busy Beaver functions that we came across in Chapter 3. We will see that we have found a use that no one thought possible, but to answer in a more general fashion, let us take an extreme position. Let us say that we had all the digits of an Ω number. How would it be useful? Well, we would have the exact halting time for every possible computer program, because even though Ω depends on the choice of Turing machine, we also know that, given Turing universality, any universal Turing machine used to calculate Ω will also calculate all the computer programs enumerated by any other universal Turing machine. In other words, having access to the infinite digits of Ω not only gives us the halting probability of a universal Turing machine, but in some fundamental way it gives us access to all of them, and under the Church–Turing thesis it also means that we have access to basic knowledge about all the computer programs that are possible. And computer programs can be anything, from a simulation of our solar system to the answer to any mathematical problem, including those for which we would know there is no answer if the computer program that encodes the problem never halts.

This is why the Ω number is also known as the infinite wisdom number. It is comparable to Jorge Luis Borges' infinite library, a repository of all possible knowledge (Fig. 4.18). However, it is also of a very different nature to the books contained in Borges' infinite library because in Borges' library, if you remember his short story, books can be conceived of as having all possible permutations of letters or words, whereas here we only have those permutations that have computational content, that are the result of a computation, and so it is a more meaningful subset of all possible statistical combinations. In other words, if you were given the choice of looking into an infinite library containing an infinite number of books from all possible permutations of words and letters or an infinite library of all possible computer programs, you should definitely choose the library of infinite computer programs, as you would more easily find answers to questions such as Fermat's last theorem. Furthermore, notice how knowing the Busy Beaver values that we explored in the previous Chapter would also give us information about the halting probability, as they would tell us that for certain computer program sizes we could decide whether they would halt or not and then calculate some digits of Ω. Hopefully you are beginning to see how everything is connected.

Of course, it is a different matter to tell which program is which, and which program encodes the question we want Chaitin's Ω to answer. However, you already know that short computer programs encode most of the meaningful objects that we care

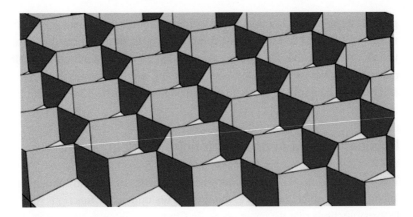

Figure 4.18 Sketch of Borges' library with an infinite number of hexagonal rooms full of bookshelves. Based on Plotted: A Literary Atlas by Andrew DeGraff. Copyright © 2015 Andrew DeGraff | Printed with permission of Zest Books, a division of Lerner Publishing Group. All rights reserved.

about, such as the mathematical constant π, compared to arbitrarily long programs, because computer programs that are arbitrarily long relative to their output may be encoding random objects that, while they may be good applications for pseudo-random generation useful in a casino, say, are hardly worth much else. So by looking at the length of computer programs relative to what they produce, we may gain a sense of what they may be encoding, as we search for interesting questions and answers.

Now you may begin to see how algorithmic complexity is related to knowledge and meaning in a very profound manner, contravening the usual portrayal of computation as incapable of dealing with deep questions about meaning and epistemology because it can only treat these concepts as trivial generalisations of classical information theory. Shannon entropy lacks the most important ingredient, precisely the concept of computation. But we also saw how even Shannon entropy is related to meaning, albeit in simpler ways.

One could think of having access to any digit of a Chaitin Ω number in finite time as having access to some sort of oracle, because one can always formulate questions in terms of whether a computer program will halt using only yes and no questions that Chaitin's Ω would be able to answer. In fact, one would have the answers to all mathematical questions, including simulations of real-world phenomena. In a sense, the Chaitin Ω oracle is similar to the computer Deep Thought in Douglas Adams's story *The Hitchhiker's Guide to the Galaxy*. In a fundamental way, asking questions using the Chaitin Ω number is like asking Deep Thought the Ultimate Question of Life, the Universe, and Everything. Like Deep Thought, Chaitin's Ω number holds all the answers to all the questions, but we would still need to figure out how to formulate these answers correctly.

But just as in this science fiction story where the computer Deep Thought gave the answer '42' to the ultimate question, any answer given by a Chaitin Ω number would be hard to understand and, in principle, impossible to follow by mechanical computation.

And again, just as in Adams' story, one would need to rely on another more powerful computer to verify the answer, which in turn may provide a more puzzling and impossible-to-follow answer. This is at the heart of Gödel's and Turing's proofs of degrees of undecidability and uncomputability. However, only knowing the first n bits of Ω would enable us to decide whether or not each program up to n bits in length ever halts, so knowing all digits would enable us to decide the halting problem of all possible computer programs. But even knowing a few bits would give us a lot of power. For example, Calude himself [53] has discovered how large the computer programs would need to be to solve mathematical problems such as Riemann's hypothesis and Goldbach's conjecture, to mention two famous problems.

When thinking of Chaitin's Ω in terms of a wisdom number containing infinite knowledge, including the answers to all questions that can be formulated as computer programs (e.g., all open mathematical problems and more), it is very interesting to find that the digits of Ω are unattainable and incompressible, meaning that there are no shortcuts to reach this knowledge. No mechanistic process can outrun or outsmart Ω because it cannot be derived by any means simpler than the sequence of bits in Ω itself. This doesn't mean that one cannot calculate a few digits of Ω for a number of cases. For example, if we knew that computer programs 0, 10, and 110 all halt (notice they are prefix-free), then we would know that the first digits of Ω are 0.111, and in turn, if we had started with 0.111 for this Ω number, we would know that the programs 0, 10, and 110 halt. In this sense, Ω encodes and maximally 'compresses' information about the halting state of all possible computer programs. Therefore, knowing Chaitin's Ω one could solve the halting problem, but knowing Chaitin's Ω we also know how to compress all possible programs, and hence because Chaitin's Ω cannot be further compressed, it is also algorithmically random.

4.12   Algorithmic Probability and the Universal Distribution

There is another measure, related to Chaitin's Ω, that quantifies the probability that a string is produced by a random computer program, that is, a computer program whose instructions are picked at random. This measure of algorithmic probability is closely related to Chaitin's Ω, but in addition to the halting probability we are also interested in the output of the computer programs that halt and, in particular, in their frequency distributions. That is, in how often a string is produced by a random computer program. Introduced by Solomonoff [43] and further expanded by Levin [44] and by Chaitin himself [51, 54] by way of his papers on the Ω number, algorithmic probability is a probability measure with the same properties as Chaitin's Ω. For example, just like Chaitin's Ω, algorithmic probability, which we will denote by the letter m, depends on the choice of reference universal Turing machine. Its value from the sum is never equal to 1, not only because there are many strings that are produced, but also because of all the computer programs that never halt. For this reason, this measure of algorithmic probability is also called a semi-probability measure, or semi-measure for short, and can formally be defined as follows:

$$m_U(s) = \sum_{p\,:\,U(p)=s} 1/2^{|p|}.$$

Any similarity to Chaitin's Ω is definitely not a coincidence. From this measure of algorithmic probability m defined for a universal Turing machine, one can approximate a Chaitin Ω for the same universal Turing machine. Thus, it is in a sense a more general measure. In contrast, Chaitin's Ω can help calculate m but it does not produce algorithmic probability itself.

However, notice that, unlike Chaitin's Ω, algorithmic probability is applied to an object, in this case a string, for which we want to estimate the probability of its being produced by a random computer program. Unlike classical probability, we are interested in how often a string can be produced algorithmically by a computer program, and this is therefore called algorithmic probability rather than simply probability. Also, just like Ω, $m(s)$ is lower semi-computable because one can numerically estimate values that are lower bounds.

In classical probability theory, we ask after the probability of an object such as a string being picked at random from a set of other strings according to a distribution. We have seen how often it is the case that with no information about the underlying distribution it is customary to assume the uniform distribution. So when talking about classical probability, we assign the same probability to every string of the same length. The classical probability of a binary string s of length n among all possible 2^n strings of the same length, is then $1/2^n$. However, this is not the case for algorithmic probability. Because if s can be produced by a short computer program, then its probability will be larger than that of a string that requires a long computer program.

Definition 4.1 The classical probability of production of a bit string s among all 2^n bit strings of size n is

$$P(s) = 1/2^n.$$

Definition 4.2 Let U be a (prefix-free from Kraft's inequality) universal Turing machine and p a program that produces s running on U, then

$$m(s) = \sum_{p\,:\,U(p)=s} 1/2^{|p|} < 1.$$

So algorithmic probability allows us to introduce a natural bias related to the underlying generating mechanisms, in this case the likelihood that a random computer program will produce a string, without having to assume probability distributions that in the most common situations we wouldn't have access to. And there would be no need to assume an almost arbitrary distribution such as the uniform distribution, which would make it hard if not impossible to differentiate subtle but important apparent properties of different objects. So, a sequence such as 11010010 would be differentiated from a sequence of only 1s or only 0s that we are sure can be generated by extremely small and simple computer programs.

Algorithmic probability induces a semi-computable probability distribution over all strings that is called the 'universal distribution'. The universal distribution has sometimes been described as miraculous in the literature because of its fundamental properties. This distribution is universal because, taking as a basis the work of Solomonoff [43], Levin [44], and, independently, Chaitin [51], proved that the measure is independent of the choice of programming language or reference universal Turing machine, in the same sense as in the invariance theorem for algorithmic complexity. In fact, these two measures, algorithmic complexity and algorithmic probability, are different sides of the same coin, and in a fundamental way they are basically the same as one can be derived from the other in a beautiful and elegant way, as we will see in the next section.

Algorithmic probability is usually regarded as a formalisation of Occam's razor because it formally applies the dictum establishing that 'among competing hypotheses, the hypothesis with the fewest assumptions should be selected'. In this case, the 'fewest assumptions' is formalised as the shortest in length as produced by a computer program. At the same time, algorithmic probability also complies with Epicurus' Principle of Multiple Explanations, which establishes that 'if several hypotheses are consistent with the data, one should retain them all' and, indeed, one can see from the definition of algorithmic probability that every computer program producing the data s is not only retained but contributes to its algorithmic probability, even though it is the shortest computer program that contributes the most and is thus the most likely to be producing the data according to algorithmic probability. Algorithmic probability is not just an interesting cherry-picked measure, it is the accepted mathematical theory of optimal inference and it imparts sense to a claim that Chaitin has made in the past, that 'comprehension is compression', because the most likely explanation for a piece of data, according to algorithmic probability, is also the most compressed.

The notion behind algorithmic probability is very intuitive, powerful, and even elegant. For example, if one wished to produce the digits of the mathematical constant π at random, one would have to try time after time until one managed to hit upon the first numbers corresponding to an initial segment of the decimal expansion of π. The probability of producing the digits of π, however, is extremely small. It is 1/10 digits raised to the number of desired digits of π to be produced, so the probability falls exponentially. For example, the classical probability of producing the very first 2,400 digits is $1/10^{2400}$.

But, if instead of shooting out random numbers to produce the digits of π, one was to randomly shoot out computer programs able to produce the digits of π, the resulting probability would be extremely different. A program that produces the digits of π would have a higher probability of being produced by a computer program. This is because concise and known formulae for π can be coded in short computer programs that generate any arbitrary number of digits of π as opposed to random objects that would have much longer programs.

4.13   The Infinite Programming Monkey

One way to understand algorithmic probability is in terms of a famous metaphor involving a monkey and a typewriter. The infinite monkey theorem, introduced by Émile

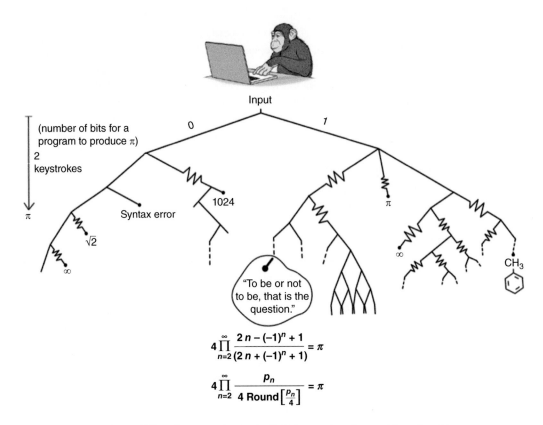

Input

(number of bits for a
program to produce π)

2
keystrokes

π

0

1

1024

π

Syntax error

$\sqrt{2}$

∞

∞

∞

CH_3

"To be or not
to be, that is the
question."

$$4\prod_{n=2}^{\infty} \frac{2n-(-1)^n+1}{(2n+(-1)^n+1)} = \pi$$

$$4\prod_{n=2}^{\infty} \frac{p_n}{4\,\mathbf{Round}\left[\frac{p_n}{4}\right]} = \pi$$

Figure 4.19 When the typewriter in the 'infinite monkey theorem' is replaced by a computer, and keystrokes represent symbols in a computer language, highly structured objects have greater probability of being generated. This probability is called 'algorithmic probability' and beautifully connects algorithmic complexity (object description length) to classical probability.

Borel in the context of randomness, is a metaphor for the probability that random events may be capable of producing structured sequences such as words. The theorem states that a monkey hitting keys at random on a typewriter can produce any text, no matter how long and sophisticated, even the complete works of William Shakespeare, given enough time.

But if instead of a typewriter one placed a monkey in front of a computer, with each key on the keyboard representing an instruction in binary for a computer program, it turns out that the probability of the monkey producing the complete works of Shakespeare, or any object of high complexity, increases dramatically, to the point that it might be likely to happen during the span of time equivalent to the age of the universe! Figure 4.19 illustrates the idea.

Indeed, according to algorithmic probability, outputs encoding information such as Shakespeare's plays, which are far from random, have a greater chance of being the output of a random computer program because the program producing them would be shorter than something producing pure randomness. In other words, the less random –

Figure 4.20 We have written a computer program so you can compare how much faster or slower a monkey on a typewriter or a computer can generate objects of different algorithmic complexity and verify that a monkey working on a computer has a greater chance of producing patterns. The program is available online at https://demonstrations.wolfram.com/InfiniteMonkeyTheorem/.

and therefore more compactly describable – the higher the probability. What algorithmic probability tells us is that it is far easier to try to reproduce something such as life, and for humans to produce something like Hamlet in, as we know, a much shorter time span than the age of the universe, and as the result of a sophisticated physical, chemical, and biological process, than it is to reproduce Hamlet from scratch letter by letter.

This small computer program illustrates how a short random computer program is capable of producing highly structured non-random outputs with much greater likelihood than classical probability theory would indicate. The exact algorithmic probability cannot really be computed, only approximated. A lower bound using Shannon entropy tells us that the probability of the programmer monkey hitting the target binary sequence is 1/(the shortest program producing the string), which cannot be shorter than the logarithm in base 2 of the string length, and should be quite close if the string is highly compressible or not random. The computer program shown in Fig. 4.20 illustrates that this is actually the case, and that the number of keystrokes is rather accurate.

4.14   The Algorithmic Coding Theorem

Recall the formal definition of algorithmic probability given by $m_U(s) = \sum_{p:U(p)=s} 1/2^{|p|}$. It can be seen that the largest term in the sum of the equation is obtained when the denominator is the smallest, that is, when the length of p is the smallest, namely the shortest length of program p in bits that produces s. But the length of the shortest program is none other than the algorithmic complexity of s. Hence, we have found a beautiful connection between the frequency of production of a string and its algorithmic complexity.

Theorem 4.1   *The algorithmic coding theorem,*

$$K(s) = -\log_2 m(s) + O(1).$$

The algorithmic coding theorem formally establishes the connection between algorithmic complexity, here denoted by K, and algorithmic probability. The theorem establishes that the algorithmic complexity of a string is proportional to the negative logarithm of its algorithmic probability. In other words, if a string is produced by many programs, then there is also a short computer program that produces the string, and if a short computer program produces a string, then it will also be more likely to be produced by more computer programs. This actually suggests a way to calculate algorithmic complexity by way of algorithmic probability, avoiding the use of the lossless compression algorithms that are so widely used to approximate algorithmic complexity. When introducing algorithmic information dynamics we will show how such a method can be implemented.

The implications in the real-world of something like algorithmic probability are very broad and fascinating if one permits oneself some speculation. For example, according to classical mechanics, the world may be an unfolding algorithmic process. Suppose that all phenomena in nature can be carried out by a Turing machine as a computation, and therefore that sub-processes are also shorter computations. Then the algorithmic probability m of a physical event s would tell us something about the likelihood of s actually happening. In fact, algorithmic probability can be used to explain the generation of structure out of randomness – order in the universe out of nothing.

4.15   Bennett's Logical Depth: A Measure of Sophistication

One criticism of algorithmic complexity is that it does not conform to our common-sense intuition of complexity. For example, when we say that something is complex we usually mean that it is quite sophisticated, not trivial, and not random. A measure of the complexity of a string can be arrived at by combining the notions of algorithmic information content and time. According to the concept of logical depth introduced by Charles Bennett [55], the complexity of a string is best defined by the time that an unfolding process takes to reproduce the string from its shortest description. The longer the time it takes, the more complex. Hence, complex objects are those which can be seen as 'containing internal evidence of a non-trivial causal history'.

Unlike algorithmic complexity, which assigns randomness its highest complexity values, logical depth assigns a low complexity or low depth to both random and trivial objects. It is thus more in keeping with our intuition of complex physical objects, because trivial and random objects are intuitively easy to produce, have no lengthy history, and they unfold very quickly.

Logical depth was originally advanced as the appropriate measure for evaluating the complexity of real-world objects such as living beings. Hence its alternative designation: physical complexity (used by Bennett himself). This is because the concept of logical depth takes into account the plausible history of an object. It combines the shortest possible description of the object with the time taken for this description to evolve to its current state. The addition of time in logical depth results in a reasonable characterisation of the organisational physical complexity of an object, which is not to be had by the application of the concept of algorithmic complexity alone.

A persuasive case for the convenience of the concept of logical depth as a measure of organised complexity, over and against algorithmic complexity by itself, is made by Bennett himself. Bennett's main motivation was actually to provide a reasonable means of measuring the physical complexity of real-world objects. Bennett provides a careful development of the notion of logical depth, taking into account near-shortest programs as well as the shortest one – hence the significance value – to arrive at a reasonably robust measure.

To understand the intuition behind logical depth, think of the following. Imagine you are given a random file in binary a thousand bits in size.

If you try to compress the file you won't be able to do so by much in comparison to, for example, compressing a file consisting of simply one thousand 1s. Now, if you measure the decompression time of the compressed random file you will find that it is not very different from the decompression time of the compressed simple file.

This is because the decompression instructions in both cases are very simple, one because the file is trivial, and the other because there are almost no decompression instructions for a random file given that we are unable to compress it by much. But in fact, you can see that there is a small difference in time between the two, perhaps because the random file does have some slightly more sophisticated decompression steps compared to the totally trivial case.

But when trying to compress a file that is neither trivial nor random, we see that just as we would have expected, the Thue–Morse sequence that is algorithmically generated and produces some non-trivial statistical patterns can be compressed by more bits than the pseudo-random sequence, but by fewer than the trivial file. It is right in the middle, as expected, and the decompression time is also longer, because now there are more decompression instructions than for the random file, and these instructions are less trivial than for the trivial file and thus require more computing time to reproduce the original Thue–Morse sequence. This is exactly the core idea of this beautiful measure of logical depth.

Formally, for a finite string Bennett's first logical depth measure is defined as follows. Let s be a string and d a significance level. A string's depth is given by the following formula:

$$D_d(s) = \min\{T(p) : (|p| - |p'| < d) \wedge (U(p) = s)\},$$

which reads as 'upper case D at significance level lower case d of an object s is the minimum time T taken by a computer program p running on a universal Turing machine U that can reproduce s and whose program length compared to the shortest computer program p' is not greater than d and $U(p)$ reproduces the object s'.

Algorithmic complexity and logical depth are therefore intimately related. The latter depends on the former because one has to first find the shortest programs producing the string and then look for the shortest times. Looking for the shortest programs is equivalent to approximating the algorithmic complexity of a string. While the compressed versions of a string are approximations of its algorithmic complexity, decompression times are approximations of the logical depth for that string.

For algorithmic complexity, the choice of universal Turing machine is bounded by an additive constant by the invariance theorem, which we covered in a previous section, while for logical depth the choices involved are bounded by a multiplicative factor. Hence the choices are more important, yet they are still somehow under control, at least theoretically.

4.16   The Use and Misuse of Lossless Compression

Notice that the same problem affects compression algorithms, widely used to approximate K. They are not exempt from the same constant problem. Lossless compression is also subject to the constant involved in the *invariance theorem*, because there is no reason to choose one compression algorithm over another.

Lossless compression algorithms have traditionally been used to approximate the Kolmogorov complexity of an object (e.g., a string) because they can provide upper bounds to K and compression is a sufficient test for non-randomness. In a similar fashion, our approximations are upper bounds based on finding a small Turing machine producing a string. Data compression can be viewed as a function that maps data onto other data using the same units or alphabet (if the translation is into different units or a larger or smaller alphabet, then the process is called an encoding).

Compression is successful if the resulting data are shorter than the original data plus the decompression instructions needed to fully reconstruct said original data. For a compression algorithm to be lossless, there must be a reverse mapping from compressed data to the original data. That is to say, the compression method must encapsulate a bijection between 'plain' and 'compressed' data because the original data and the compressed data should be in the same units. By a simple counting argument, lossless data compression algorithms cannot guarantee compression for all input data sets because there will be some inputs that do not get smaller when processed by the compression algorithm, and for any lossless data compression algorithm that makes at least one file smaller, there will be at least one file that it makes larger. Strings of data of length N or shorter are clearly a strict superset of the sequences of length $N - 1$ or shorter. It follows, therefore, that there are more data strings of length N or shorter than there are data strings of length $N - 1$ or shorter. And it follows from the *pigeonhole principle* that it is not possible to map every sequence of length N or shorter to a unique sequence of length $N - 1$ or shorter. Therefore, there is no single algorithm that reduces the size of all data.

Data compression can be viewed as a function that maps data onto other data using the same units or alphabet (if the translation is into different units or a larger or smaller alphabet, then the process is called a 're-encoding' or simply a 'translation'). Compression is successful if the resulting data are shorter than the original data plus the decompression instructions needed to fully reconstruct said original data. For a compression algorithm to be lossless, there must be a reverse mapping from compressed data to the original data. That is to say, the compression method must encapsulate a bijection between 'plain' and 'compressed' data, because the original data and the compressed data should be in the same units.

One of the more time consuming steps of implementations of, for example, LZ77 compression (one of the most popular), is the search for the longest string match. Most lossless compression implementations are based upon the LZ algorithm. The classical LZ77 and LZ78 algorithms enact a greedy parsing of the input data. That is, at each step, they take the longest dictionary phrase that is a prefix of the currently unparsed string suffix. LZ algorithms are said to be 'universal' because, assuming unbounded memory (arbitrary sliding window length), they asymptotically approximate the (infinite) entropy rate of the generating source [56]. Not only does lossless compression fail to provide any estimation of the algorithmic complexity of small objects [57, 58], it is also no more closely related to algorithmic complexity than Shannon entropy itself [59], being only capable of exploiting statistical regularities (if the observer has no other method to update/infer the probability distribution) [60].

The greatest limitation of popular lossless compression algorithms, in light of algorithmic complexity, is that their implementations only exploit statistical regularities (repetitions up to the size of the sliding window length). Thus, in effect, no general lossless compression algorithm does better than provide the Shannon entropy rate (see Section 5.3) of the objects it compresses. It is then obvious that other possible methods for approximating K are not only desirable but necessary, especially methods that can, at least in principle, and more crucially in practice, detect algorithmic features in data that statistical approaches such as entropy and to some extent compression would miss.

4.17   Discussion Questions

1. What is the main difference between randomness and pseudo randomness?
2. Are all random strings Borel normal?
3. Does the knowledge about the source of the randomness play a role in the selection of the method to measure it?
4. Should one specify a programming language and the particular universal Turing machine on which these computer programs would run for the calculation of algorithmic complexity? Why?
5. In which sense is the infinite wisdom number is comparable to Jorge Luis Borges' infinite library, and in which way it is different?
6. How are algorithmic complexity and algorithmic probability connected?
7. What is the greatest limitation of popular lossless compression algorithms? Can you show it with an example?

5   The Coding Theorem Method *(CTM)*

Chapter Summary

The coding theorem method (CTM) is an alternative to popular compression algorithms such as Lempel–Ziv (LZ). It enables the approximation of algorithmic complexity via algorithmic probability. CTM does not rely upon purely statistical principles, as do popular lossless compression algorithms such as LZ and LZW (which are designed to find statistical regularities and are thus more closely related to classical information theory than to algorithmic complexity). The aim of CTM is to take the probabilistic content that other approaches pump into their generative models out of the generative models explaining natural or artificial phenomena. Statistical and probabilistic approaches have tended to conflate the probabilistic content belonging to a model with the probabilistic content of the phenomena they set out to explain. For example, when modelling the outcome of throwing a dice, what a statistical model quantifies is the degree of uncertainty of the observer, an uncertainty external to the process itself. It thus quantifies a property of the observer, not the process the dice is subjected to. Throws of the dice are deterministic according to the laws of classical mechanics, which are known to govern the trajectory of the dice (quantum fluctuations are believed to be probabilistic, but at short distances they wouldn't have any effect). The probabilistic content of the model describing the outcome of dice throwing is therefore alien to the actual process. An algorithmic probability-like approach such as CTM is significantly different. It produces a set of deterministic models describing the dice trajectory and outcome without recourse to probability. It is in the distribution of possible computable models explaining the dice that probability is introduced; probability is not inherent in the individual models themselves. These models can be tested as mechanistic (step-by-step) descriptions of the process, irrespective of their outcome predictions. Moreover, they represent an alternative to black-box approaches, since each model can be followed step-by-step, with model states corresponding to constructive (e.g., physical) states, not to random variables that do not allow for state-to-state correspondence between model and data.

5.1   Correlation versus Causation

Shannon entropy can only quantify a subset of non-random features in an object, the subset represented by statistical regularities, what we usually know as statistical patterns. Let's consider the following examples:

1. Thue–Morse sequence 01101001100101101001011001101001
2. Segment of π in binary 11001001000011111101101010100010

The above objects have either been proven or are suspected to be Borel normal, meaning they are of maximal entropy. So despite their high algorithmic content they will appear random to entropy if no information about their deterministic source is available (as is the case when tossing a coin, which is the result of a deterministic process, though we model it using a random variable).

For the same reason, we can see that graph entropy is not robust. Let's define in very general terms the Shannon entropy of a graph: $H(G) = - \sum_i^{|G|} P(G_i) \log_2 P(G_i)$, where G_i is a 'counting' property i of G. As shown, H is dependent on the choice of i. Once i is chosen, H can be replaced by a counting function of the form $f(G, i) \sim H(G_i)$ such that H is no longer anything special, simply a counting function of the number of times i occurs in the enumeration G_i. Entropy is therefore not a robust measure, because unlike K (invariance theorem), H is not invariant to different descriptions of the same object. H is not a robust measure of randomness.

To illustrate the limitations of entropy we created a very simple graph (Fig. 5.1). Let $1 \rightarrow 2$ be a starting graph G connecting a node with label 1 to a node with label 2. If a node with label n has degree n, we call it a core node; otherwise, we call it

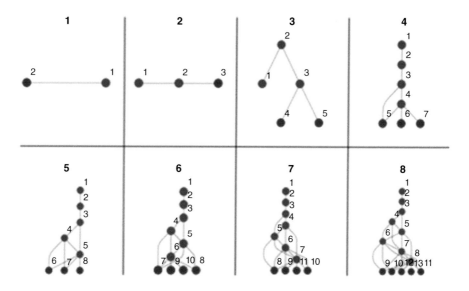

Figure 5.1 A recursive graph that has its lowest and highest entropy at the same time, depending on the way the graph is described.

a supportive node. Iteratively make the supportive node with the lowest label a core node by adding outgoing edges to immediately following, potentially new, nodes. The degree sequence d of the labelled nodes $d = 1, 2, 3, 4, \ldots, n$ is the Champernowne constant in base 10, a transcendental real whose decimal expansion is Borel normal. The sequence of number of edges is a function of core and supportive nodes, defined by $[1/r] + [2/r] + \ldots + [n/r]$, where $r = (1 + \sqrt{5})/2$ is the golden ratio and $[\ ]$ the floor function (sequence A183136 in the OnLine Encyclopedia of Integer Sequences [https://oeis.org/]), whose values are: 1, 2, 4, 7, 10, 14, 18, 23, 29, 35, 42, 50, 58, 67, 76, 86, 97, 108, 120, 132, 145, ...

So from the ZK graph we draw contradictory conclusions: The sparse adjacency of ZK has an entropy rate 0 at the limit, so it is not random; the degree sequence of ZK, however, has maximal entropy (1, 2, 3, ...), so it is random.

From both descriptions the graph can be unequivocally reconstructed. This means that the descriptions are lossless (and deterministic); they neither add or delete any information. Reconstructing an object from its entropy values means reconstructing its statistical properties, not the object itself, unlike the program for K.

The ZK graph can be constructed recursively for any number of nodes n by nesting the AddEdges[ ] function as follows:

```
1  EdgeList /@ NestList[AddEdges, Graph[{1 -> 2}], n]
```

where

```
1   AddEdges[graph_] :=
2    EdgeAdd[
3     graph,
4     Rule @@@ Distribute[
5      {Max[VertexDegree[graph]] + 1,
6       Table[
7        i,
8        {i,
9         Max[VertexDegree[graph]] + 2,
10        2 (Max[VertexDegree[graph]] + 1) -
11         VertexDegree[graph, Max[VertexDegree[graph]] + 1]}]},
12     List]]
```

Unlike popular lossless compression algorithms, which are guaranteed not to be able to characterise objects such as s, the CTM can, in principle, do so. This is because when running all possible computer programs up to the size of s in bits, there is a non-zero probability that the CTM will find such a program if it exists (i.e., if s is not algorithmically random, which we know this particular s is not).

5.2   Relevance of Algorithmic Complexity in the Practice of Science

(Un)Computability mediates in the challenge of causality by way of the algorithmic information theory:

$$K_U(s) = \min\{|p|, U(p) = s\},$$

s:	Observation
min:	Most likely mechanism according to Occam's razor
$|p|$:	Generating mechanism
U:	Constructor
$U(p) = s$:	Compatible with the observation

The Coding theorem establishes the connection between $m(s)$ and $K(s)$:

$$m(x) = \sum_{p:U(p)=x} 1/2^{|p|}.$$

$$K(s) = -\log_2 m(s) + O(1).$$

Essentially, the CTM uses the fact that the more frequent a string is, the lower the Kolmogorov complexity it has; and strings of lower frequency have higher Kolmogorov complexity:

$$D(n, 2)(s) = \frac{\text{\# of times that a machine } (n, 2) \text{ produces } s}{\text{\# of machines in } (n, 2)},$$

$$CTM(s) \sim -\log_2 D(n, 2)(s).$$

If any part of the whole system (samples) is of high $m(x)$ and low $K(x)$, then that part can be generated by mechanistic/algorithmic means and thus is causal. The lower BDM, the more causal. For example, consider 2 strings 10111 and 10111. A tighter upper bound of their algorithmic complexity K is not: $K(U(p) = 10111) + K(U(p) = 10111)$ but $K(U(p) = 10111$ 2 times) because, clearly, $K(U(p) = 10111) + K(U(p) = 10111) > K(U(p) = 10111$ 2 times).

The coding theorem method involves approximating the algorithmic complexity of an object by running every possible program, from the shortest to the longest, and counting the number of times that a program produces every string of the object.

The length of the computer program will be an upper bound of the algorithmic complexity of the object, following the coding theorem (Eq. (4.3)) and a (potentially) compressed version of the data itself (the shortest program found), for a given computer language or 'reference' universal Turing machine (UTM). This guarantees discovery of the shortest program in the given language or reference UTM but entails an exhaustive search over the set of countable infinite computer programs that can be written in such a language. A program of length n has an asymptotic probability close to 1 of halting in time 2^n [61], making this procedure exponentially expensive, even assuming that all programs halt or that programs are assumed never to halt after a specified time, with those that do not being discarded.

As shown in [57] and [58], an exhaustive search can be carried out for a small-enough number of computer programs (more specifically, Turing machines) for which the halting time is known because of the Busy Beaver problem [36]. This problem sets us the task of finding the Turing machine of fixed size (states and symbols) that runs longer than any other machine of the same size. Values are known for Turing machines with 2 symbols and up to 4 states that can be used to stop a resource-bounded exploration, that is, by discarding any machine taking more steps than the Busy Beaver values. For longer strings, we also proceed with an informed runtime cut-off below the theoretical 2^n optimal runtime that guarantees an asymptotic drop of non-halting machines [61] but above the value needed to capture most strings with a specific degree of accuracy, as demonstrated in [62].

The CTM [57, 58] is a bottom-up approach to algorithmic complexity and, unlike common implementations of lossless compression algorithms, the main motivation of CTM is to find algorithmic features in data that are beyond the range of application of Shannon entropy and popular lossless compression algorithms [59], not merely statistical regularities.

CTM is rooted in the relation [57, 58] established by algorithmic probability between the frequency of production of a string from a random program and its algorithmic complexity as described by Eq. (4.3). Essentially, it uses the fact that the more frequent a string is, the lower its Kolmogorov complexity, with strings of lower frequency having a higher Kolmogorov complexity. The advantage of using algorithmic probability to approximate K by application of the coding theorem (4.3) is that $m(s)$ produces reasonable approximations of K based on an average frequency of production, which retrieves values even for small objects.

Let (t, k) denote the set of all Turing machines with t states and k symbols using the Busy Beaver formalism [36], and let T be a Turing machine in (t, k) with empty input. Then the empirical output distribution $D(t, k)$ for a sequence s produced by a given $T \in (t, k)$ yields an estimation of the *algorithmic probability* of s, $D(t, k)(s)$ defined by:

$$D(t,k)(s) = \frac{|\{T \in (t,k) : T \ produces \ s\}|}{|\{T \in (t,k) : T \ halts \ \}|}. \tag{5.1}$$

For small values of t and k, $D(t, k)$ is computable for values of the Busy Beaver problem that are known. The Busy Beaver problem [36] is the problem of finding the t-state, k-symbol Turing machine that writes a maximum number of non-blank symbols before halting, starting from an empty tape, or the Turing machine that performs a maximum number of steps before halting, having started on an initially blank tape. For $t = 4$ and $k = 2$, for example, the Busy Beaver machine has maximum runtime $S(t) = 107$ [63], from which one can deduce that if a Turing machine with 4 states and 2 symbols running on a blank tape hasn't halted after 107 steps, then it will never halt. This is how D was initially calculated – by using known Busy Beaver values. However, because of the undecidability of the halting problem, the Busy Beaver problem is only computable for small t, k values [36]. Nevertheless, one can continue approximating D for a greater number of states (and colours), proceeding by sampling, as described in [57, 58], with an informed runtime based on both theoretical and numerical results.

Notice that $0 < D(t, k)(s) < 1$; $D(t, k)(s)$ is thus said to be a semi-measure, just as $m(s)$ is.

Now we can introduce a measure of complexity that is heavily reliant upon *algorithmic probability* $m(s)$, as follows:

Table 5.1 Empirical distributions calculated from rulespace $(t;k)$. Letter codes: F full space, S sample, $R(t;k)$ reduced enumeration. Time is given in seconds (s), hours (h), and days (d).

(t,k)	Calculation	Number of Machines	Time
$(2,2)$	$F- (6 \text{ steps})$	$\lvert R(2,2) \rvert = 2,000$	0.01 s
$(3,2)$	$F- (21)$	$\lvert R(3,2) \rvert = 2,151,296$	8 s
$(4,2)$	$F- (107)$	$\lvert R(4,2) \rvert = 3,673,320,192$	4 h
$(4,2)_{2D}$	$F_{2D}- (1,500)$	$\lvert R(4,2)_{2D} \rvert = 315,140,100,864$	252 d
$(4,4)$	$S\ (2,000)$	334×10^9	62 d
$(4,5)$	$S\ (2,000)$	214×10^9	44 d
$(4,6)$	$S\ (2,000)$	180×10^9	41 d
$(4,9)$	$S\ (4,000)$	200×10^9	75 d
$(4,10)$	$S\ (4,000)$	201×10^9	87 d
$(5,2)$	$F- (500)$	$\lvert R(5,2) \rvert = 9,658,153,742,336$	450 d
$(5,2)_{2D}$	$S_{2D}\ (2,000)$	$1,291 \times 10^9$	1,970 d

Let (t,k) be the space of all t-state k-symbol Turing machines, $t,k > 1$, and $D(t,k)(s) = $ the function assigned to every finite binary string s. Then,

$$\text{CTM}(s,t,k) = -\log_b D(t,k)(s), \tag{5.2}$$

where b is the number of symbols in the alphabet (traditionally 2 for binary objects, which we will take as understood hereafter). That is, the more frequently a string is produced, the lower its Kolmogorov complexity, with the converse also being true.

Table 5.1 shows the rule spaces of Turing machines that were explored, from which empirical algorithmic probability distributions were sampled and estimated.

We will designate as *base string*, *base matrix*, or *base tensor* the objects of size l for which CTM values were calculated such that the full set of k^l objects have CTM evaluations. In other words, the base object is the maximum granularity of application of CTM. Table 5.1 provides figures relating to the number of base objects calculated.

Validations of CTM undertaken before show the correspondence between CTM values and the exact number of instructions used by Turing machines when running to calculate CTM [64, Fig. 1 and Table 1] to produce each string, i.e., direct K complexity values for this model of computation (as opposed to CTM using algorithmic probability and the coding theorem) under the chosen model of computation [36]. The correspondence in values found between K as directly calculated, and CTM arrived at by way of frequency of production was near perfect.

Sections 7.1.2 and 7.2, and Figs. 10, 11, 12, and 15 in [65] support the agreements in correlation using different rule spaces of Turing machines and different computing models altogether (cellular automata). Section 7.1.1 of the same paper provides a first comparison to lossless compression. The sections 'Agreement in probability' and 'Agreement in rank' provide further material comparing rule space (5,2) to the rule space (4,3) previously calculated in [57]. The section 'Robustness' in [58] provides evidence relating to the behaviour of the *invariance theorem constant* for a standard model of Turing machines [36].

5.3   Building upon Block Entropy

The entropy H of a discrete random variable s with possible values $s_1, \ldots, s_n$ and probability distribution $P(s)$ is defined as

$$H(s) = - \sum_{i=1}^{n} P(s_i) \log_2 P(s_i).$$

In the case of $P(s_i) = 0$ for a given i, the value of the corresponding summand $0 \log_2(0)$ is taken to be 0.

It is natural to ask how random a string appears when blocks of finite length are considered. For example, the string $01010101\ldots01$ is periodic, but for the smallest granularity (1 bit) or 1-symbol block, the sequence has maximal entropy, because the number of 0s and 1s is the same, assuming a uniform probability distribution for all strings of the same finite length. Only for longer blocks, of length 2 bits, can the string be found to be regular, identifying the smallest entropy value for which the granularity is at its minimum.

When dealing with a given string s, assumed to originate from a stationary stochastic source with known probability density for each symbol, the following function H_l gives what is variously denominated as block entropy and is Shannon entropy over blocks (or subsequences of s) of length l. That is,

$$H_l(s) = - \sum_{b \in \text{blocks}} P_l(b) \log_2 P_l(b),$$

where blocks is the set resulting from the decomposition of s into substrings or blocks of size l and $P_l(b)$ is the probability of obtaining the combination of n symbols corresponding to the block b. For infinite strings assumed to originate from a stationary source, the *entropy rate* of s can be defined as the limit

$$\lim_{l \to \infty} \frac{1}{l} \sum_{|s'|=l} H_l(s'),$$

where $|s'| = l$ indicates that we are considering all strings generated of length l. For a fixed string, we can think of the normalised block entropy value, where l better captures the periodicity of s.

Entropy was originally conceived by Shannon as a measure of information transmitted over a stochastic communication channel with known alphabets, and it establishes hard limits for maximum lossless compression rates. For instance, Shannon coding (and Shannon–Fano coding) sorts the symbols of an alphabet according to their probabilities, assigning smaller binary self-delimited sequences to symbols that appear more frequently. Such methods form the base of many, if not most, commonly used compression algorithms.

Given its utility in data compression, entropy is often used as a measure of the information contained in a finite string $s = s_1 s_2 \ldots x \ldots s_k$. Let's consider the *natural distribution*, the uniform distribution that makes the least number of assumptions but which does consider every possibility equally likely – hence its uniformity. Suggested

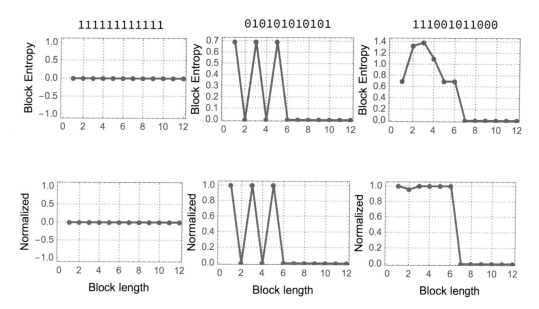

Figure 5.2 The best version of Shannon entropy can be rewritten as a function of variable block length where the minimum value best captures the (possible) periodicity of a string, here illustrated with three strings of length 12, regular, periodic, and random-looking, respectively. Because blocks larger than $n/2$ would in effect be single blocks and therefore have an entropy equal to 0, the largest possible block is $n/2$. The normalised version (bottom) divides the entropy value for the relevant block size by the largest possible number of blocks for that size and alphabet (here, binary). From H. Zenil, S. Hernández-Orozco, N.A. Kiani, F. Soler-Toscano, A. Rueda-Toicen. A Decomposition Method for Global Evaluation of Shannon Entropy and Local Estimations of Algorithmic Complexity. *Entropy* 20(8), 605, 2018.

by the set of symbols in s and the string length, the *natural distribution* of s is the distribution defined by $P(x) = \frac{n_x}{|s|}$, where n_x is the number of times the object x occurs in s (at least one to be considered) and the respective entropy function H_l. If we consider blocks of size $n >> l \geq 2$ and the string $s = 01010101\ldots01$, where n is the length of the string, then s can be compressed in a considerably smaller number of bits than a statistically random sequence of the same length and, correspondingly, has a lower H_l value. However, entropy with the natural distribution suggested by the object, or any other computable distribution, is a computable function, and is therefore an imperfect approximation of algorithmic complexity.

The best possible version of a measure based upon entropy can be reached by partitioning an object into blocks of increasing size (up to half the length of the object) in order for Shannon entropy to capture any periodic statistical regularity. Fig. 5.2 illustrates the way in which such a measure operates on three different strings.

However, no matter how sophisticated a version or variation of an entropic measure is used, it will fail to characterise algorithmic aspects of data that are not random and escape entropy if no knowledge about the source is available. Fig. 5.3 shows how algorithmic probability/complexity can find such patterns and ultimately characterise any of them, including statistical ones, thereby offering a generalisation of and a complement to the application of entropy by itself.

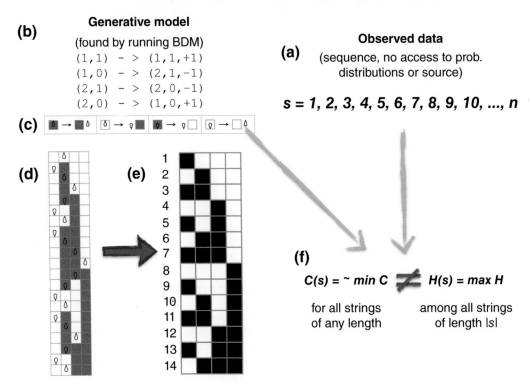

Figure 5.3 (a) Observed data, a sequence of successive positive natural numbers. (b) The transition table of the Turing machine found by running all possible small Turing machines. (c) The same transition table in visual form. (d) The space-time evolution of the Turing machine starting from an empty tape. (e) Space-time evolution of the Turing machine implementing a binary counter, taking as halting criterion the leftmost position of the original Turing machine head as depicted in (c) (states are arrows). (e) This small computer program that our CTM and BDM methods find (see Chapter 6) means that the sequence in (a) is not algorithmically random, because the program represents a succinct generative causal model for any arbitrary length, though Shannon entropy would have assigned it a maximal randomness among all strings of the same length (in the absence of other knowledge about the source) despite its highly algorithmic non-random, structured nature. Entropy alone – which without access to probability distributions can only spot statistical regularities – cannot find this kind of generative model, demonstrating the low randomness of an algorithmic sequence. From H. Zenil, S. Hernández-Orozco, N.A. Kiani, F. Soler-Toscano, A. Rueda-Toicen. A Decomposition Method for Global Evaluation of Shannon Entropy and Local Estimations of Algorithmic Complexity. *Entropy* 20(8), 605, 2018.

BDM builds upon block entropy's decomposition approach, using algorithmic complexity methods to obtain and combine building blocks. The result is a complexity measure that, as shown in Section 6.4, approaches K in the best case and behaves like entropy in the worst case (Section 6.10), outperforming H_l in various scenarios. First we introduce the algorithm that demarcates the building blocks of BDM, which are local estimations of algorithmic complexity. Specific examples of objects that not even

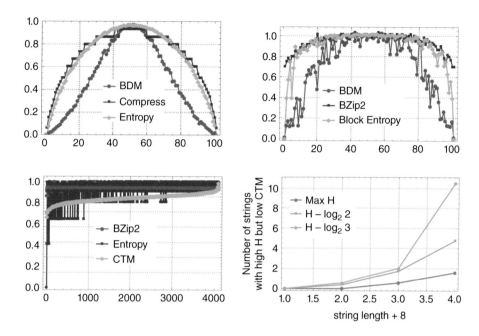

Figure 5.4 Strings that are assigned lower randomness than that estimated by entropy. (Top left) Comparison between values of entropy, compression (`Compress[]`), and BDM over a sample of 100 strings of length 10,000 generated from a binary random variable following a Bernoulli distribution and normalised by maximal complexity values. Entropy follows a Bernoulli distribution and, unlike compression, which follows entropy, BDM values produce clear convex-shaped gaps on each side, assigning lower complexity to some strings compared to both entropy and compression. (Top right) The results confirmed using a popular lossless compression algorithm BZip2 (and also confirmed, though not reported, with LZMA) on 100 random strings of 100 bits each (BZip2 is slower than Compress but achieves greater compression). (Bottom left) The $CTM_{low}(s) - H_{high}(s)$ gap between near-maximal entropy and low algorithmic complexity grows and is consistent along different string lengths, here from 8 to 12 bits. This gap is the one exploited by BDM and extended over longer strings by it, which gives it the algorithmic edge over entropy and compression. (Bottom right) When strings are sorted by CTM, one notices that BZip2 collapses most strings to minimal compressibility. Over all $2^{12} = 4,096$ possible binary strings of length 12, entropy only produces six different entropy values, but CTM is much more fine-grained, and this is extended to the longer strings by BDM, which succeeds in identifying strings of lower algorithmic complexity that have near-maximal entropy and therefore no statistical regularities. Examples of such strings are in Section 6.8. From H. Zenil, S. Hernández-Orozco, N.A. Kiani, F. Soler-Toscano, A. Rueda-Toicen. A Decomposition Method for Global Evaluation of Shannon Entropy and Local Estimations of Algorithmic Complexity. *Entropy* 20(8), 605, 2018.

block entropy can characterise are found in Section 6.8, showing how our methods are a significant improvement over any measure based upon entropy and traditional statistics (see Fig. 5.4).

For example, the following two strings were assigned near maximal complexity but they were found to have low algorithmic complexity by CTM/BDM, given that we were able to find not just one small Turing machine that reproduces them but indeed many Turing machines producing them upon halting. Which means that, by the coding theorem, they are of low algorithmic complexity: 001010110101, 001101011010 (and their negations and reversions). These strings display nothing particularly special and in a sense they look typically random. Yet we were expecting this – to find strings that would appear random but are actually not algorithmically random. These strings would have been assigned higher randomness by entropy and popular lossless compression algorithms, but are assigned lower randomness by our methods, which thus represent a real advantage over methods that can only exploit statistical regularities (Fig. 5.4).

6   The Block Decomposition Method *(BDM)*

Chapter Summary

In this chapter we will introduce the block decomposition method (BDM). One way to view BDM is as a weighted version of Shannon's entropy that injects algorithmic randomness into the original formulation of classical information, and is therefore able to distinguish statistical randomness from algorithmic randomness. This is not a trivial task, since these weights are actually uncomputable, though approximations are possible using CTM, which we covered in the previous chapter. We will learn that the distinction between statistical and algorithmic randomness is a key distinction because (1) an object would be characterised as maximally 'disordered' by statistical approaches such as Shannon entropy, unless one had prior information that it had been generated by a deterministic process, which of course defeats the purpose of using the measure (in science this is the rule rather than the exception: one investigates objects of unknown nature) and (2) the distinction is of great empirical value in scientific practice. For example, if one wished to quantify human memory, it would not be necessary to make a subject learn $s = 12345678...$ digit by digit in order to be able to reproduce the sequence. This illustrates the algorithmic nature of even the simplest human cognitive abilities, whereas measures of statistical randomness like Shannon entropy would characterise s as random. The sequence s may seem special, but in fact most sequences are of this type; they won't have any statistical regularity but would not be algorithmically random. Another example is the digits of a mathematical constant such as π. There are many more numbers like π than there are numbers having only statistical patterns. CTM and BDM have found many applications in psychometrics and cognition due to this advantage (see Part III). In this chapter you will learn that what BDM does is to extend the power of CTM to quantify algorithmic randomness by implementing a divide-and-conquer approach whereby data are decomposed into pieces small enough so that an exhaustive computational search can find the set of all computable models able to generate said data. The sequence of small generators then supports the larger piece of data and provides insight into the algorithmic properties of the latter. Small pieces of data supported by even smaller programs than themselves are called causal patches in the context of algorithmic information dynamics (AID), as they are the result of a causal relationship between the set of these computable models and the observed data, whereas models of about the same length as the data are not causally explained by a shorter generating mechanism and are therefore considered random and not causally supported. A sequence of computer programs smaller than their respective matched patches constitutes a sufficient test for non-randomness, and is therefore informative as regards the causal content of a large piece of data, as its components can be explained by underlying computable models shorter than the data itself.

Capturing the 'complexity' of an object, for purposes such as classification and object profiling, is one of the most fundamental challenges in science. This is so because one has to either choose a computable measure (e.g., Shannon entropy) that is not invariant to object descriptions and probability distributions [66] and lacks an *invariance theorem* – forcing one to make an arbitrary choice of a particular feature of interest among several possible others – or else estimate values of an uncomputable function when applying a 'universal' measure of complexity that is invariant to object description (such as *algorithmic complexity*). This latter drawback has led to computable variants and the development of time- and resource-bounded algorithmic complexity/probability that is finitely computable [67, 68, 69, 70].

A good introduction and list of references is provided in [71]. Here, we study a measure that lies half-way between two universally used measures that enables the action of both at different scales by dividing data into smaller pieces, for which the halting problem besetting an uncomputable function can be partially circumvented in exchange for a huge calculation based upon the concept of algorithmic probability. The calculation can, however, be precomputed, and hence re-used in future applications, thereby constituting a strategy for efficient estimations – bounded by Shannon entropy and by algorithmic (Kolmogorov–Chaitin) complexity – in exchange for a loss of accuracy.

In the past, lossless compression algorithms dominated the landscape of applications of algorithmic complexity. When researchers have chosen to use lossless compression algorithms for reasonably long strings, the method has proven to be of value (see, e.g., [72]). Their successful application has had to do with the fact that compressibility is a sufficient test for non-algorithmic randomness (though the converse is not true). However, popular implementations of lossless compression algorithms are based upon estimations of *entropy* [59], and are therefore no more closely related to algorithmic complexity than is Shannon entropy in and of itself. They can only account for statistical regularities and not for algorithmic ones, though accounting for algorithmic regularities ought to be crucial, since these regularities represent the main advantage of using algorithmic complexity.

One of the main difficulties with computable measures of complexity such as Shannon entropy is that they are not robust enough [60, 66]. For example, they are not invariant to different descriptions of the same object – unlike algorithmic complexity, where the so-called *invariance theorem* guarantees the invariance of an object's algorithmic complexity. This is due to the fact that one can always translate a lossless description into any other lossless description simply with a program of a fixed length, in effect merely adding a constant. Computability theorists are not much concerned with the relatively negligible differences between evaluations of Kolmogorov complexity, which are owed to the use of different descriptive frameworks (e.g., different programming languages), yet these differences are fundamental in applications of algorithmic complexity.

Here, we study a block decomposition method (BDM) that is meant to extend the power of the so-called *coding theorem method* (CTM). Applications of CTM include image classification [65] and visual cognition [73, 74, 75], among other applications in cognitive science. In these applications, other complexity measures, including entropy and lossless compressibility, have been outperformed by CTM. Graph complexity is another subject of active research [76, 77, 78, 79, 80]. The method presented here has made a contribution to this subject by proposing robust measures of algorithmic graph complexity [81, 82].

6.1   Being Greedy, Divide and Conquer!

Because finding the program that reproduces a large object is computationally very expensive and ultimately uncomputable, one can aim at finding short programs that reproduce small fragments of the original object, parts that together compose the larger object. This is precisely what the BDM does.

BDM is divided into two parts. On the one hand, approximations to K are performed by CTM, with the values obtained then being applied in $O(1)$ by exchanging time for memory in the entries of a precomputed look-up table for small strings. This diminishes its precision as a function of object size (string length) unless a new iteration of CTM is precomputed. On the other hand, BDM decomposes the original data into fragments for which CTM provides an estimation, and then puts the values together based upon classical information theory.

BDM is thus a hybrid complexity measure that applies Shannon entropy over the long range while providing local estimations of algorithmic complexity. It is meant to improve upon Shannon entropy, which in practice is reduced to finding statistical regularities, and to extend the power of CTM. It proceeds by decomposing objects into smaller pieces for which algorithmic complexity approximations have been numerically estimated using CTM, then reconstructing an approximation of the Kolmogorov complexity for the larger object by adding the complexity of the individual components of the object, according to the rules of information theory. For example, if s is an object and $10s$ is a repetition of s ten times smaller, upper bounds can be achieved by approximating $K(s) + \log_2(10)$ rather than $K(10s)$ because we know that repetitions have a very low Kolmogorov complexity, given that one can describe repetitions with a short algorithm.

Unlike popular lossless compression algorithms that are guaranteed to be unable to characterise objects such as 12345678..., CTM can, in principle, do so, because when running all possible computer programs up to the size of s in bits, there is a non-zero probability that CTM will find such a program if it exists (i.e., if s is not algorithmically random, which we know this particular s is not).

6.2   Properties of BDM

Based on a method explored in [57, 58], BDM takes advantage of the powerful relationship established by algorithmic probability between the frequency of a string produced by a random program running on a *(prefix-free)* UTM and the string's Kolmogorov complexity. The chief advantage of this method is that it deals with small objects with ease, and it has shown stability in the face of changes of formalism, producing reasonable Kolmogorov complexity approximations.

BDM must be combined with CTM if it is to scale up properly and behave optimally for upper-bounded estimations of K. BDM + CTM is universal in the sense that it is guaranteed to converge to K given the invariance theorem and, as we will prove later, if CTM no longer runs, then BDM alone approximates the Shannon entropy of a finite object.

Like compression algorithms, BDM is subject to a trade-off. Compression algorithms face a trade-off between compression power and compression/decompression speed.

6.3   *l*-overlapping String Block Decomposition

Let us fix values for t and k and let $D(t, k)$ be the frequency distribution constructed from running all the Turing machines with n states and k symbols. Following Eq. (5.1), we have it that $- \log D$ is an approximation of K (denoted by CTM). We define the BDM of a string or finite sequence s as,

$$\text{BDM}(s, l, m) = \sum_i \text{CTM}(s^i, m, k) + \log(n_i), \qquad (6.1)$$

where n_i is the multiplicity of s^i, and s^i the subsequence i after decomposition of s into subsequences s^i, each of length l, with a possible remainder sequence $y < |l|$ if $|s|$ is not a multiple of the decomposition length l.

The parameter m goes from 1 to the maximum string length produced by CTM, where $m = l$ means that there is no overlapping, inducing a partition of s. Thus, m is an overlapping parameter when $m < l$, for which we will investigate its impact on BDM (in general, the smaller m, the greater the overestimation of BDM).

The parameter m is needed because of the remainder. If $|s|$ is not a multiple of the decomposition length l, then the option is to either ignore the remainder in the calculation of BDM or define a sliding window with overlapping $m - l$.

The choice of t and k for CTM in BDM depends only on the available resources for running CTM, which involves running the entire (t, k) space of Turing machines with t symbols and k states.

BDM approximates K in the following way: if p_i is the minimum program that generates each base string s^i, then $\text{CTM}(s^i) \approx |p_i|$ and we can define a unique program q that runs each p_i, obtaining all the building blocks. How many times each block is present in s can be given in $\log n_i$ bits. Therefore, BDM is the sum of the information needed to describe the decomposition of s in base strings. How close this sum comes to K is explored in Section 6.4.

The definition of BDM is interesting because one can plug in other algorithmic distributions, even computable ones, approximating any measure of algorithmic complexity, even if it is not the one defined by Kolmogorov–Chaitin, such as the one defined by Calude et al. [83] based upon finite-state automata. BDM thus allows the combination of measures of classical information theory and algorithmic complexity.

For example, for binary strings we can use $t = 2$ and $k = 2$ to produce the empirical output distribution $(2, 2)$ of all machines with 2 symbols and 2 states by which all strings of size $l = 12$ are produced, with the exception of two (one string and its complement). But we assign them values $\max \{\text{CTM}(y, 2, 2) + r : |y| = 12\}$ where e is different from zero because the missing strings were not generated in $(2, 2)$ and therefore have a greater algorithmic random complexity than any other string produced in $(2, 2)$ of the same length. Then, for $l = 12$ and $m = 1$, $\text{BDM}(s, l, m)$ decomposes $s = 010101010101010101$ of length $|s| = 18$ into the following sub-sequences:

$$010101010101$$
$$101010101010$$
$$010101010101$$
$$101010101010$$
$$010101010101$$
$$101010101010$$
$$010101010101$$

with 010101010101 having multiplicity 4 and 101010101010 a multiplicity of 3.

We then get the CTM values for these sequences:

$$\text{CTM}(010101010101, 2, 2) = 26.99073,$$
$$\text{CTM}(101010101010, 2, 2) = 26.99073.$$

To calculate BDM, we then take the sum of the CTM values plus the sum of the $\log_b$ of the multiplicities, with $b = 2$ because the string alphabet is 2, the same as the number of symbols in the set of Turing machines producing the strings. Thus,

$$\log_2(3) + \log_2(4) + 26.99 + 26.99 = 57.566.$$

To ask after the likelihood of an array, we can consider a two-dimensional Turing machine. The block decomposition method can then be extended to objects beyond unidimensional ones such as strings, e.g., arrays representing bitmaps such as images, or graphs (by way of their adjacency matrices). We would first need CTM values for two- and w-dimensional objects that we call base objects (e.g., base strings or base matrices).

A popular example of a two-dimensional tape Turing machine is Langton's ant [84]. Another way to see this approach is to take the BDM as a way of deploying all possible two-dimensional deterministic Turing machines of small size in order to reconstruct the adjacency matrix of a graph from scratch (or smaller pieces that fully reconstruct it). Then, as with the *coding theorem method* (above), the algorithmic complexity of the adjacency matrix of the graph can be estimated via the frequency with which it is produced from running random programs on the (prefix-free) two-dimensional Turing machine. More specifically,

$$BDM(X, \{x_i\}) = \sum_{(r_i, n_i) \in \text{Adj}(X)_{\{x_i\}}} CTM(r_i) + \log(n_i), \qquad (6.2)$$

where the set $\text{Adj}(X)_{\{x_i\}}$ is composed of the pairs (r, n), r is an element of the decomposition of X (as specified by a *partition* $\{x_i\}$, where x_i is a submatrix of X) in different sub-arrays of size up to $d_1 \times \ldots \times d_w$ (where w is the dimension of the object) that we call a *base matrix* (because CTM values were obtained for them) and n is the multiplicity of each component. $CTM(r)$ is a computable approximation from below of the algorithmic information complexity of r, $K(r)$, as obtained by applying the coding theorem method to w-dimensional Turing machines. In other words, $\{r_i\}$ is the set of *base objects*.

Because string block decomposition is a special case of matrix block decomposition, and square matrix block decomposition is a special case of w-block decomposition for objects of w dimensions, let us describe the way in which BDM deals with boundaries on square matrices, for which we can assume CTM values are known, and that we call base strings or base matrices.

Figure 6.1 shows that the number of permutations is a function of the complexity of the original object, with the number of permutations growing in proportion to the original object's entropy – because the number of different resulting blocks determines the number of different n objects to distribute among the size of the original object (e.g. 3 among 3 in Fig. 6.1 (top) or only two different 4×4 blocks in Fig. 6.1 [bottom]). This means that the non-overlapping version of BDM is not invariant vis-à-vis the variation of the entropy of the object, on account of which it has a different impact on the error introduced in the estimation of the algorithmic complexity of the object. Thus, non-overlapping objects of low complexity will have little impact, but with random objects, non-overlapping increases inaccuracy. Overlapping decomposition solves this particular permutation issue by decreasing the number of possible permutations in order to avoid trivial assignment of the same BDM values. However, overlapping has the undesired effect of systematically overestimating values of algorithmic complexity by counting almost every object of size n, $n - 1$ times, hence overestimating at a rate of about $n(n - 1)$ for high complexity objects of which the block multiplicity will be low, and by $n \log(n)$ for low complexity objects.

Applications to graph theory [81], image classification [73] and human behavioural complexity have been produced in the last few years [74, 75].

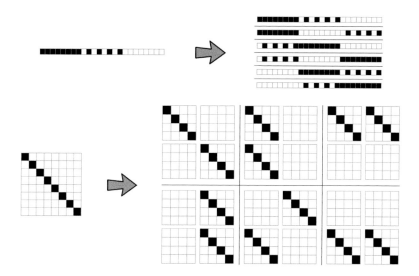

Figure 6.1 Non-overlapping BDM calculations are invariant to block permutations (reshuffling base strings and matrices), even when these permutations may have different complexities due to the reorganisation of the blocks that can produce statistical or algorithmic patterns. For example, starting from a string of size 24 (top) or an array of size 8×8 (bottom), with decomposition length $l = 8$ for strings and decomposition $l = 4 \times 4$ block size for the array, all six permutations for the string and all six permutations for the array have the same BDM value, regardless of the shuffling procedure.

6.4   BDM Upper and Lower Absolute Bounds

In what follows we show the hybrid nature of the measure. We do this by setting lower and upper bounds to BDM in terms of the algorithmic complexity $K(X)$, the partition size, and the approximation error of CTM, such that these bounds are tighter in direct relation to smaller partitions and more accurate approximations of K. These bounds are independent of the partition strategy defined by $\{x_i\}$.

Proposition 6.1   *Let BDM be the function defined in Eq. (6.2) and let X be an array of dimension w. Then $K(X) \leq \mathrm{BDM}(X, \{x_i\}) + O\left(\log^2 |A|\right) + \epsilon$ and $\mathrm{BDM}(X, \{x_i\}) \leq | \mathrm{Adj}(X)_{\{x_i\}} | K(X) + O\left(|\mathrm{Adj}(X)_{\{x_i\}}| \log |\mathrm{Adj}(X)_{\{x_i\}}|\right) - \epsilon$, where A is a set composed of all possible ways of accommodating the elements of $\mathrm{Adj}(X)_{\{x_i\}}$ in an array of dimension w, and ϵ is the sum of errors for the approximation of CTM over all the sub-arrays used.*

Proof   Let $\mathrm{Adj}(X)_{\{x_i\}} = \{(r_1, n_1), ..., (r_k, n_k)\}$ and $\{p_j\}$, $\{t_j\}$ be the sequences of programs for the reference prefix-free UTM U such that, for each $(r_j, n_j) \in \mathrm{Adj}(X)_{\{x_i\}}$, we have $U(p_j) = r_j$, $U(t_j) = n_j$, $K(r_j) = |p_j|$, and $|t_j| \leq 2 \log(n_j) + c$. Let ϵ_j be a positive constant such that $\mathrm{CTM}(r_j) + \epsilon_j = K(r_j)$; this is the error for each sub-array. Let ϵ be the sum of all the errors.

For the first inequality we can construct a program q_w, whose description only depends on w, such that, given a description of the set $\mathrm{Adj}(X)_{\{x_i\}}$ and an index l, it enumerates all the ways of accommodating the elements in the set and returns the array corresponding to the position given by l.

Note that $|l|$, $|\operatorname{Adj}(X)_{\{x_i\}}|$, and all n_js are of the order of $\log|A|$. Therefore,

$$U(q_w p_1 t_1 ... p_k t_k l) = X$$

and

$$K(X) \leq |q_w p_1 t_1 ... p_k t_k l|$$

$$\leq |q_w| + \sum_1^k (|p_j| + |t_j|) + |l|$$

$$\leq \mathrm{BDM}(X, \{x_i\}) + \epsilon + |q_w|$$
$$+ (\log|A| + c)\,|\operatorname{Adj}(X)_{\{x_i\}}| + O(\log|A|)$$

$$\leq \mathrm{BDM}(X, \{x_i\}) + O(\log^2|A|) + \epsilon,$$

which gives us the inequality.

Now, let q_X be the smallest program that generates X. For the second inequality we can describe a program $q_{\{x_i\}}$, which, given a description of X and the index j, constructs the set $\operatorname{Adj}(X)_{\{x_i\}}$ and returns r_j, i.e., $U(q_{\{x_i\}} q_X j) = r_j$. Note that each $|j|$ is of the order of $\log|\operatorname{Adj}(X)_{\{x_i\}}|$. Therefore, for each j we have

$$\mathrm{CTM}(r_j) + \epsilon_j = |p_j| \leq |q_{\{x_i\}}| + |q_X| + O\left(\log|\operatorname{Adj}(X)_{\{x_i\}}|\right)$$

and

$$\mathrm{CTM}(r_j) + \epsilon_j + \log(n_j) \leq |q_{\{x_i\}}| + |q_X| + O\left(\log|\operatorname{Adj}(X)_{\{x_i\}}|\right) + \log(n_j).$$

Finally, by adding all the terms over the js we find the second inequality:

$$\mathrm{BDM}(X, \{x_i\}) + \epsilon \leq |\operatorname{Adj}(X)_{\{x_i\}}|\left(|q_X| + |q_{\{x_i\}}| + \log(n_j) + O\left(\log|\operatorname{Adj}(X)_{\{x_i\}}|\right)\right)$$

$$\leq |\operatorname{Adj}(X)_{\{x_i\}}|\,K(X) + O\left(|\operatorname{Adj}(X)_{\{x_i\}}|\log|\operatorname{Adj}(X)_{\{x_i\}}|\right).$$

$\square$

Corollary 6.2   *If the partition defined by $\{x_i\}$ is small, that is, if $|\operatorname{Adj}(X)_{\{x_i\}}|$ is close to 1, then $\mathrm{BDM}(X, \{x_i\}) \approx K(X)$.*

Proof   Given the inequalities presented in Proposition 6.1, we have that

$$K(X) - O\left(\log^2|A|\right) - \epsilon \leq \mathrm{BDM}(X, \{x_i\})$$

and

$$\mathrm{BDM}(X, \{x_i\}) \leq |\operatorname{Adj}(X)_{\{x_i\}}|\,K(X) + O\left(|\operatorname{Adj}(X)_{\{x_i\}}|\log|\operatorname{Adj}(X)_{\{x_i\}}|\right) - \epsilon,$$

which at the limit leads to $K(X) - \epsilon \leq \mathrm{BDM}(X) \leq K(X) - \epsilon$ and $\mathrm{BDM}(X) = K(X) - \epsilon$. From [58], we can say that the error rate ϵ is small, and that by the invariance theorem it will converge towards a constant value.

$\square$

6.5   Dealing with Object Boundaries

Because partitioning an object – a string, array or tensor – leads to boundary leftovers that are not multiples of the partition length, the only two options for taking into consideration such boundaries in the estimation of the algorithmic complexity of the entire object are to either estimate the complexity of the leftovers or to define a sliding window allowing overlapping in order to include the leftovers in some of the block partitions. The former implies mixing object dimensions that may be incompatible (e.g., CTM complexity based on one-dimensional TMs versus CTM based on higher dimensional TMs). Here, we explore these strategies for dealing with object boundaries. We introduce a strategy for partition minimisation and *base object* size maximisation that we will illustrate for two-dimensionality. The strategies are intended to overcome under- or over-fitting complexity estimations that are attributable to conventions, not just technical limitations (e.g., uncomputability and intractability).

Notice that some of the explorations in this section may give the impression that we are introducing and using ad-hoc methods to deal with object boundaries. However, this is not the case. What we will do in this section is explore all the conceivable ways in which we can estimate K according to BDM, taking into consideration boundaries that may require special treatment when they are not length-wise multiples of the partition lengths resulting from the decomposition of the data after BDM. Moreover, we show that in all cases, the results are robust, because the errors found are convergent and can thus be corrected. Thus, any apparently ad-hoc condition has little to no implications for the calculation of BDM at the limit, and only a limited impact at the beginning.

6.6   Recursive BDM

In Section 6.4, we showed that using smaller partitions for BDM yields more accurate approximations of the algorithmic complexity K. However, the computational costs of calculating CTM are high. We have compiled an exhaustive database for square matrices of size up to 4×4. Therefore, it is in our best interest to find a method to minimise the partition of a given matrix into squares of size up to $d \times d = l$ for a given l.

The strategy consists in taking the biggest base matrix multiple of $d \times d$ on one corner and dividing it into adjacent square submatrices of the given size. Then we group the remaining cells into two submatrices and apply the same procedure, but now for $(d-1) \times (d-1)$. We continue dividing into submatrices of size 1×1.

Let X be a matrix of size $m \times n$ with $m, n \geq d$. Let's denote by

$$\text{quad} = \{\text{UL}, \text{LL}, \text{DR}, \text{LR}\}$$

the set of quadrants on a matrix and by quad^d the set of vectors of quadrants of dimension l. We define a function $\text{part}(X, d, q_i)$, where $\langle q_1, \ldots, q_d \rangle \in \text{quad}^d$, as follows:

$$\text{part}(X, l, q_i) = \max(X, d, q_i)$$
$$\cup \, \text{part}(\text{resL}(X, d, q_i), d - 1, q_{i+1})$$
$$\cup \, \text{part}(\text{resR}(X, d, q_i), d - 1, q_{i+1})$$
$$\cup \, \text{part}(\text{resLR}(X, d, q_i), d - 1, q_{i+1}),$$

where $\max(X, d, q_i)$ is the largest set of adjacent submatrices of size $d \times d$ that can be clustered in the corner corresponding to the quadrant q_i, $\text{resR}(X, d - 1, q_i)$ is the submatrix composed of all the adjacent rightmost cells that could not fit on $\max(X, d, q_i)$ and are not part of the leftmost cells, $\text{resL}(X, d - 1, q_i)$ is an analogue for the leftmost cells, and $\text{resLR}(X, d - 1, q_i)$ is the submatrix composed of the cells belonging to the rightmost and leftmost cells. We call the last three submatrices *residual matrices*.

By symmetry, the number of matrices generated by the function is invariant with respect to any vector of quadrants $\langle q_1, \dots, q_d \rangle$. However, the final BDM value can (and will) vary according to the partition chosen. Nevertheless, with this strategy we can evaluate all the possible BDM values for a given partition size and choose the partition that yields the minimum value or the maximum value, or else compute the average for all possible partitions.

The partition strategy described can easily be generalised and applied to strings (one dimension) and tensors (objects of n-dimensions).

6.7   Periodic Boundary Conditions

One way to avoid having remainder matrices (from strings to tensors) of different sizes is to embed a matrix in a topological torus (see Fig. 6.2 bottom) such that no more object borders are found. Then let X be a square matrix of arbitrary size m. We screen the matrix X for all possible combinations to minimise the number of partitions maximising block size. We then take the combination of the smallest BDM for fixed *base matrix* size d and we repeat for $d - 1$ until we have added all the components of the decomposed X. This procedure will, however, overestimate the complexity values of all objects (in unequal fashion along the complexity spectra), but will remain bounded, as we will show in Section 6.9.

Without loss of generality, the strategy can be applied to strings (one dimension) and tensors (any larger number of dimensions, e.g. greater than two), the former embedded in a cylinder while tensors can be embedded in n-dimensional tori (see Fig. 6.2).

6.8   BDM versus Shannon Entropy

Let us address the task of quantifying how many strings with maximum entropy rate are actually algorithmically compressible, i.e., have low algorithmic complexity. That is, how many strings are actually algorithmically (as opposed to simply statistically) compressible but are not compressed by popular lossless compression algorithms, which

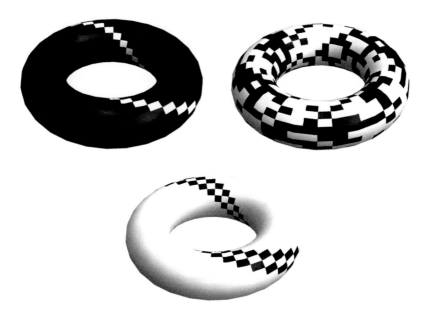

Figure 6.2 One way to deal with the decomposition of n-dimensional tensors is to embed them in an n-dimensional torus ($n = 2$ in the case of the one depicted here), making the borders cyclic or periodic by joining the borders of the object. Depicted here are three examples of graph canonical adjacency matrices embedded in a two-dimensional torus that preserves the object complexity on the surface: a complete graph, a cycle graph, and an Erdös–Rényi graph with edge density 0.5, all of size 20 nodes and free of self-loops. Avoiding borders has the desired effect of producing no residual matrices after the block decomposition with overlapping. From H. Zenil, S. Hernández-Orozco, N.A. Kiani, F. Soler-Toscano, A. Rueda-Toicen. A Decomposition Method for Global Evaluation of Shannon Entropy and Local Estimations of Algorithmic Complexity. *Entropy* 20(8), 605, 2018.

are statistical (entropy rate) estimators [59]. We know that most strings have both maximal entropy (most strings look equally statistically disordered, a fact that is the foundation of thermodynamics) and maximal algorithmic complexity (according to a pigeonhole argument, most binary strings cannot be matched to shorter computer programs as these are also binary strings). But the gap between those with maximal entropy and low algorithmic randomness diverges, and is infinite at the limit (for an unbounded string sequence). That is, there is an infinite number of sequences that have maximal entropy but low algorithmic complexity.

The promise of BDM is that, unlike compression, it does identify some cases of strings with maximum entropy that actually have low algorithmic complexity. BDM assigns lower complexity to more strings than entropy, as expected. Unlike entropy and implementations of lossless compression algorithms, BDM recognises some strings that have no statistical regularities but have algorithmic content that makes them algorithmically compressible.

Examples of strings with lower randomness than that assigned by entropy and block entropy are 101010010101 (and its complement) or the Thue-Morse sequence

011010011001 . . .

(or its complement) obtained by starting with 0 and successively appending the Boolean complement [85], the first with low CTM $=$ 29 and the Thue–Morse with CTM $=$ 33.13 (the maximum CTM value in the subset is 37.4 for the last string in this table). The Morse sequence is uniformly recurrent without being periodic, not even in its further reaches, so it will continue to have high entropy and high block entropy.

CTM and BDM as functions of the object's size (and therefore the size of the Turing machine rule space that has to be explored) have the following time complexity:

- CTM is uncomputable but for decidable cases runs in exponential time.
- Non-overlapping string BDM and LD runs in linear time, and n^d polynomial time for d-dimensional objects.
- Overlapping BDM runs in ns time, with m the overlapping offset.
- Full overlapping with $m = 1$ runs in 2^n polynomial time as a function of the number of overlapping elements n.
- Smooth BDM runs in linear time.
- Mutual information BDM runs in exponential time for strings and d exponential for dimension d.

So how does this translate into the real profiling power of recursive strings and sequences that are of low algorithmic complexity but appear random to classical information theory? Figures 6.3, and 6.4 provide real examples showing how BDM can outperform the best versions of Shannon entropy.

Figure 6.3 shows the randomness estimation of two known low algorithmic complexity objects and CTM to BDM transitions of the mathematical constant π and the Thue–Morse sequence, to both of which numerical estimations by CTM assign lower randomness than that suggested by entropy and its best version, block entropy. We expect to find that CTM does much better at characterising the low algorithmic randomness of a sequence like the Thue–Morse sequence (beyond the fact that it is not Borel normal [85]), given that every part of the sequence is algorithmically obtained from another part of the sequence (by logical negation or the substitution system $0 \rightarrow 01, 1 \rightarrow 10$ starting from 0) while the digits of π have been shown to be independent (at least in powers of 2) from each other [86] and only algorithmic in the way they are produced from any of the many known generating formulae.

One of the most recently discovered formulae producing any digits of any segment of the mathematical constant π (in base 2^k), is given by a very short symbolic summation [86]: $\sum_{n=1}^{\infty}(4/(8n + 1) - 2/(8n + 4) - 1/(8n + 5) - 1/(8n + 6))/k^n$. The probability of producing such a string of 80 ASCII characters (less than 1K bits) is $1/256^{83} \times f$, or $1.30642 \times 10^{-200} \times f$ if typing 'random formulae' by chance, where f is a multiplying factor quantifying the number of other formulae of fixed (small) size that can also produce π and of which many are known from the work of Vieta, Leibniz, Wallis, Euler, Ramanujan, and others. In theory, classical probability is exponentially divergent from the much higher algorithmic probability $1/2^n$, that is, the classical probability of producing an initial segment of π (in binary) of length n. Good sources of formulas producing digits of π can be found at the Online Encyclopedia of Integer Sequences (OEIS) (https://oeis.org/A000796) listing more than 50 references,

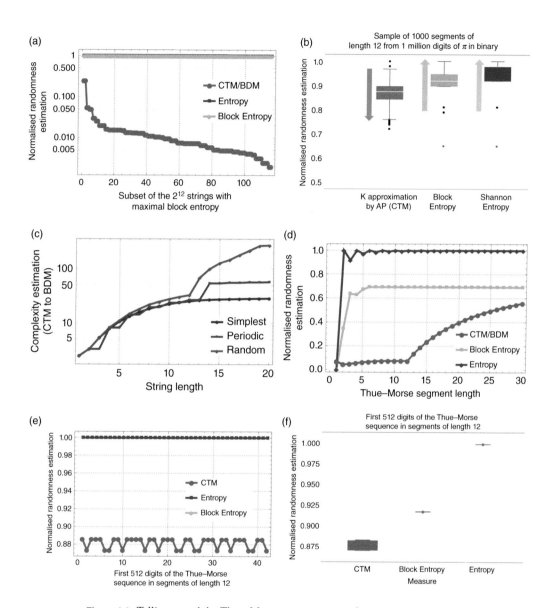

Figure 6.3 Telling π and the Thue–Morse sequence apart from truly (algorithmically) random sequences. CTM assigns significantly lower randomness (b, d, e, f) to known low algorithmic complexity objects. (b) If absolutely Borel normal (as strongly suspected and statistically demonstrated to a specific degree of confidence), π's entropy and block entropy are asymptotically approximate to 1, while by the invariance theorem of algorithmic complexity, CTM asymptotically approximates 0. Smooth transitions between CTM and BDM are also shown (c,d) as a function of string complexity. Other smooth transition functions of BDM are explored and introduced in Section 6.12. From H. Zenil, S. Hernández-Orozco, N.A. Kiani, F. Soler-Toscano, A. Rueda-Toicen. A Decomposition Method for Global Evaluation of Shannon Entropy and Local Estimations of Algorithmic Complexity. *Entropy* 20(8), 605, 2018.

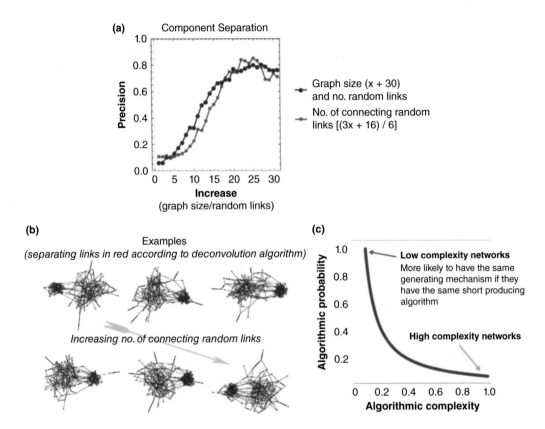

Figure 6.4 Example of causal deconvolution by algorithmic complexity based on the detection of each element contribution at different algorithmic information regimes in an example of network separation with different topology (random versus small world) connected by random links. From H. Zenil, N.A. Kiani, A. Zea, J. Tegnér. Causal Deconvolution by Algorithmic Generative Models. *Nature Machine Intelligence*, vol 1, pages 58–66, 2019.

and at Wolfram MathWorld, listing around a hundred (http://mathworld.wolfram.com/PiFormulas.html).

Unlike classical probability, algorithmic probability quantifies the likelihood of production of the object (Fig. 6.3, and Fig. 6.4) by indirect algorithmic/recursive means rather than by direct production (the typical analogy is writing on a computer equipped with a language compiler program versus writing on a typewriter).

To the authors' knowledge, no other numerical method is known that suggests the low algorithmic randomness of statistically random-looking constants such as π and the Thue–Morse sequence (Fig. 6.3, and Fig. 6.4) from the perspective of an observer lacking access to prior information, probability distributions, or knowledge about the nature of the source (i.e., a priori deterministic).

There are two sides of BDM. On the one hand, it is based on a non-computable function, but once approximations are computed we build a lookup table of values that make BDM computable. Lossless compression is also computable but it is taken as able to make estimations of an uncomputable function like K because it can provide

Table 6.1 Summary of ranges of application and scalability of CTM and all versions of BDM. Here, d stands for the dimension of the object.

	short strings < 100 bits	long strings > 100 bits	scalability
Lossless compression	$\times$	✓	$O(n)$
Coding Theorem Method (CTM)	✓	$\times$	$O(\exp)$ to $O(\infty)$
Non-overlapping BDM	✓	✓	$O(n)$
Full-overlapping Recursive BDM	✓	✓	$O(n^{d-1})$
Full-overlapping Smooth BDM	✓	✓	$O(n^{d-1})$
Smooth add col BDM	✓	✓	$O(n)$

upper bounds and estimate K from above, just as we do with CTM (and thus BDM) by exhibiting a short Turing machine capable of reproducing the data/string.

Table 6.1 summarises the range of applications, with CTM and BDM pre-eminent in that they can more efficiently deal with short, medium, and long sequences and other objects such as graphs, images, networks, and higher dimensionality objects.

6.9 Error Estimations

One can estimate the error in different calculations of BDM, regardless of the error estimations of CTM (quantified in [57, 58]), in order to calculate their departure and deviation both from granular entropy and algorithmic complexity, for which we know the lower and upper bounds (Fig. 6.5). For example, a maximum upper bound for binary strings is the length of the strings themselves. This is because no string can have an algorithmic complexity greater than its length, simply because the shortest computer program (in bits) to produce the string is the string itself.

In the calculation of BDM, when an object's size is not a multiple of the base object of size d, boundaries of size $< d$ will be produced, and there are various ways of dealing with them to arrive at a more accurate calculation of an object that is not a multiple of the base. First, we will estimate the error introduced by ignoring the boundaries or dealing with them in various ways, and then we will offer alternatives to take into consideration in the final estimation of their complexity.

If a matrix X of size $k \times j$ is not a multiple of the base matrix of size $d \times d$, it can be divided into a set of decomposed blocks of size $d \times d$, and R, L, T, and B residual matrices on the right, left, top, and bottom boundaries of M, all of smaller size than d.

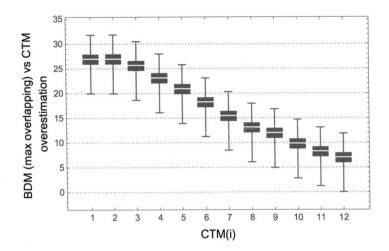

Figure 6.5 Box plot showing the error introduced by BDM quantified by CTM. The sigmoid appearance has to do with the fact that we actually have exact values for CTM up to bitstrings of length 12 and so BDM(12) = CTM(*i*) for *i* = 12, but the slope of the curve gives an indication of the errors when assuming that BDM has only access to CTM(*i*) for *i* < 12 versus the actual CTM(12). This means that the error grows linearly as a function of CTM and of the string length, and the accuracy degrades smoothly and slowly towards entropy if CTM is not updated. From H. Zenil, S. Hernández-Orozco, N.A. Kiani, F. Soler-Toscano, A. Rueda-Toicen. A Decomposition Method for Global Evaluation of Shannon Entropy and Local Estimations of Algorithmic Complexity. *Entropy* 20(8), 605, 2018.

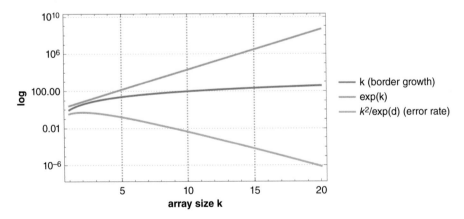

Figure 6.6 Error rate for two-dimensional arrays. With no loss of generalisation, the error rate for *n*-dimensional tensors $\lim_{d \to \infty} \frac{k^n}{n^k} = 0$ is convergent and thus negligible, even for the discontinuities, disregarded in this plot, which are introduced by some BDM versions, such as non-overlapping blocks and discontinuities related to trimming the boundary condition. From H. Zenil, S. Hernández-Orozco, N.A. Kiani, F. Soler-Toscano, A. Rueda-Toicen. A Decomposition Method for Global Evaluation of Shannon Entropy and Local Estimations of Algorithmic Complexity. *Entropy* 20(8), 605, 2018.

Then boundaries *R*, *L*, *T* and *B* can be dealt with in the following way:

- Trimming boundary condition: *R*, *L*, *T*, and *B* are ignored, then BDM(*X*) = BDM(*X*, *R*, *L*, *T*, *B*), with the undesired effect of general underestimation

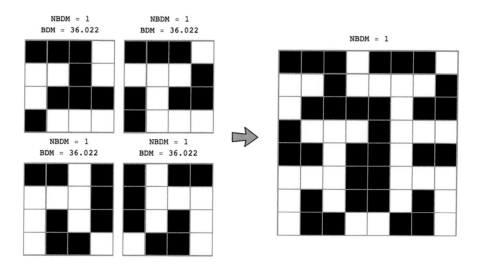

Figure 6.7 NBDM assigns a maximum value of 1 to any base matrix with the highest CTM or any matrix constructed out of base matrices. In this case, the four base matrices on the left are those with the highest CTM in the space of all base matrices of the same size, while the matrix to the left is assigned the highest value because it is built out of the maximum complexity base matrices. From H. Zenil, F. Soler-Toscano, J.-P. Delahaye and N. Gauvrit. Two-Dimensional Kolmogorov Complexity and Validation of the Coding Theorem Method by Compressibility. *PeerJ Computer Science*, 1:e23, 2015.

for objects not multiples of d. The error introduced (see Fig. 6.6) is bounded between 0 (for matrices divisible by d) and $k^2/\exp(k)$, where k is the size of X. The error is thus convergent $\left(\exp(k)\right.$ grows much faster than k^2) and can therefore be corrected, and is negligible as a function of array size, as shown in Fig. 6.6.

- Cyclic boundary condition (Fig. 6.7 bottom): The matrix is mapped onto the surface of a torus such that there are no more boundaries and the application of the overlapping BDM version takes into consideration every part of the object. This will produce an over-estimation of the complexity of the object but will, for the most part, respect the ranking order of estimations if the same overlapping values are used with maximum overestimation $d - 1 \times \max\{\text{CTM}(b) \mid b \in X\}$, where $K(b)$ is the maximum CTM value among all base matrices b in X after the decomposition of X.
- Full overlapping recursive decomposition: X is decomposed into $(d - 1)^2$ base matrices of size $d \times d$ by traversing X with a sliding square block of size d. This will produce a polynomial overestimation in the size of the object of up to $(d - 1)^2$, but if consistently applied it will, for the most part, preserve ranking.
- Adding low complexity rows and columns (we call this 'add col'):
  If a matrix of interest is not a multiple of the size of the base matrices, we add rows and columns until completion to the next multiple of the base matrix, then we correct the final result by subtracting the borders that were artificially added.

The BDM error rate (see top of Fig. 6.2) is the discrepancy in the sum of the complexity of the missed borders, which is an additive value of, at most, polynomial growth. The error is not uniform and linear for objects of different complexity. For a tensor of d

dimensions, with all 1s as entries, the error is bounded by $\log(k^d)$ for objects with low algorithmic randomness and by k^d/d^k for objects with high algorithmic randomness.

Ultimately there is no optimal strategy for making the error disappear, but in some cases the error can be better estimated and corrected (Fig. 6.9) and all cases are convergent, hence asymptotically negligible, and in all cases complexity ranking is preserved and under- and over-estimations bounded.

6.10   BDM Worst-Case Convergence towards Shannon Entropy

Let $\{x_i\}$ be a partition of X defined as in the previous sections for a fixed d. Then the Shannon entropy of X for the partition $\{x_i\}$ is given by:

$$\mathrm{H}_{\{x_i\}}(X) = - \sum_{(r_j, n_j) \in \mathrm{Adj}(X)_{\{x_i\}}} \frac{n_j}{|\{x_i\}|} \log\left(\frac{n_j}{|\{x_i\}|}\right), \qquad (6.3)$$

where $\mathrm{P}(r_j) = \frac{n_j}{|\{x_i\}|}$ and the array r_j is taken as a symbol itself. The following proposition establishes the asymptotic relationship between $\mathrm{H}_{\{x_i\}}$ and BDM.

Proposition 6.3   *Let M be a two-dimensional matrix and $\{x_i\}$ a partition strategy with elements of maximum size $d \times d$. Then,*

$$\left|\mathrm{BDM}_{\{x_i\}}(X) - \mathrm{H}_{\{x_i\}}(X)\right| \le O(\log(|\{x_i\}|)).$$

Proof   First, we note that $\sum n_j = |\{x_i\}|$ and, given that the set of matrices of size $d \times d$ is finite and so is the maximum value for $\mathrm{CTM}(r_j)$, there exists a constant c_d such that $\left|\mathrm{Adj}(X)_{\{x_i\}}\right| \mathrm{CTM}(r_j) < c_d$. Therefore,

$$\mathrm{BDM}_{\{x_i\}}(X) - \mathrm{H}_{\{x_i\}}(X) = \sum\left(\mathrm{CTM}(r_j) + \log(n_j) + \frac{n_j}{|\{x_i\}|}\log\left(\frac{n_j}{|\{x_i\}|}\right)\right)$$

$$\le c_d + \sum\left(\log(n_j) + \frac{n_j}{|\{x_i\}|}\log\left(\frac{n_j}{|\{x_i\}|}\right)\right)$$

$$= c_d + \sum\left(\log(n_j) - \frac{n_j}{|\{x_i\}|}\log\left(\frac{|\{x_i\}|}{n_j}\right)\right)$$

$$= c_d + \frac{1}{|\{x_i\}|}\sum\left(|\{x_i\}|\log(n_j) - n_j\log\left(\frac{|\{x_i\}|}{n_j}\right)\right)$$

$$= c_d + \frac{1}{|\{x_i\}|}\sum\log\left(\frac{n_j^{|\{x_i\}|+n_j}}{|\{x_i\}|^{n_j}}\right).$$

Now, let's recall that the sum of n_j's is bounded by $|\{x_i\}|$. Therefore, there exists c'_d such that

$$\frac{1}{|\{x_i\}|}\sum\log\left(\frac{n_j^{|\{x_i\}|+n_j}}{|\{x_i\}|^{n_j}}\right) \le \frac{c_d}{|\{x_i\}|}\log\left(\frac{|\{x_i\}|^{|\{x_i\}|+c'_d|\{x_i\}|}}{|\{x_i\}|^{c'_d|\{x_i\}|}}\right)$$

$$= \frac{c_d}{|\{x_i\}|}\log\left(|\{x_i\}|^{|\{x_i\}|}\right)$$

$$= c_d\log(|\{x_i\}|). \qquad \square$$

Now, it is important to note that the previous proof sets the limit in terms of the constant c_d, the minimum value of which is defined in terms of matrices for which the CTM value has been computed. The smaller this number is, the tighter is the bound set by Proposition 6.3. Therefore, in the worst case, that is, when CTM has been computed for a comparatively small number of matrices, or the larger base matrix has a low algorithmic complexity, the behaviour of BDM is similar to entropy. In the best case, when CTM is updated by any means, BDM approximates algorithmic complexity (Corollary 6.2).

Furthermore, we can think of $c_d \log(|\{x_i\}|)$ as a measure of the deficit in information incurred by entropy and BDM relative to each other. Entropy is missing the number of base objects needed in order to arrive at an approximation of the compression length of M, while BDM is missing the position of each base symbol. And giving more information to both measures won't necessarily yield a better approximation to K.

6.11   Normalised BDM

A normalised version of BDM is useful for applications in which a maximal value of complexity is known or desired for comparison purposes. The chief advantage of a normalised measure is that it enables a comparison among objects of different sizes, without allowing size to dominate the measure. This will be useful in comparing arrays and objects of different sizes. First, for a square array of size $n \times n$, we define

$$\text{MinBDM}(n)_{d \times d} = \lfloor n/d \rfloor + \min_{x \in M_d(\{0,1\})} \text{CTM}(x), \qquad (6.4)$$

where $M_d(\{0,1\})$ is the set of binary matrices of size $d \times d$. For any n, $\text{MinBDM}(n)_{d \times d}$ returns the minimum value of Eq. (6.4) for square matrices of size n, so it is the minimum BDM value for matrices with n nodes. It corresponds to an adjacency matrix composed of repetitions of the least complex $d \times d$ square. It is the all-1 or all-0 entries matrix, because $0_{d,d}$ and $1_{d,d}$ are the least complex square base matrices (hence the most compressible) of size d.

Secondly, for the maximum complexity, Eq. (6.4) returns the highest value when the result of dividing the adjacency matrix into the $d \times d$ base matrices contains the highest possible number of different matrices (to increase the sum of the right terms in Eq. (6.4)) and the repetitions (if necessary) are homogeneously distributed along those squares (to increase the sum of the left terms in Eq. (6.4)), which should be the most complex ones in $M_d(\{0,1\})$. For $n, d \in \mathbb{N}$, we define a function

$$f_{n,d} : M_d(\{0,1\}) \longmapsto \mathbb{N}$$

that verifies

$$\sum_{r \in M_d(\{0,1\})} f_{n,d}(r) = \lfloor n/d \rfloor^2, \qquad (6.5)$$

$$\max_{r \in M_d(\{0,1\})} f_{n,d}(r) \leq 1 + \min_{r \in M_d(\{0,1\})} f_{n,d}(r), \qquad (6.6)$$

$$\text{CTM}(r_i) > \text{CTM}(r_j) \implies f_{n,d}(r_i) \geq f_{n,d}(r_j). \qquad (6.7)$$

The value $f_{n,d}(r)$ indicates the number of occurrences of $r \in M_d(\{0,1\})$ in the decomposition into $d \times d$ squares of the most complex square array of size $n \times n$. Condition (6.5) establishes that the total number of component squares is $\lfloor n/d \rfloor^2$. Condition (6.6) reduces the square repetitions as much as possible, so as to increase the number of differently composed squares by as much as possible and distribute them homogeneously. Finally, Eq. (6.7) ensures that the most complex squares are the best represented. Then, we define

$$\text{MaxBDM}(n)_{d \times d} = \sum_{\substack{r \in M_d(\{0,1\}), \\ f_{n,d}(r) > 0}} \log_2(f_{n,d}(r)) + \text{CTM}(r).$$

Finally, the normalised BDM value of an array X is:

Given a square matrix X of size n, $\text{NBDM}(X)_d$ is defined as

$$\frac{\text{CTM}(X) - \text{MinBDM}(n)_{d \times d}}{\text{MaxBDM}(n)_{d \times d} - \text{MinBDM}(n)_{d \times d}}. \tag{6.8}$$

In this way, we take the complexity of an array X to have a normalised value that is not dependent on the size of X but rather on the relative complexity of X with respect to other arrays of the same size. Figure 6.7 provides an example of high complexity for purposes of illustration. The use of $\text{MinBDM}(n)_{d \times d}$ in the normalisation is relevant. Note that the growth of $\text{MinBDM}(n)_{d \times d}$ is linear with n, and the growth of $\text{MaxBDM}(n)_{d \times d}$ exponential. This means that for high complexity matrices, the result of normalising by using just $\text{CTM}(X)/\text{MaxBDM}(n)_{d \times d}$ would be similar to $\text{NBDM}(X)_d$. But it would not work for low complexity arrays, as when the complexity of X is close to the minimum, the value of $\text{CTM}(X)/\text{MaxBDM}(n)_{d \times d}$ drops exponentially with n. For example, the normalised complexity of an empty array (all 0s) would drop exponentially in size. To avoid this, Eq. (6.8) considers not only the maximum but also the minimum.

Notice the heuristic character of $f_{n,d}$. It is designed to ensure a quick computation of $\text{MaxBDM}(n)_{d \times d}$, and the distribution of complexities of squares of size $d \in \{3,4\}$ in $D(5,2)$ ensures that $\text{MaxBDM}(n)_{d \times d}$ is actually the maximum complexity of a square matrix of size n, but for other distributions it could work in a different way. For example, condition (6.6) assumes that the complexities of the elements in $M_d(\{0,1\})$ are similar. This is the case for $d \in \{3,4\}$ in $D(5,2)$, but it may not be true for other distributions. But at any rate it offers a way of comparing the complexities of different arrays independent of their size.

6.12　CTM to BDM Transition

How BDM scales CTM remains a question, as does the rate at which BDM loses the algorithmic estimations provided by CTM. Also unknown is what the transition between CTM and CTM + BDM looks like, especially in the case of applications

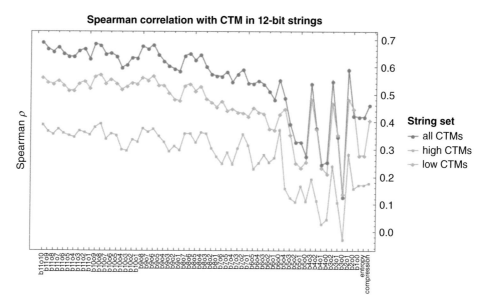

Figure 6.8 Spearman correlation coefficients (ρ) between CTM and BDM of all possible block sizes and overlap lengths for 12 bit strings, compared with the correlation between CTM and Shannon entropy, and the correlation between CTM and compression length (shown at the rightmost edge of the plot) in blue. The ρ coefficients for the 2,048 strings below and above the median CTM value are shown in green and orange, respectively. BDM block size and overlap increases to the left. Compression length was obtained using *Mathematica*'s Compress[] function. All values were normalised as described in Section 6.11.

involving objects of medium size between the range of application of CTM (e.g., 10 to 20 bit strings) and larger objects (e.g., longer sequences in the hundreds of bits).

We perform a Spearman correlation analysis to test the strength of a monotonic relationship between CTM values and BDM values with various block sizes and block overlap configurations in all 12 bit strings. We also test the strength of this relationship with CTM on Shannon entropy and compression length.

Figure 6.8 shows the agreement between BDM and CTM for strings for which we have exact CTM values, against which BDM was tested. The results indicate an agreement between CTM and BDM in a variety of configurations, thereby justifying BDM as an extension of the range of application of CTM to longer strings (and to larger objects in general).

In the set of all 12 bit strings, the correlation is maximal when block size = 11 and overlap = 10 (b11o10, $\rho = 0.69$); Shannon entropy has $\rho = 1$ with BDM when strings are divided into blocks of size = 1 and overlap = 0 (b1o0, $\rho = 0.42$), as is expected, given what is described in Section 6.10.

The Spearman rank test performed on the first 4,096 binary strings has p-values $< 1 \times 10^{-15}$, while the Spearman rank test on the 2,048 strings with CTM below the median has p-values $< 1 \times 10^{-9}$. Finally, the Spearman rank test on the 2,048 strings with CTM value above the median has p-values $< 1 \times 10^{-5}$ in all cases except those corresponding to b4o1, b4o0, and b3o0, where $\rho < 0.03$ and $0.045 \leq p\text{-value} \geq 0.25$. The lower ρ coefficients in above-median CTM strings indicates that there is a greater difficulty in estimating the algorithmic complexity of highly irregular strings through either BDM, entropy, or compression length than in detecting their regularity. Figure 6.8 shows that for block size > 6, the Spearman ρ of BDM is always higher than the correlation of CTM with either Shannon entropy or compression length. Some block configurations of size <6 (e.g., b2o1) also have higher ρ than both Shannon entropy and compression.

While BDM approximates the descriptive power of CTM and extends it over a larger range, we prove in Section 6.4 that BDM approximates Shannon entropy if base objects are no longer generated with CTM, but if CTM approximates algorithmic complexity, then BDM does.

Smooth BDM (and 'add col')

An alternative method for increasing accuracy while decreasing computational cost involves the use of a weighted function as penalisation parameter in BDM. Let the base matrix size be 4×4. We first partition the matrix into submatrices of the matrix base size 4×4. If the matrix size is not divisible by 4, we (1) use a smooth BDM with full overlap boundary condition (we call this method simply 'smooth' BDM) or (2) add an artificial low complexity boundary to 'complete' the matrix to the next multiple of 4 and apply 'smooth' (we call this approach 'add col' in future sections).

When using the BDM full overlap boundary condition, we screen the entire matrix by moving a sliding square of size 4×4 over it (as it is done for 'recursive BDM'). When adding artificial low complexity boundaries, we only calculate non-overlapping submatrices of size 4×4, because the expanded matrix of interest is a multiple of 4. These artificial low complexity boundaries are columns and rows of single symbols (zeroes or ones). We then correct the final result by subtracting the information added to the boundaries from $\log(|R|) + \log(|C|)$.

To prevent the undesired introduction of false patterns in the 'completion' process (add col), we use the minimum BDM of the extended matrix for both cases (columns and rows of zeroes and ones denoted by $BDM_1(X)$ and $BDM_0(X)$, respectively).

In both cases, to distinguish the occurrence of rare and thus highly complex patterns, we assign weights to each base matrix based on the probability of seeing each pattern, denoted by W_i, where i is the index of the base matrix. We thereby effectively 'smooth' the transition to decide matrix similarity, unlike the previous versions of BDM, which count multiples of equal matrices. Thus the main difference introduced in the 'smooth' version of BDM is the penalisation by base matrix (statistical) similarity rather than only perfect base matrix match.

To simplify notation, in what follows let us denote the adjacency matrix $\text{Adj}(X)$ of a matrix M simply as M. The *smooth* version of BDM is then calculated as follows:

$$BDM(X) = \min(BDM_0, BDM_1), \tag{6.9}$$

$$BDM_f(X) = \sum_{(r_i, n_i) \in \text{Adj}(X)_{\{x_i\}}} BDM(r_i) \times W_i + \log(n_i). \tag{6.10}$$

Weighted Smooth BDM with Mutual Information

The smooth BDM version assigns a weight to each base matrix depending on its statistical likelihood, which is equivalent to assigning a weight based on the entropy of the base matrix over the distribution of all base matrices of size 4×4. An equivalent version that is computationally more expensive is arrived at using classical mutual information by measuring the statistical similarity between base matrices precomputed by mutual information.

Mutual information is a measure of the statistical dependence of a random variable X on a random variable Y in the joint distribution of X and Y relative to the joint distribution of X and Y under an assumption of independence. If $\text{MI}(X, Y) = 0$, then X and Y are statistically independent, but if the knowledge of X fully determines Y, $\text{MI}(X, Y) = 1$, then X and Y are not independent. Because MI is symmetric $\text{MI}(X, Y) = \text{MI}(Y, X)$; if $\text{MI}(X, Y) = 1$, then knowing all about Y also implies knowing all about X. In one of its multiple versions, MI of X and Y can be defined as

$$\text{MI}(X, Y) = \text{H}(X) - \text{H}(X|Y), \tag{6.11}$$

where $\text{H}(X)$ is the Shannon entropy of X and $\text{H}(X|Y)$ the conditional Shannon entropy of X given Y.

In this way, statistically similar base matrices are not counted as requiring two completely different computer programs, one for each base matrix, but rather a slightly modified computer program producing two similar matrices accounting mostly for one and for the statistical difference of the other. More precisely, BDM can be defined by

$$\text{BDM}(X) = \sum_{(r_i, n_i) \in \text{Adj}(X)_{\{x_i\}}} \text{MIBDM}(r_i) + \log n_i, \tag{6.12}$$

where MIBDM is defined by

$$\text{MIBDM}(r_i) = \min \Big\{ \text{MI}(r_i, r_j)\, \text{CTM}(r_i) + (1 - \text{MI}(r_j, r_i))\, \text{CTM}(r_j), \tag{6.13}$$

$$\text{MI}(r_i, r_j)\, \text{CTM}(r_j) + (1 - \text{MI}(r_j, r_i))\, \text{CTM}(r_i) \Big\},$$

and where $\text{MI}(r_i, r_j)$ is a weight for each CTM value of each base matrix such that j is the index of the matrix that maximises MI (or maximises statistical similarity) over the distribution of all the base matrices such that $\text{MI}(r_i, r_j) \geq \text{MI}(r_i, r_k)$ for all $k \in \{1, \ldots, N\}$, $N = |\text{Adj}(X)_{\{x_i\}}|$.

However, this approach requires $N \times N$ comparisons $\text{MI}(r_i, r_j)$ between all base matrices r with indices $i \in \{1, \ldots, N\}$ and $j \in \{1, \ldots, N\}$.

Notice that because MI is symmetric, then $\text{MI}(r_i, r_j) = \text{MI}(r_j, r_i)$, but the min in Eq. (6.13) is because we look for the minimum CTM value (i.e., the length of the shortest program) for the two cases in which one base matrix is the one helping define the statistical similarities of the other and vice versa.

6.13   Testing BDM and Boundary Condition Strategies

A test for both CTM and BDM can be carried out using objects that have different representations or may look very different but are in fact algorithmically related. First, we will prove some theorems relating to the algorithmic complexity of dual and cospectral graphs, and then we will perform numerical experiments to see if CTM and BDM perform as theoretically expected.

A dual graph of a planar graph G is a graph that has a vertex corresponding to each face of G, and an edge joining two neighbouring faces for each edge in G. If G' is a dual graph of G, then $A(G') = A(G)$, making the calculation of the Kolmogorov complexity of graphs and their dual graphs interesting – because of the correlation between Kolmogorov complexity and $A(G')$, which should be the same for $A(G)$. One should also expect the estimated complexity values of graphs to be the same as those of their dual graphs, because the description length of the dual graph generating program is $O(1)$.

Cospectral graphs, also called *isospectral* graphs, are graphs that share the same graph spectrum. The set of graph eigenvalues of the adjacency matrix is called the spectrum $Spec(G)$ of the graph G. This cospectrality test for complexity estimations is interesting because two non-isomorphic graphs can share the same spectrum.

We have demonstrated that isomorphic graphs have similar complexity as a function of graph automorphism group size [81]. We have also provided definitions for the algorithmic complexity of labelled and unlabelled graphs based on the automorphism group [82]. In the Appendix we prove several theorems and corollaries establishing the theoretical expectation that dual and cospectral graphs have similar algorithmic complexity values, and so we have theoretical expectations of numerical tests with BDM to compare with.

The compression lengths and BDM values in Table 6.2 are obtained from the adjacency matrices of 113 dual graphs and 193 cospectral graphs from *Mathematica*'s `GraphData[]` repository [87]. Graphs and their dual graphs were found by BDM to have estimated algorithmic complexities close to each other. While entropy and entropy rate do not perform well in any test compared to the other measures, compression retrieves similar values for cospectral graphs as compared to BDM, but it is outperformed by BDM on the duality test. The best BDM version for duals was different from that for cospectrals. For the duality test, the smooth, fully overlapping

Table 6.2 Spearman ρ values of various BDM versions tested on dual and cospectral graphs that theoretically have the same algorithmic complexity up to a (small) constant.

	Non-overlapping BDM	Fully overlapping recursive BDM	Smooth fully overlapping BDM	Smooth add row or column BDM
Duality test	0.874	0.783	0.935	0.931
Cospectrality test	0.943	0.933	0.9305	0.931

version of BDM outperforms all others, but for cospectrality, overlapping recursive BDM outperforms all others.

In [81], we showed that BDM behaves in agreement with the theory with respect to the algorithmic complexity of graphs and the size of the automorphism group to which they belong. This is because the algorithmic complexity $K(G)$ of G is effectively a tight upper bound on $K(Aut(G))$.

7   Graph and Tensor Complexity

Networks, which are used extensively in science and engineering, are often complex when representing static and dynamic data, where edges are relations among objects or events. It is therefore of fundamental importance to address the challenge of quantifying their complexity and their information content if we are to understand and eventually act upon the objects they represent in meaningful ways. The ability of a computational model-based analysis of an object to implement a complexity or information-theoretic measure, as shown in [59], is key to understanding the object as well as the capabilities and limitations of the model. Widely popular implementations of lossless compression algorithms used to estimate algorithmic information content [71], such as those based on the Lempel–Ziv (LZ) algorithm, can effectively be implemented using finite state automata (FSA) [88]. However, this means that they do not possess sufficient computational power to characterise all the features in data [60]. To be able to capture all possible computable (recognisable by a computer) properties, the full power of compression implicit in algorithmic complexity is needed, as well as the computational power equivalent to a universal Turing machine, not currently present in popular implementations of lossless compression such as LZ. FSAs will therefore only be capable of capturing statistical properties, some basic algorithmic features at the level of regular languages, and so on; other languages of higher or lower power will cover only a partial subset of all the possible properties that data such as networks display. The widespread use of popular implementations of lossless compression algorithms to approximate algorithmic complexity is thus, in practice, only a very minor improvement over classical Shannon information indices [59], and can only capture statistical regularities at their respective computational powers; they miss relevant algorithmic properties.

In this chapter we briefly survey some of the literature related to information-theoretic approaches to network complexity, but, more importantly, we stress some of the limitations of current approaches to applications of both classical information theory and algorithmic complexity in the realm of graphs and networks. In particular, we highlight the fragility of entropy in its requirements and its dependency on associated mass probability distributions and the important limitations of lossless compression as used to estimate algorithmic information. Finally, we survey novel directions that represent attempts to overcome some of these shortcomings.

7.1   Notation, Metrics, and Properties of Graphs and Networks

To ensure common ground, in this section we briefly reprise some common definitions and properties of graphs and complex networks. A vertex labelling V of a graph $G = (V, E)$ is a function of the set of vertices V relative to a set of labels, different for each vertex. A graph with such a mapping function is called a labelled graph. Otherwise, the graph is said to be unlabelled. $|V(G)|$ and $|E(G)|$ will denote the vertex and edge/link count of G.

Graphs G and H are said to be *isomorphic* if there is a bijection between the vertex sets of G and H, $\lambda: V(G) \to V(H)$, such that any two vertices $u, v \in G$ are adjacent in G if and only if $\lambda(u)$ and $\lambda(v)$ are adjacent in H. When G and H are the same graph, the bijection is referred to as an *automorphism* of G. The adjacency matrix $A(G)$ of a graph G is not invariant under *graph relabellings*. Figure 7.1 illustrates two adjacency matrices for isomorphic graphs.

The number of links per node constitutes a key characteristic of a graph. When all nodes have the same number of links, the graph is said to be *regular*. The *degree* of a node v, denoted by $d(v)$, is the number of links to other nodes.

A *canonical form* of G is a labelled graph $Canon(G)$ that is isomorphic to G, such that every graph that is isomorphic to G has the same canonical form as G. An advantage of $Canon(G)$ is that, unlike $A(G)$, $A(Canon(G))$ is a graph invariant of $Canon(G)$ [89].

A popular type of graph that has been studied because of its use as a fundamental random baseline is the *Erdős–Rényi (ER)* graph [13, 14, 90]. Here, vertices are

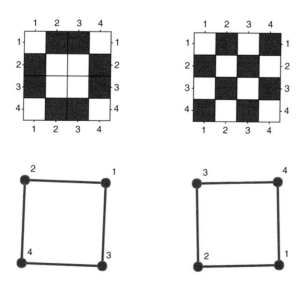

Figure 7.1 The adjacency matrix is not an invariant description of an unlabelled graph. Two isomorphic graphs can have two different adjacency matrix representations. This translates into the fact that the graphs can be relabelled. However, similar graphs have adjacency matrices with similar algorithmic information content, as proven in [81].

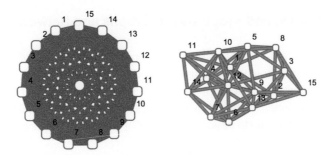

Figure 7.2 From simple to random graphs. The graphs are ordered on the basis of estimations of their algorithmic complexity (K). $K(G) \sim \log_2 |V(G)| = \log_2 15 \sim 3.9$ bits when a graph is simple (left) and is highly compressible. In contrast, a random graph (right) with the same number of nodes and number of links requires more information to be specified, because there is no simple rule connecting the nodes and therefore $K(G) \sim |E(G)| = 15$ in bits, i.e., the ends of each edge have to be specified (so a tighter bound would be $2|E(G)| \sim 30$ for an *ER* graph of edge density ~ 0.5).

randomly and independently connected by links using a fixed prescribed probability (also called *edge density*) (see Fig. 7.2 for a comparison between a regular graph and a random graph of the same size). The probability of vertices being connected is referred to as the *edge probability*. The main characteristic of random graphs is that all nodes have approximately the same number of links, equal to the average number of links per node. An *ER* graph $G(n,p)$ is a graph of size n constructed by connecting nodes randomly with probability p independent of every other edge. Usually, edge-independent *ER* graphs are assumed to be non-recursive (i.e. truly random), but *ER* graphs can be constructed recursively with, for example, pseudo-random algorithms. Here, it is assumed that *ER* graphs are non-recursive, as theoretical comparisons and bounds hold only in the non-recursive case. For numerical estimations, however, a pseudo-random edge connection algorithm is used, in keeping with common practice.

The so-called *small-world* graph describes many empirical networks where most vertices are separated by a relatively small number of edges. A network is considered to be a *small-world* graph G if the average graph distance D grows no faster than the log of the number of nodes: $D \sim \log |V(G)|$. Many networks are *scale-free*, meaning that their degrees are size independent, in the sense that the empirical degree distribution is independent of the size of the graph up to a logarithmic term. That is, the proportion of vertices with degree k is proportional to γk^τ for some $\tau > 1$ and constant γ. In other words, many empirical networks display a power law degree distribution.

7.2   Classical Information Theory

Information theory originated in the need to quantify fundamental limits on signal processing, such as communicating, storing, and compressing data. Shannon's concept of information entropy quantifies the average number of bits needed to store or communicate a message. Shannon's entropy determines that one cannot store (and therefore communicate) a symbol with n different symbols in less than $\log(n)$ bits. In this sense, Shannon's entropy determines a lower limit below which no message can be further compressed, not even in principle. A complementary viewpoint on Shannon's information theory would be to consider it a measure quantifying the *uncertainty* involved in predicting the value of a random variable. For example, specifying the outcome of a fair coin flip (two equally likely outcomes) requires one bit at a time because the results are independent of each other, each result therefore having maximal entropy. Things begin to get interesting when the coin is not fair. If one considers a coin with heads on both sides, then the tossing experiment always results in heads, and the message will always be 1, with absolute certainty.

For an ensemble $X(R, P(x_i))$, where R is the set of possible outcomes (the random variable), $n = |R|$ and $P(x_i)$ is the probability of an outcome in R. The Shannon information content or entropy of X is then given by

$$H(X) = - \sum_{i=1}^{n} P(x_i) \log_2 P(x_i).$$

Thus, calculating $H(X)$ requires the mass distribution probability of ensemble X. Here, we wish to note that using Shannon entropy entails a choice regarding the level of granularity of the analysis. This follows from it being a metric requiring the counting of discrete elements or events. For example, consider the bit string 01010101010101, which clearly has a regular pattern. But the Shannon entropy of the string at the level of single bits is maximal, as at this level of granularity the string contains the same number of 1s and 0s. Shifting perspective, the string is clearly regular when two-bit blocks are taken as basic units, an instance in which the string has minimal complexity because it contains only one symbol (01) from among four possible ones (00, 01, 10, 11). One strategy to mitigate this problem is to take into consideration all possible 'granularities' (we call this *block entropy*), from length 1 to n, where n is the length of the sequence. This measure is related to what's also called *predictive information* or *excess entropy* (the differences among the entropies for consecutive block sizes). However, such an approach comes with a computational price tag. To compute the block entropy is prohibitively expensive, as compared with fixing the block size at n, as it entails producing all possible overlapping $\binom{i}{n}$ substrings for all $i \in \{1, \ldots, n\}$.

In conclusion, characterising the complexity or information in a network requires a specification of the level of granularity of the analysis. However, since this happens to be what we set out to discover in the first place, for a given complex network, we are faced with having to assume what we are trying to discover. The block entropy 'solution' is to run the analysis across all levels of granularity, which evidently is not a scalable approach. This motivates the search for a more unbiased metric, a challenging task indeed.

7.3   Classical Information and Entropy of Graphs

One of the major challenges in modern physics is to provide proper and suitable representations of network systems for use in fields ranging from physics [91] to chemistry [92]. A common problem is the description of order parameters with which to characterise the '*complexity of a network*'. Here, we note that the issue of order parameters is closely related to the selection of the level of granularity of the analysis, as discussed in Section 7.2. One common conceptual solution is to perform a selection of the kind of alphabet being used in the analysis of complex networks. Selection of level of granularity or of order parameters are examples of ways to meet this challenge. A complementary approach involves using a set of predefined measures to describe a complex network. For example, graph complexity has traditionally been characterised using graph-theoretic measures such as degree distribution, clustering coefficient, edge density, and community or modular structure. In all these cases there are numerous algorithms available that can compute these properties of networks provided one preselects the feature that each of these different graph-theoretic indices will be used to measure.

More recently, networks have also been characterised using classical information theory. One complication of this approach is the interdependence of many graph-theoretic properties, which makes measures more sophisticated than single-property measurements difficult to come by [93]. A common practice is to generate graphs that have a certain specific property while being random in all other respects, the rationale being to assess whether or not the property in question is typical among an ensemble of graphs with otherwise seemingly different properties. We have recently advanced methods to improve upon this idea underlying the so-called 'principle of maximum entropy' (Maxent) by way of approximating algorithmic complexity [94].

Indeed, approaches using measures based on Shannon entropy that claim to quantify the information content of a network [95] as an indication of its 'typicality' are based on an assumption about which ensembles are associated as per the entropy evaluation, the idea behind *Maxent* being that the more statistically random, the more typical. The claim is that one can construct a 'null model' that captures some aspects of a network (e.g., graphs that have the same degree distribution) and see how different the network is from the null model as regards particular features, such as clustering coefficient, graph distance, or other features of interest. The procedure aims at generating an intuition about an ensemble of graphs that are assumed to have been sampled uniformly at random from the set of all graphs with the same property in order to determine if such a property occurs with high or low probability. If the graph is not significantly different, statistically, from the null model, then the graph is said to be as 'simple' as the null model; otherwise, the measure is said to be a lower bound on the 'complexity' of the graph, as an indication of its random as opposed to causal nature. Yet, to construct a proper null model is far from trivial, since as a rule one does not know what properties are present in the network.

Some applications of entropy apply to node and graph degree distributions. For example, a method to estimate upper and lower bounds for extremal node degrees was recently proposed in [96] as a measure of relative entropy, calculated from the graph edge probability matrix and largest eigenvalues. Entropy has also been applied to other graph features, such as functions of their adjacency matrices [97], and to distance and Laplacian matrices [79].

A recent example is the computation of the Shannon entropy of adjacency matrices to discover CRISPR candidate regions, as a method for transforming DNA sequences into graphs [98]. A survey contrasting adjacency matrix–based (walk) entropies and other entropies (e.g., based on degree sequence) is offered in [97]. The study finds that adjacency-based entropies are more robust vis-à-vis graph size and are correlated to graph algebraic properties, as these are also based on the adjacency matrix (e.g., graph spectrum). However, these walk entropy approaches are designed for static or fixed graphs. For time-variant and evolving graphs, other measures have been proposed [99], based on the calculation and change of graph spectral properties.

In estimating the complexity of objects, in particular of graphs, it is common practice to rely on graph- and information-theoretic measures. Here, using integer sequences with properties such as Borel normality, we explain how these measures are not independent of the way in which an object, such as a graph, can be described or observed. From observations that can reconstruct the same graph and are therefore essentially translations of the same description, we will see that when applying a computable measure such as Shannon entropy, not only is it necessary to pre-select a feature of interest where there is one, and to make an arbitrary selection where there is not, but also that more general features can be significantly misrepresented by computable measures such as entropy and entropy rate. Therefore, recursive and non-recursive (uncomputable) graphs and graph constructions based on these integer sequences have been introduced, whose different lossless descriptions have disparate entropy values, thereby enabling the study and exploration of a measure's range of applications and demonstrating the weaknesses of computable measures of complexity.

One way to describe a network is to use the notion of a node degree sequence of a graph. Clearly, when formulated in this manner Shannon entropy can be used to characterise the node degree sequence of a graph by analogy with strings. This use of entropy was first introduced in [100]. Similar approaches have been investigated and adopted in characterising chemical graphs and networks by the computational systems biology community [101]. Yet, a notion of coarse graining is needed, specifically, using the idea of a layered computation of the graph degree distribution, such as, for example, a sphere covering. Such an approach represents a hierarchical application of entropy, which can be considered a version of graph traversal entropy rate. Assessing molecular complexity is naturally of considerable interest in chemistry. Here, Shannon entropy has been used to quantify the entropy associated with the degree sequence of graphs reflecting molecular structure. If the network is unlabelled, then a description using the degree distribution is invariant to relabellings. Hence, the degree distribution is not a lossless representation of a labelled network. This is valid whenever the node labels are not relevant. In contrast, when dealing with time-dependent or temporal networks, the degree distribution cannot be used for analysing or reconstructing the network from data, as the labels of the nodes, the time-stamps, contain the critical information.

We note that the concept of entropy rate cannot be directly applied to the degree distribution. The reason is that the node degree sequence has no inherent order, because any label numbering will be arbitrary. It therefore follows that Shannon entropy is not invariant vis-à-vis the language description of a network. This is in line with the previous discussion about the level of granularity, or, on the same note, order parameters. Note that a labelled or unlabelled network has a flat degree distribution and, therefore, the lowest Shannon entropy for degree sequence and adjacency matrix.

7.4   Fragility of Computable Measures such as Entropy

Common statistical and computable measures such as Shannon entropy can easily be proven not to be robust and to require arbitrary choices such as coarse graining at multiple levels. Dependent on underlying mass distributions, and mostly quantifying how removed the assumptions of the premises are from real-world applications, Shannon entropy has proved to have limited use in dealing with complexity, information content, and, ultimately, causation and temporal information. To further our intuition as to why this is the case and assess how to progress beyond it, let us first consider the formal basis of Shannon entropy. In a graph G we define the (Shannon) entropy as

$$\mathrm{H}(A(G)) = - \sum_{i=1}^{n} \mathrm{P}(A(x_i)) \log_2 \mathrm{P}(A(x_i)),$$

where G is the random variable with n possible outcomes (all possible adjacency matrices of size $|V(G)|$). For example, a completely disconnected graph G with all adjacency matrix entries equal to zero has entropy $\mathrm{H}(A(G)) = 0$, since the number of different symbols in the adjacency matrix is 1. However, if the frequency of 1s and 0s differs in $A(G)$, then $\mathrm{H}(A(G)) \neq 0$. In general, we will use block entropy to detect additional graph regularities. The idea is to gloss over the adjacency matrix at different and greater resolutions. We note, however, that in calculating the unlabelled block entropy of a graph one has to consider all possible adjacency matrix representations for all possible labellings. Therefore, the block entropy of a graph is computed as

$$\mathrm{H}(G) = \min\{\mathrm{H}(A(g_L))|G_L \in L(G)\},$$

where $L(G)$ is the group of all possible labellings of G.

Other entropy-based measures of network elements are available. Yet as a rule, they require that an observer focus on a particular element or property of a graph, such as the adjacency matrix, degree sequence, or number of bifurcations. Notably, they do not all converge in entropy, thus illustrating that Shannon entropy is not invariant vis-à-vis different descriptions of the same object. This is in contrast to algorithmic complexity, which has the ability to characterise any general or universal property of a graph or network [66]. Indeed, in [66], we introduced a graph that is generated recursively by a small computer program of (small) fixed length. Yet, when looking at its degree sequence we see that it tends to maximal entropy, and when looking at the adjacency matrix it tends to zero entropy at the limit, thus displaying divergent values for the same object when assuming different mass probability distributions – when assuming the uniform distribution to characterise an underlying ensemble comprising all possible adjacency matrices of increasing size, or when assuming all possible degree sequences. In a follow-up paper by an independent group, our techniques were used to find other high-entropy graphs generated recursively (and thus actually of low randomness) [102].

The proposed refinement of the so-called principle of maximum entropy – Maxent – based on algorithmic complexity demonstrates (formally and numerically) that not all *ER* networks are random [94], and that methods based on algorithmic complexity can, to some extent, distinguish random from pseu*do*-random *ER* networks, calling into question the ability of classical Maxent to compare objects against their most randomised versions.

7.5   Towards the Algorithmic Complexity of a Graph

A graph with low entropy has low algorithmic complexity because the statistical regularities found in the graph can be used in a computer program to generate it. However, a graph with high entropy can have high or low algorithmic complexity [66], and the number of high entropy but low algorithmic complexity graphs will diverge [103]. This means that entropy overestimates randomness because it can only characterise traditional statistical properties and not algorithmic randomness, as can be directly inferred from the works of Kolmogorov [38] and Martin-Löf [41], and as investigated both theoretically and numerically in [59].

Using an entropy-based metric also poses the challenge of selecting an appropriate level of scale, or coarse graining: the problem of knowing which level of description is most relevant to the task of identifying order parameters of interest for a given system to minimise its estimated randomness [103], including the selection of a feature of interest as per our discussion of the way in which entropy requires the definition of a variable that belongs to an ensemble with an associated mass probability distribution. It would be optimal if we knew in advance which properties we were looking for prior to beginning our search for them. But how then would we get out of this vicious circle, chasing our own tail in the general case when no feature can be chosen beforehand, as is often, if not always, the case in real-world scenarios?

During recent decades we have benefited from pioneering work on the mathematics of algorithmic complexity originating from Kolmogorov [38], Chaitin [39], Solomonoff [43], Levin [44], and Martin-Löf [41], among others. Their framework offers the opportunity, in principle, to analyse complex objects in an unbiased manner from first mathematical principles. That is, from the accepted mathematical definition of randomness by way of algorithmic randomness, which diverges from the definition of pseu*do*-randomness, based on classical information theory, that is so widely used in practice. Yet, conventionally, this approach has been hampered by the notion that since algorithmic randomness is not a computable property, then it must be of limited to no practical value.

7.6   BDM and Algorithmic Probability

To illustrate how BDM is based on and driven by algorithmic probability, let us consider the mathematical constant π. Under the assumption of its absolute Borel normality, the digits in the n-ary expansion of π appear randomly distributed, and with no knowledge of the deterministic source and nature of π as produced by short mathematical formulae, we ask how an entropy versus an algorithmic metric performs. First, the Shannon entropy rate (assuming the uniform distribution along all integer sequences of N digits) of the N first digits of π, in any base, would suggest maximum randomness at the limit. However, without access to or without making assumptions about the probability distribution, approximations to algorithmic probability would assign π a high probability, and thus the lowest complexity per the coding theorem. This illustrates the fundamental difference between the two approaches, as they characterise the object as being strikingly different with regard to their complexity.

Just as with π, it has been proven that certain graphs can be artificially constructed to exemplify any level of Shannon entropy [66, 102], and thus any estimation of statistical randomness, without changing their algorithmic complexity. Such graphs are recursively generated, thus of a deterministic nature, and consequently at odds with the entropy values derived from them when their generating sources and mass probability distributions are unknown. While this is no surprise, it is a fact that has been overlooked in the wide application of Shannon entropy to all manner of objects, in particular to graphs and networks, as if it could be made meaningful or robust without having to pre-select a feature via choice of associated probability distributions [66].

7.7   Approximations of Graph Algorithmic Complexity

It is pertinent to ask how well algorithmic complexity can be estimated, i.e., how well different graphs can be distinguished. For example, if we have two graphs of the same size, can we tell whether they are regular, complex, or random graphs? This was first demonstrated in [81] to be feasible for graphs of the same size and, by extension, when they grew asymptotically. Here, K was calculated using the BDM as a compression algorithm. It assigned low algorithmic complexity to regular graphs, medium complexity to complex networks following Watts and Strogatz [4] or Albert and Barabási [104] algorithms, and higher algorithmic complexity to random networks. This is what we expect theoretically, since random graphs are the most algorithmically complex. Note that all long binary strings are algorithmically random, and approximately all random unlabelled graphs are algorithmically random [105]. Hence, using algorithmic complexity we can prove that the number of unlabelled graphs is a function of their randomness deficiency, which translates into establishing numerically how distant they are from the maximum value of $K(G)$, in line with recent proposals to generalise Maxent to algorithmic randomness [94].

As noted above, the *coding theorem method* (CTM) [57, 58] provides the means for approximation via the frequency of a string. Why is this so? The underlying mathematics originates in the relation specified by algorithmic probability between the frequency of production of a string from a random program and its algorithmic complexity. It is also therefore denoted as the algorithmic *coding theorem*, in contrast to a well-known coding theorem in classical information theory. Essentially, the numerical approximation hinges on the fact that the more frequently a string (or object) occurs, the lower its algorithmic complexity. Conversely, strings with a lower frequency have higher algorithmic complexity.

The way to implement a compression algorithm at the level of Turing machines, unlike popular compression algorithms based on Shannon entropy, is to go through all possible compression schemes. This is equivalent to traversing all possible programs that generate a piece of data, which is exactly what the CTM algorithm does.

In [81], numerical evidence was presented supporting the theoretical assumption that the algorithmic complexity of an unlabelled graph would not differ dramatically from any of its labelled versions. This can be understood in terms of the observation that there is a small computer program of fixed size that determines the order of the labelling proportional to the size of the isomorphism group. The size of the isomorphism group makes a difference. Given a large isomorphism group, the labelled networks have more equivalent descriptions, which follow from their symmetries. These can therefore, according to algorithmic probability, be of lower algorithmic complexity.

7.8 Reconstructing *K* of Graphs from Local Patterns

For any given network, it is of considerable interest to ask whether it possesses any specific features or patterns. In addressing this challenge we note the gulf between a pre-selected level of coarse graining versus an 'unbiased' algorithmic complexity search for patterns. In what has been referred to as network biology, there is a large body of work originating in pioneering efforts to discover motifs in networks [106]. Here, there is a analogy to patterns in the DNA string, i.e., motifs determining the structures of proteins. In the context of networks, several abundant motifs, such as feed-forward and feedback circuits, have been discovered. Yet, higher order patterns have been much more difficult to detect due to the exponential increase in the number of possible motifs. Hence, a brute force counting strategy is difficult to implement in practice. Yet, these attempts hinge upon a search for predefined patterns of a certain size (small number of nodes), which differ relative to the random null model. Using the algorithmic complexity framework offers a complementary view of this important problem.

Here, we instead determine the algorithmic complexity of a graph. This translates into considering how often the adjacency matrix of a motif is generated by a random Turing machine on a 2-dimensional array, also called a *turmite* or *Langton's ant* [84]. Hence, an accounting procedure is performed using Turing machines with a view to approximating the algorithmic complexity of the structures identified. This technique is referred to as the *block decomposition method* (BDM), introduced in [81] and [103]. The BDM technique requires a partition of the adjacency matrix corresponding to the graph into smaller matrices. With these in hand, we numerically calculate the corresponding algorithmic probability by running a large set of small 2-dimensional deterministic Turing machines, and then – by applying the algorithmic coding theorem discussed above – its algorithmic complexity.

Following such a divide-and-conquer strategy, we can then approximate the overall complexity of the original adjacency matrix by the sum of the complexity of its parts. Note that we have to take into account a logarithmic penalisation for repetition, given that n repetitions of the same object only add $\log n$ to its overall complexity, as one can simply describe a repetition in terms of the multiplicity of the first occurrence. Technically, this translates into the algorithmic complexity of a labelled graph G by means of BDM being defined as follows:

$$K_{\text{BDM}}(G,d) = \sum_{(r_u,n_u)\in A(G)_{d\times d}} \log_2(n_u) + K_m(r_u), \qquad (7.1)$$

where $K_m(r_u)$ is the approximation of the algorithmic complexity of the sub-arrays r_u arrived at by using the algorithmic coding theorem, while $A(G)_{d\times d}$ represents the set with elements (r_u,n_u), obtained by decomposing the adjacency matrix of G into non-overlapping squares, i.e., the block matrix of size d by d. In each (r_u,n_u) pair, r_u is one such square and n_u its multiplicity (number of occurrences). From now on, $K_{\text{BDM}}(g,d=4)$ will be denoted simply by $K(G)$, but it should be taken as an approximation of $K(G)$ unless otherwise stated (e.g., when taking the theoretical true $K(G)$ value). Once the CTM is calculated, BDM can be implemented as a look-up table, and hence runs efficiently in linear time for non-overlapping fixed size submatrices.

Like the notion of block entropy (see Section 7.5), the algorithmic complexity of a graph G is given by

$$K'(G) = \min\{K(A(G_L))|G_L \in L(G)\},$$

where $L(G)$ is the group of all possible labellings of G, whereas G_L is a particular labelling. Note that $K(G)$ provides a choice for graph canonisation, since it uses the adjacency matrix of G having the lowest algorithmic complexity. Here we combine G with the smallest lexicographical representation when the adjacency matrix is concatenated by rows. This is by way of dealing with the fact that G does not have to be unique. Next, one may ask how this relates to results obtained using a more standard search for motifs, as has been employed in network biology, discussed above. Specifically, is the BDM approach able to recover known network motifs? To this end we use sub-arrays of the adjacency matrix in order to ensure that network motifs (over-represented graphs), used in biology and proven to classify superfamilies of networks [12, 106], are taken into consideration in the BDM calculation. This demonstrates that BDM alone classifies and identifies the same superfamilies of networks [107] as classical network motifs – discussed above – were able to identify.

7.9   Group-Theoretic Robustness of Algorithmic Graph Complexity

How robust is the measure of algorithmic complexity? Here, we review this question by contrasting the computation using unlabelled versus labelled graphs. In short, the metric is robust up to an additive constant. Let's consider this issue in more detail. First, regular graphs have been shown to have low values of K, whereas random graphs have high estimated values of K. This has been shown by actually performing the non-trivial calculation of unlabelled complexity, namely K'. Furthermore, graphs with a larger set of automorphisms have lower K values compared to graphs with a smaller set of automorphisms [81]. Now, an important question is how accurate a labelled estimation of $K(G)$ is with respect to the unlabelled $K'(G)$. This is a valid concern and a useful question to ask since in the general case the calculation of $K(G)$ is computationally cheap compared to $K'(G)$, which carries an exponential overhead. Perhaps surprisingly, the difference $|K(G) - K'(G)|$ is bounded by a constant. Indeed, as first suggested in [82], there exists an algorithm denoted by α of fixed length (bit-size) $|\alpha|$ such that all $L(G)$ relabellings of G can be computed. This is doable using a brute-force scheme, e.g., by producing all the indicated adjacency matrix rows and their associated permutations of columns. It therefore follows that $|K(G) - K(G_L)| < |\alpha|$ for any relabelled graph G_L of G. In other words, $K(G_L) = K'(G) + |\alpha|$, where $|\alpha|$ is independent of G. We wish to note here that even if the time complexity of α is commonly believed to not be in the **P** class, this is not a relevant observation. It is sufficient for the proof to go through that an α exists and is of finite size. We can therefore safely deduce that an estimation of the unlabelled $K'(G)$ obtained by piggy-backing on the estimate of a labelled $K(G_L)$ is indeed an accurate asymptotic approximation. The brute-force schema is likely the shortest program description capable of producing all graph relabellings, and therefore the best choice to minimise α.

This result is both relevant and useful in practice. First, we can accurately estimate $K_L(G)$ through $K(G)$ for any lossless representation of G up to an additive term. Yet, as noted above, the existence of a finite entity does not readily inform us about the convergence rate of $K(G)$ to $K_L(G)$. Interestingly, numerical estimations demonstrate that the convergence rate is fast. For example, the median of the BDM estimations of all the isomorphic graphs of the graph in Fig. 7.1 is 31.7, with a standard deviation of 0.72. However, when generating a graph, the BDM median is 27.26 and the standard deviation 2.93, clearly indicating a statistical difference. But more importantly, the probability of a random graph having a large automorphism group count is low, as shown in [81]. These observations are consistent with what we would expect of the algorithmic probability of a random graph – a low frequency of production as a result of running a *turmite* Turing machine. Here, and in [81], we have also shown that graphs, their formal duals, and their co-spectral versions have similar algorithmic complexity values, as estimated using algorithmic probability (BDM). This means that, in practice, the convergence is not only guaranteed but also sufficiently rapid.

7.10   *K*(*G*) Is not Graph Invariant but Is Highly Informative

Following the discussion of robustness in the previous section, here we address what could be referred to as sensitivity. Namely, if two graphs end up with similar $K(G)$, does it follow that they are isomorphic? Let us explain why this is not the case. $K(G)$ can readily be computed via an approximation up to a bounded error that vanishes asymptotically with the increase in the size of the graph. Note, however, that $K(G)$ does not uniquely determine G. This is evident from the fact that two non-isomorphic graphs G and H can have $K(G) = K(H)$. More precisely, the algorithmic coding theorem provides an estimation of how often this occurs, and it is also related to a simple pigeonhole argument. Indeed, if G or H are algorithmically (Kolmogorov–Chaitin) random graphs, then the probability that $K(G) = K(H)$ grows exponentially. If G and H are complex, then their algorithmic probability $\sim 1/2^K(G)$ and $\sim 1/2^K(H)$, respectively, are small and are located in the tail of the algorithmic probability distribution, also referred to as the *universal distribution* or Levin's *semi-measure* distribution. This ranges over a very tiny interval of (maximal) algorithmic complexity, hence increasing the chance of collision of the values, i.e., $K(G) = K(H)$.

In [82], we utilised theoretical and experimental estimates of the algorithmic complexity of trivial/simple (denoted here by S) and random Erdős–Rényi (ER) graphs. Regular graphs, such as completely disconnected or complete graphs, have an algorithmic complexity $K(S) = \log |V(S)|$. ER graphs have maximal complexity, so any other complex network is upper bounded by $K(ER)$ graphs. Finally, note that the algorithmic complexity of a Barabási–Albert (BA) network is low because it is based on a recursive procedure while preserving an element of randomness, since the generative model comes equipped with an attachment probability.

In [82], theoretical and numerical estimations of algorithmic information content for a range of theoretical and real-world networks is provided. Here, Table 7.1 offers a larger picture, summarising theoretical expectations of the asymptotic behaviour of K for different graphs and networks. Notice that ER' represents a graph that satisfies the definition of an ER graph, but its edges are not independent because ER' is recursively generated by, for example, a pseu*do*-random number generator (PRNG). This ability to distinguish randomness from pseu*do*-randomness and tell ER' from ER is what the introduction of algorithmic complexity to the study of graphs and networks has to offer.

To test numerical approximations of these theoretical estimations, a wide range of tools and experiments have been devised, one of the most conclusive with regard to graphs being the one performed on dual and co-spectral graphs. We know that graphs

Table 7.1 Theoretical calculations of K for different network topologies for $0 \leq p \leq 1$. Clearly, minimum values are for fully connected, fully disconnected, and recursive graphs, while maximum K is reached for edge-independent ER graphs with edge density $p = 0.5$ and a fixed number of nodes for which $K(ER) \sim \binom{|V(ER)|}{2}/2$. For WS graphs, p is the rewiring probability.

Type of graph/network	Asymptotic expected behaviour
Empty/Complete (E)	$K(E) \sim \log |V(E)|$
Regular recursive (R) (e.g., cycles, stars)	$K(R) \sim \log |V(R)|$
Barabási–Albert (BA)	$K(BA) \sim |V(BA)| + c$
Watts–Strogatz (WS)	$\lim_{p \to 0} K(WS) \sim K(R)$
	$\lim_{p \to 1} K(WS) \sim K(ER)$ or $K(ER')$
Algorithmic random Erdős–Rényi (ER)	$K(ER) \sim \frac{n(n-1)}{16p|p-1|}$
Pseu*do*-random Erdős–Rényi (ER')	$K(ER') \sim K(S)$

and their duals must have about the same algorithmic complexity because there is a computer program of (small) fixed size that can transform any graph into its dual by simple replacement of edges for nodes and nodes for edges. While for duals it was clear that our methods were numerically sound, the co-spectrality test confirmed the robustness of our measures. For both tests the methods outperformed other approaches, such as compression and Shannon entropy, as reported in [103].

8 Algorithmic Information *Dynamics* (AID)

Chapter Summary

Algorithmic information dynamics (AID) is an algorithmic probabilistic framework for model generation, an alternative and a complement to other approaches to experimental inference and causal discovery such as statistical methods and classical information theory, and is of relevance to areas such as computational mechanics and program synthesis. Unlike other methods such as Bayesian networks, AID does not rely on graphical models or classical probability distributions, or at least the models it generates do not rely on these. AID is the result of combining algorithmic information theory (AIT) on the one hand, and, on the other, perturbation/intervention analysis, which involves subjecting an open system to induced or naturally occurring interventions. AID is the body of foundations and methods that makes algorithmic information theory more suitable for scientific application in areas ranging from genetics to cognition, and constitutes a true alternative to popular lossless compression algorithms such as LZW – which are entropy estimators – in estimating algorithmic complexity, as well as to other probabilistic approaches to scientific discovery and model inference. AID connects various fields of science and areas of active research such as logical inference, reasoning, causality, complexity, probability, and dynamics. It provides the foundations, framework, and numerical methods with which to systematically produce computable hypotheses as candidate generative (mechanistic) models of natural or artificial phenomena. AID provides the tools with which to explore the space of computable models. It is based on the theory of algorithmic complexity, and, more to the point, on algorithmic probability.

In a nutshell, AID involves the exploration of the effects that perturbations of systems and data (such as strings or networks) have on their underlying computable models. For example, when a random bit-flip is performed on the string pattern 0000000... in position N, such a perturbation will have a significant effect with respect to the length of the original shortest program P consisting of a loop printing N number of 0s, because P would need to account for the 1 inserted at a random position X between 1 and N. If, however, the string is random, a random change will have no effect, because every bit already required its own description. When it comes to deletion as a perturbation, random deletions have similar effects in both simple and random objects, amounting to a term upper bounded by $\log(X)$, with X the number of elements deleted. Some applications to well-known discrete dynamical systems will help us illustrate how algorithmic dynamics can be applied to questions of causation and inverse problems.

8.1 A Calculus of Algorithmic Change

At the core of reprogrammability analysis is a causal calculus as introduced in [108], based upon the change in complexity of a system subject to perturbations, particularly the direction (sign) and magnitude of the change in algorithmic information content C between two states of the same object, such as objects G and G', which for purposes of illustration can be graphs with a set of nodes $V(G)$ and a set of edges $E(G)$.

The dynamics of a graph can then be defined as transitions between different states, and one can always ask after the potential causal relationship between G and G'. In other words, what possible underlying minimal computer program can explain G' as evolving over discrete time from state G?

For graphs, we can allow the operation of edge e removal from G denoted by $G \backslash e$, where the difference $|C(G) - C(G \backslash e)|$ is an estimation of the shared algorithmic mutual information of G and $G \backslash e$ or the *algorithmic information dynamics* (or *algorithmic dynamics* for short) for evolving time-dependent systems (e.g., if G' evolves from G after t steps). If e does not contribute to the description of G, then $|C(G) - C(G \backslash e)| \sim \log_2 |V(G)|$, where $|V(G)|$ is the node count of G, i.e., the algorithmic dynamic difference will be very small and, at most, a function of graph size, and thus the relationship between G and G' can be said to be causal and not random, as G' can be derived from G with at most $\log_2$ bits. If, however, $|C(G) - C(G \backslash e)| > \log_2 |V(G)|$ bits, then G and $G \backslash e$ do not share causal information, and the removal of e results in a loss. In contrast, if $C(G) - C(G \backslash e) > n$, then e cannot be explained by G alone and nor is it algorithmically not contained in/derived from G, and it is therefore a fundamental part of the description of G, with e as a generative causal mechanism in G; or else it is not part of G but has to be explained independently, e.g., as noise. Whether it is noise or part of the generating mechanism of G depends on the relative magnitude of n with respect to $C(G)$ and on the original causal content of G itself. If G is random, then the effect of e will be small in either case, but if G is richly causal and has a very small generating program, then e as noise will have a greater impact on G than would removing e from the description of an already short description of G. However, if $|C(G) - C(G \backslash e)| \leq \log_2 |V(G)|$, where $|V(G)|$ is the vertex count of G, then e is contained in the algorithmic description of G and can be recovered from G itself (e.g., by running the program from a previous step until it produces G with e from $G \backslash e$).

8.2   AID and Causality

As we have seen, the theory of computability is the quintessential theory of mechanistic behaviour.

Exemplified by the work of Judea Pearl and Dana Mackenzie on causality, and based on perturbation analysis, one can see how AID contributes to the field of causality by exploiting the theory of computation and introducing the concepts of simulation, perturbation analysis and algorithmic probability.

One can formulate a cause-effect question in the language of the pseudo-calculus, from herein referred to as *do*-calculus, as a probability linking a random variable and an intervention P($L|do(D)$). In a typical example given by Pearl himself [109], if L refers to the human lifespan and $do(D)$ the use of some drug D, the probability P of D having an effect on L is quantified by using classical probability to calculate P. What AID does is to substitute AP for P, the algorithmic probability that D exerts an effect on L. The chief advantage of so doing is that not only does one obtain a probability for the dependency (and direction) between D and L, but also a set of candidate models explaining the connection, with no recourse to classical probability in the inference or description of each individual model. AP would then impose a natural, non-uniform algorithmic probability distribution over the space of candidate models connecting D and L, based on estimations of the universal distribution, which in effect introduces a simplicity bias favouring shorter models, as opposed to algorithmically random ones: models which have already been found to explain the causal connection between D and L. AID thus removes the need for classical probability distributions and, more importantly, produces a set of generative models no longer derived from traditional statistics (e.g., regression, correlation) which even the *do*-calculus uses to determine the probability of a dependency, thereby falling back on the very methods it set out to circumvent. While the *do*-calculus is a major improvement in the area of causal discovery, AID complements it, offering a strategy for leaving classical probability behind altogether and moving away from methods that confound correlation and causation. Another key difference between the original *do*-calculus and the algorithmic probability calculus is that the *do*-calculus makes a distinction between P($L|do(D)$) and P($L|D$), but in AID both hypotheses, AP($L|D$) and AP($D|L$), can be tested independently, as they have different meanings under algorithmic probability. AP($L|D$) means that L is the generative mechanism of D, and AP($D|L$) means that D is the generative mechanism of L. Each of them would trigger a different explorative process under AID. The former would prompt us to look for the set of smaller to larger computable models denoted by L that can explain D, while the latter would have us look for all the generative models $\{D\}$ that generate L. Intuitively, this means that if, for example, one wanted to explain how a barometer falling (X) could be the cause of a storm, one would likely not find many shorter generative mechanisms that explain the occurrence of the storm (Y), while there would be shorter computable models for the reverse, for the barometer falling being an effect and not a cause. Clearly, AP($X|Y$) and AP($Y|X$) are not equivalent. In the language of AID, which is more in conformity with typical notations in graph and set theory, the first example would be written as AP($L\backslash D$)

or C($L \backslash D$), where C is the algorithmic complexity of L with intervention D (in the context of AID, this is usually a deletion to a model that includes D), and the second example as AP($X \backslash Y$) or C($X \backslash Y$). Alternatively, AP($L \backslash \{D\}$) would be L with a set of interventions $\{D\}$. According to Pearl [109], the causal ladder reaching up to human-grade reasoning consists of three levels. The first is that of pattern detection, covered by traditional statistics and based for a long time on regression and correlation; the second is interventions of the type his *do*-calculus suggests and which AID allows, and the third is that of counterfactuals, or the power to imagine what would happen if conditions were different. While the *do*-calculus offers no replacement for regression and correlation in the last mile of the calculations, AID covers all three levels of causality and provides the methodological framework within which to address each without any recourse to classical probability and to regression or correlation. While interventions are possible in AID, both interventions and counterfactuals are covered by the underlying simulation of all possible computable hypotheses (hence all 'if' questions) up to the length of the original observation, and methods such as CTM and BDM allow the estimation of processes at all these three levels by decomposing the original problem into smaller problems, with the added advantage that AID ranks the counterfactuals naturally by likely algorithmic probability, based on the universal distribution as a prior.

8.3 Transforming Graphs towards and away from Randomness

While we have said that a network representation is not required for the application of AID, AID is particularly interesting in network theory and can help us illustrate various concepts in our study of evolving dynamic systems.

Let's consider a complete graph. In a complete graph (Fig. 8.1), the removal of any single node leads to a logarithmic reduction in its algorithmic complexity, but the removal of any single edge leads to an increase in randomness. The former because the result is simply another complete graph of smaller size, and the latter because the deleted link would need to be described after the description of the complete graph itself.

If a graph is evolving deterministically over time, its algorithmic complexity remains (almost) constant up to a logarithmic term as a function of time, because its generating mechanism is still the same, but if its evolution is non-deterministic and possible changes to its elements are assumed to be uniformly distributed, then a full perturbation analysis can simulate its next state, basically by exhaustively applying all possible changes one step at a time. In this case, any node/edge perturbation of a simple graph has a very different effect than performing the same interventions on a random graph.

The purpose of algorithmic information dynamics is to trace in detail the changes in algorithmic probability – estimated by local observations – produced by natural or induced perturbations in evolving open complex systems. This is possible even for partial data that may look discrepant but have the same underlying source. For in the real world, we usually have only partial access to a system's underlying generating mechanism, yet from partial observations algorithmic models can be derived, and their likelihood of being the producers of the phenomena observed estimated.

8.3.1   Sequence Models

For snapshots of a system, such as a static network, calculating the AID of every element, say of a graph, yields a list of edges and vertices ranked by their algorithmic contribution, with each item explaining a part of the original system followed by the system after perturbation. The list of BDM values for each case is in itself a computable candidate model of the set of smaller candidate computable models (better approximated by CTM) explaining the larger piece of data.

One can then see how the list of observations or perturbations in, e.g., an evolving network, represents a sequence of interventions ranked by their effects on the underlying models describing the difference between the original object and the object after perturbation, and provides a vector of candidate models quantifying each change. One can use these sequences to reconstruct objects and dynamical systems.

8.4   Reconstruction of Space-Time diagrams of Dynamical Systems

Algorithmic information dynamics can be used to reconstruct dynamical systems. The method proceeds as follows:

1. Make an observation X.
2. If observations are unsorted, try all order combinations to find a combination that maximises algorithmic probability, thus favouring the simplest configuration, in line with Occam's razor.
3. Then start perturbing each observation (the order is irrelevant).
4. Find the perturbation whose deletion had a neutral impact on the complexity of the candidate model (found by CTM/BDM). Neutral impact is defined as the impact $\sim \log(N)$, where N is the number of observations.
5. Once a a perturbation having a neutral effect is found, remove it from the dataset and return to step 2, continuing until no more neutral perturbations are left. What remains is likely noise or signals from the outside that the largest generative model of the major observation component cannot explain without a significant increase in size.
6. Reconstruct by stacking all the observations. Then the order (internal time) of the dynamical system is given by the stacking order from top to bottom (i.e., in reverse order). We sometimes call this process 'peeling back' a dynamical system.

The optimal method would actually require us to apply all powerset perturbations, not just single rows but all 2-row perturbations, and so on. This would give us an even better order but it would require exponential running time of course, and so we settled first on single row perturbation. But we explain how the powerset increases accuracy.

This is a partial answer to the question of why several regions are neutral. Notice that neutrality is not the only guide, though it is certainly the main one. In general, the later the step, the less perturbing, because you can still recover the first part without model modification, and the earlier the perturbation, the more likely (in non-trivial systems with high integrated information) it is to be more disruptive. When you have two regions that are equally neutral, then you would have two new model candidates instead of one.

8.5 Reprogrammability

We introduced measures of reprogrammability (denoted by $P_r(G)$ and $PA(G)$ in [108] that can differentiate between elements that may move an object or a system G closer to randomness $\sigma_N(G)$ or away from randomness $\sigma_P(G)$, as follows:

- Relative (re)programmability: $P_r(G) := \mathrm{MAD}(\sigma(G)))/n$ or 0 if $n = 0$, where $n := \max\{|\sigma(G)|\}$ measures the shape of $\sigma_P(G)$ and how it deviates from other distributions (e.g., uniform or normal).
- Absolute (re)programmability: $PA(G) := |S(\sigma_P(G)) - S(\sigma_N(G))|/m$, where $m := \max(S(\sigma_P(G)), S(\sigma_N(G)))$ and S is an interpolation function. This measure of reprogrammability captures not only the shape of $\sigma_P(G)$ but also the sign of $\sigma_P(G)$ above and below $x = 0$.

For a complete graph, all nodes and all edges should have the same algorithmic-information contribution, and thus $\sigma(G)$ can be analytically derived (a flat uniform distribution $x = \log_2 |V(G)|$ with $|V(G)|$ the node count of G). Thus all the nodes of a complete graph are 'slightly' positive (or more precisely, neutral, if they are 'positive' by only $\log_2 |V(G)|$).

Another way to illustrate the phenomenon of asymmetric reprogrammability is by considering networks as computer programs (Fig. 8.1) (or as produced by computer programs). The algorithmic complexity K of a complete graph k grows by its number of nodes because the generating mechanism is of the form connect all N nodes, where $N = |V(k)|$. In contrast, the algorithmic complexity of a random Erdös–Rényi (E–R) graph with edge density ~ 0.5 grows by the number of edges $|E(\text{E–R})|$, because to reproduce a random graph from scratch the sender would need to specify every edge connection, as there is no way to compress the description.

As depicted in Fig. 8.1, removing a node n from a complete graph k produces another complete, albeit smaller graph. Thus the generating program for both k and $k' = k\backslash n$ is also the relationship between k and k', which is causal. In contrast, if an edge e is removed from k, the generating program of $k'' = k\backslash e$ requires the specification of e and the resulting generating program of $C(k'') > C(k)$, moving k towards randomness for every e randomly removed.

On the one hand, moving a complete graph towards randomness (see Figs. 8.1 and 8.2) requires random changes if we are only interested in reproducing the statistical properties of the random graph, that is, its degree distribution, which requires

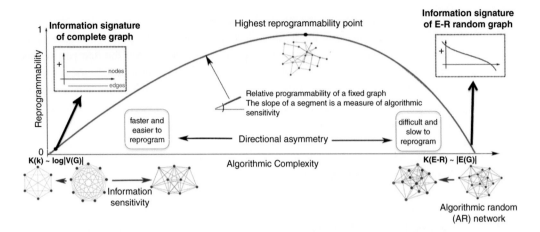

Figure 8.1 Networks as programs. All real-world networks lie between the extreme cases of being as simple as a complete graph whose algorithmic complexity K is minimal and grows by only $\log|V(k)|$, and a random (also statistically random and thus E–R) graph whose algorithmic complexity is maximal and grows by its number of edges $|E(\text{E–R})|$. If we ask what it takes to change the program producing k so that it produces E–R and vice versa, in a random graph any single node or edge removal does not entail a major change in the program size of its generating program, which is similar in size to the graph itself, i.e., $|E(G)|$.

(1)`Print[1] 200 times`	(2) `Print[s]`
`1111111111111111111`	`00110010010010110011`
`1111111111111111111`	`01100101110010011111`
`1111111111111111111`	`00110101000110011101`
`1111111111111111111`	`11001010100010011100`
`1111111111111111111`	`10000101000011011100`
`1111111111111111111`	`11000101000110010100`
`1111111111111111111`	`11010111000110110011`
`1111111111111111111`	`01111100000001010110`
`1111111111111111111`	`00010110110011010111`
`1111111111111111111`	`01011010100010111101`

$s =$ (for column 2)

Figure 8.2 Thermodynamic reprogrammability. (Left) The programs producing simple versus random data have different reprogrammability properties. If repurposed to generate programs to print blocks of 0s, we only need a single intervention in the generative program of (1), changing 1 to 0 inside the **Print** instruction indicating that 200 0s be printed instead of 1s. (Right) In contrast, asking a program that prints a random binary string s to print only 0s will require on average $|s|/2$ interventions to manually change every bit 1 to 0. Random perturbations can be seen as the exploration of the possible paths through which an object may evolve over time. Thus uniform random perturbations provide a picture of the set of possible future states. This means that, in both cases, the asymmetric cost of moving random to simple and simple to random from a purely algorithmic perspective is also relevant in the case of naturally evolving systems.

no previous knowledge. On the other hand, we see how a random graph can also be easily rewired to be a complete graph by simply adding the edges needed to make it complete. However, if the complete graph is required to exactly reproduce a specific random graph and not just its statistical properties, then one would need to have full knowledge of the specific random graph and apply specific changes to the complete graph, making the process slow and requiring a lot of knowledge. Yet, moving the random graph toward completeness still requires the same effort as before because the complete graph is unique, given the fixed number of nodes. Nevertheless, changing a simple graph such as the complete graph (see Fig. 8.1) by edge removal has a greater impact on its algorithmic complexity than performing the same operation on a random graph. Specifically, if S is a simple graph and R a random one, then we have that $C(S) - C(S \backslash e) > C(R) - C(R \backslash e)$, i.e., the rate of change from S to (a non-specific) R is greater than in the other direction, thus imposing a thermodynamic-like asymmetry related to the difficulty in reprogramming one object into another according to its initial program-size complexity. The asymmetric axis where the highest reprogrammability point can be found is exactly the point at which $C(S) - C(S \backslash e) = C(R) - C(R \backslash e)$ for a specific S and R.

It is clear then how analysing the contribution of each element to the object, as shown in Fig. 8.1, has the potential to reveal the algorithmic nature of the original object, and how difficult it is to reprogram the underlying generative computer program in order to produce a different output/graph.

A thermodynamic-like effect can be found in the (re)programmability capabilities of an object. Changing random networks by edge removal is significantly more difficult than moving simple networks toward randomness. For random graphs, there are only a few elements, if any, that can be used to move them slowly towards simplicity, as shown in Fig. 8.1. In contrast, a larger number of elements can move a simple network faster towards randomness. This relationship, captured by the reprogrammability rate for S simple versus R random graphs, induces a thermodynamic-like asymmetry based on algorithmic complexity and reprogrammability [108].

Figure 8.1 shows how, without loss of generality, the reprogramming capability of networks as computer programs produces an asymmetry imposed by algorithmic complexity and reminiscent of traditional thermodynamics as based on classical probability. A maximally random network has only positive (blue) elements (Fig. 8.1) because there exists no perturbation that can increase the randomness of the network either by removing a node or an edge, as it is already random (and thus non-deterministic). Thus, changing its (near) minimal program-size length by edge or node removal is slow. However, a simple graph may have elements that push its program-size length toward randomness. In each extreme case (simple vs random), the distribution of sorted elements capable of pushing in each direction is shown in the form of what we call 'signatures', both for edge and node removal. The highest reprogrammability point is the place where a graph has as many elements to push it in one direction as in the other.

8.6   Practice and Discussion Questions

1. Kolmorogov–Chaitin complexity is not computable, but can be approximated:
   (a) From below, meaning it is upper semi-computable
   (b) From above, meaning it is lower semi-computable
   (c) By averaging upper and lower bounds
2. Graph-theoretic measures such as the degree distribution, clustering coefficient, edge density, and community or modular structure are examples of:
   (a) AID-based measures
   (b) Entropy-based measures
   (c) Computable measures
   (d) All of the above
3. Creating a graph step-by-step using a formula means:
   (a) That the graph is deterministic and has a low algorithmic complexity value because a computer program of fixed length can generate it
   (b) That the graph is complex because it is described as a step-by-step process
4. Popular lossless compression algorithms such as LZW (gzip, etc) is an example of a measure:
   (a) Closely related to algorithmic complexity
   (b) Supposedly related to algorithmic complexity but in fact closer to Shannon entropy
5. In the simplest description of the algorithmic (Kolmorogov–Chaitin) complexity $K(s) = \min\{|p| : U(p) = s\}$, what do the following roughly mean?

   $|p|$:
   (a) The size of the number of programs we are going to explore
   (b) The average size of all programs we are going to explore
   (c) The size of a program

   $U(p)$:
   (d) A machine that only accepts p
   (e) A universal Turing machine that is running p
   (f) A universal Turing machine that produced p as its final output

   s:
   (g) The chunk of data we observe as the result of some process
   (h) The input of the program p
   (i) A substring of the longest possible string p can produce

   min:
   (j) We only use the smallest program that satisfies this, which is much like Occam's razor
   (k) The local minimum length of all possible outputs of this Turing machine running p
   (l) Only use the programs that create the minimal number of unique strings

6. In a very general sense, the coding theorem method (CTM):
   (a) Runs all possible small programs on universal Turing machines and sees what kinds of strings they produce. The frequency of seeing your string, out of this set, is very roughly the CTM.
   (b) Runs all possible small programs up to a certain size based on the Turing machine formalism and sees what kinds of strings they produce. The frequency of seeing your string, out of this set, is very roughly the CTM.

7. The CTM uses Turing machines because [choose all that apply]:
   (a) It is the most fundamental computing model
   (b) It does not matter what model is used as long as it is universal because it means one can cover the space of all computer programs no matter the specific computing model
   (c) Turing machines can emulate any other Turing-complete model, including cellular automata or neural networks so the model is irrelevant but the Turing model is precise enough and largely well-understood enough to use it as the baseline

8. There are two ways we avoid the Halting problem by using CTM. The first way is an approximation where:
   (a) We avoid using universal Turing machines to run programs and instead use cellular automata
   (b) Programs are too large in general to run all the way, so we only run each program for a random number of steps

9. The second way we avoid the Halting problem is the following: When we run a program, we are never sure if the program is finished or not because we can't guess how long it will run until it halts. So (choose all that apply):
   (a) For larger programs for which Busy Beaver values are not known we make educated guesses based on theoretical foundations to stop the machines and consider only those that had halted before that time as producing an output, all others are ignored
   (b) We calculate all Busy Beaver programs beforehand and say that if the program is running longer than the Busy Beaver equivalent, it will never stop
   (c) We run the program after a fixed number of time steps larger than the Busy Beaver values and if it is not done by then, we skip it because we know it will never halt

10. Why is CTM very difficult for large strings?
    (a) Because the entropy values are very high for large strings and calculating the compression for each long string starts to take a long time
    (b) Because to explore all possible programs up to a certain size that output a large string would take thousands of years

11. If p refers to the input of the universal Turing machine U (i.e., the initial tape) then shouldn't a realistic estimate of the universal distribution $m(s)$ include runs of the machines in (t, k) with t the number of states and k the number symbols, with non-empty initial tapes?

   (a) Yes, because otherwise the universal distribution would be incomplete
   (b) No, because Turing showed as part of his universality results that one can hardcode any input in the transition table of a Turing machine effectively showing how software and hardware can be exchanged so running all Turing machines with empty input eventually also covers all Turing machines with some input

12. What is the block decomposition method in a nutshell?
   (a) Square-like patterns in matrices are approximated under simple programs and can be thrown out
   (b) It takes various similar parts of a string and calculates all the possible programs that could have produced any similar string
   (c) If we have large strings, we can break them down into smaller ones, then calculate the CTM for each piece and add the resulting K approximate values in a clever way

13. When we apply CTM and BDM methods to graphs, what graph representation do we apply them to?
   (a) The adjacency matrix
   (b) The edge weight distribution
   (c) The degree distribution

14. A 2D matrix can be made from:
   (a) A 1D Turing machine
   (b) A 2D Turing machine
   (c) A 3D Turing machine

15. Based on what you know about algorithmic complexity so far, do you expect a graph's complexity to change much when you change the node's labels?
   (a) Yes, since the generating mechanism is different if one considers the node labels
   (b) Yes, since the labels are different
   (c) No, since the generating mechanism behind the graph is the same

16. Would the values for algorithmic complexity change for identical strings if one string was made by a 2D Turing machine and the other by a 1D Turing machine?
   (a) No, the ranked correlation turns out to be exactly the same across many different strings
   (b) Yes, because the generating mechanism behind each string is different

17. We expect fully connected graphs and fully disconnected graphs to be:
   (a) Algorithmically complex, depending on the number of nodes
   (b) Algorithmically as simple as possible
   (c) No different from random if the generating mechanism is the same

18. If a string is truly algorithmically random, then changing part of the string will make the string:
   (a) More random
   (b) Less random
   (c) Neither

19. In a graph, we can remove an edge or a node to see if the change makes the graph more or less algorithmically random. If a graph is made entirely of nodes and edges whose removal is neutral to the complexity of the resulting graph, then the graph is:
    (a) Random
    (b) Complex
    (c) Simple

20. In a graph, we can remove an edge or a node to see if the change makes the graph more or less algorithmically random. If a graph is made entirely of nodes whose removal is neutral to the complexity of the resulting graph, and edges whose removal moves the graph towards randomness, then the graph is:
    (a) Simple
    (b) Random
    (c) Complex

21. A graph with nodes and edges that are only mostly going to move the graph drastically towards or away from randomness after their removal is:
    (a) Complex
    (b) Simple
    (c) Random

Part III

Applications

9   From Theory to Practice

If, by now, you are convinced of the potential of AID methods, how do they compare against others when it comes to real applications? One has to first understand that it would be unfair to unthinkingly compare AID to traditional methods used in current applications for the following reasons:

1. Fitting a curve. As you know, given X data points, you can fit a polynomial of degree X that goes through every point, giving you 100% accuracy. This is **exactly** what happens with current machine learning techniques. Yet generalising from the polynomial taken as a model makes absolutely no sense. Machine learning techniques are not designed to produce actual models of data, but to fit data at all costs. The purpose is completely different, and any comparison to AID would make little sense. One case literally involves cheating (even if it proves useful for classification purposes), while the other (AID) represents one of our best shots at doing things properly, and, despite its limitations, produces results.

2. Machine learning tests are designed to overestimate their own performance. This is because tests are designed to classify digits or faces, say, with a cross-entropy cost function. If the test were instead to produce a scientific model explaining the data (which is the true purpose of science, not classification or fitting), then they fail 100% of the time, versus a non-zero success probability for our methods based on AIT, which can be improved by 100% as a function of computational resources (whereas no matter what computational resources the other methods have at their disposal, they will never produce a model). Models from machine learning are traditionally models deduced by scientists interpreting the data, so they, rather than the machine learning or probabilistic technique used, are doing 100% of the work to extract/infer a model from data; data does not thus hold all the information but the pair data + scientist. AID tries to help the scientist infer models in a more objective manner.

3. We can do better in many areas. 1) Producing models. Even with small samples, statistical methods produce 0 models, and other equation-driven approaches to model generation require a lot of human invention and intervention. 2) AID can deal with small objects. Neither entropy nor compression can properly deal with the complexity of small objects, which are key to, e.g., perturbation analysis. Entropy will only count symbols and will be hypersensitive, while popular lossless compression algorithms will be unstable and insensitive. 3) We have shown in several papers

that our method does better than other methods, in particular Shannon entropy. See for instance [110], where we show that we can reproduce the known cognitive capabilities of humans over their lifespan by asking people to generate randomness, something that could not be reproduced by entropy because it produced contradictory results for different tests. This is consistent with what we have been saying and demonstrating, i.e., that for all objects, such as the ZK graph, any computable approach such as machine learning or entropy requires a preselected feature of interest and is not robust in the face of changes of object description, even if the descriptions are of exactly the same object. 4) Making progress on perturbation analysis. Since other methods are unable to deal with small changes as per (2), and since we need to move away from correlation, CTM and BDM are among the few currently available practical tools capable of contributing even if their power may be limited.

4. Applications such as steering complexity. There are several papers showing cases in which something along these lines could not have been accomplished without our tools, e.g., steering a dynamical system (e.g. [108]).

5. Along the same lines, we show how our measures reduce the dimensions of data (in practice) better than the other two best methods [111].

The bottom line is that no method that is independent of algorithmic probability is equipped to update itself, and thus, even if better numerical methods for estimating K and related measures are found, AID remains valid. Indeed, only AP as exploited in AID can deal with the universal nature of arbitrary data, and is thus deeply linked to 'meaning', where 'meaning' amounts to access to the candidate algorithmic models that can explain the data over and above statistical pattern recognition. This means that there are two ways to proceed. One of them provides never-ending incremental improvements for understanding an observation and making generalisations and causal predictions, and the other, the most commonly used one, creates an illusion of understanding but is ill-equipped to do anything that the first method can do, that is, anything beyond statistical characterisation, and cannot be improved upon. No machine learning technique, including deep learning, is designed to do anything beyond 'curve-fitting' without human intervention in the way of hyper-parameter selection, data labelling, etc. It is only because they appear to us as black boxes that we believe them to be doing something much more sophisticated.

The current situation is that scientists are required to devote all their effort and resources to dealing with data, gathering, cleansing, and (over-)fitting, with little to no effort left to spare for the design and choice of methods suited to their goals. For example, if the purpose is to infer causality, understand causation, or discover candidate generative models, statistical approaches should not dominate the inference methodology. The first thing we do as scientists is to gather data and apply the easiest method available, namely a machine learning approach or a computable measure such as Shannon entropy. Both may seem sound, and even mysterious, which gives a misleading air of sophistication to this way of proceeding. When people realise that such statistical methods are weak for delivering their goals, they sometimes switch to algorithmic complexity but settle for popular lossless compression algorithms such as LZW [112] to estimate it. But this is yet another black box that misleads and usually only supplies a single number. A compressed executable file may indeed be a model for the data, but 1) it is cryptic, as nobody opens the file or has the tools to understand it (making it a quintessential black box) and 2) popular lossless compression algorithms (such as LZW) cannot do better than Shannon entropy rate estimators, because just as with machine and deep learning, one cannot expect lossless compression algorithms to do something they were not designed to do, such as find algorithmic/mechanistic models as opposed to merely highlighting statistical patterns, that machine learning or variations of Shannon entropy can supply [112]. So we are left with few options in computational mechanics (CM). Most of them have been proposed in very abstract and theoretical terms. On the one hand we have those such as minimum description length (MDL) that are based, again, on popular lossless compression algorithms, and on the other hand, approaches such as Crutchfield's epsilon machines or Hutter's AIXI that can be complemented with our numerical methods, without the need for techniques related to popular lossless compression algorithms or weaker models of computation (such as Markov chains). This leaves us with only one method that is truly closer to the principles of algorithmic information and has been shown to be applicable and to produce interesting results, i.e., the methods based on CTM, BDM, and AID, that other methods are not equiped to deal with, not only because they do not implement principles of algorithmic information beyond the statistical ones offered by Shannon entropy alone but also because they are not sensitive enough to deal with, e.g., perturbation analysis, as shown in this application to causal discovery in machine intelligence [113].

9.1   Information Biology

One of the aims of our research is to exploit these ideas in order to try to reprogram living systems, so as to make them do things we would like them to do, for in the end this is the whole idea behind programming something.

In biology, the greatest challenge is the prediction of behaviour and shape ('shape' determines function in biology). Examples are protein folding or predicting whether immune cells will differentiate in one direction rather than another. However, in recent decades we have come to understand that complex systems such as the immune system cannot be thoroughly analysed and fully understood as regards its components and function by using 'reductionist' approaches, and we have moved to the science of systems, as in systems and computational biology. We have nevertheless failed to make as radical a move as we ought and apply the tools of complexity science in a holistic way. And reprogramming cells to fight complex diseases (see Table 9.1) is one of our ultimate goals.

Table 9.1 Properties of simple versus complex diseases. They cannot be considered in isolation. For example, the presence of many interacting elements does not automatically imply a complex disease (system); it is the way they interact and the multiple factors involved that make for a complex and therefore unpredictable system, even if it is fully deterministic (which may or may not be the case for complex diseases).

Simple diseases	Complex diseases
- Single and isolated causes - Simple gene expression patterns - Focalised effects - High predictability	- Multiple disjoint and joint causes - Many interacting elements - External factors - Low predictability
Examples: Monogenic (single gene) diseases (e.g. Cystic fibrosis), chromosomal abnormalities, Thalassaemia, Haemophilia, Huntington's disease	**Examples:** Multiple sclerosis, Alzheimer's, Parkinson's, most cancers

9.2 Complex versus Simple in the Real World

Complex diseases (see Table 9.1) require a complex debugging system, and the immune system can be viewed as playing just this role. The immune system is the highly complex and dynamic counterpart of many complex (and simple) diseases, and the most common approach to understanding it has been via evaluating its individual components. However, as with computer programs, this kind of approach cannot always expose the way in which an immune system works under a range of possible circumstances, particularly those leading to complex diseases such as autoimmune conditions and cancer. One instance where our knowledge will come up against limitations imposed by the systems approach itself is the process by which the immune system responds to infectious agents by activating innate inflammatory reactions and dictating adaptive immune responses (the immune system is split into two branches: the innate and the adaptive immune systems). A similar example is how cancer systems biology has emerged to address the increasing challenge of cancer as a complex, multifactorial disease, but has failed to move away once and for all from the archaic classification of cancer by tissue of origin.

9.3 How Natural Selection Reprograms Life

At some point early in the process of replicating (copying) biological information for cell growth, which is particularly necessary for multicellular organisms, the process reaches the critical juncture of having to deal with errors and redundancy, which can be quantified by the noisy-channel coding theorem determining the degree of redundancy needed to be able to deal with different degrees of noise.

If this balance is not reached, no copying process would be able to convey the information necessary for organisms to reproduce. The noisy-channel coding theorem establishes that for any given degree of noise in a communication channel, it is still possible to communicate (digital) data nearly error-free up to a computable maximum rate by introducing redundancy into the channel. This also suggests why nature has chosen a clearly digital code for life in the form of nucleic acids (e.g., DNA and RNA). On the other hand, if the balance is successfully struck, that is, if the accuracy of the replication process is greater than the incidence of error, those errors that do occur could be quantified using classical information theory.

While the nature of the variations (of which some can be identified as errors) can be attributed to noise, the correction is simply a mathematical consequence. If errors prevail, both in the primary source and replicants, then the cells and organisms have a greater chance of dying from these variations. On the other hand, new variations may confer survival or reproductive advantages in changing environmental conditions, and these will be positively selected during evolution. Natural selection hence effectively reprograms nature, cells, and organisms. Of course this does not explain how the whole process began, the origin of life being still an open question.

Approached as computer programs, one can explain how certain patterns, e.g., the information content in molecules such as DNA or RNA, may have been produced in the first place. There is also the problem of the uncaused first computer program, given that the process generating information and eventually computation required computation in the first place (e.g., the very first laws in our universe).

Hacking strategies have been in place in biological systems since the beginning. Viruses for example, are unable to replicate by themselves, but they breach the cell membrane of the host and release their DNA content into the cell nucleus so that it may be replicated. Cells are the basic units of life because of this property: they are the smallest units that fully replicate themselves. No simpler unit is able to replicate in full like the cell, even though a precursor of RNA is suspected to have been capable of replication, and RNA itself is possibly the first self-replicating molecule, because it can both store information, like DNA, and also fold, like proteins, and thus serves as a building block with which to build more complex structures.

Viruses are clearly (as per the consensus among biologists) non-living structures of encapsulated DNA that evolve by undergoing a process of natural selection, and are evidence that non-living matter may be subject to the same process. Though they may appear to do so, viruses have no aim or will of their own, either as individuals or in the aggregate. Indeed, evolution by natural selection is a mathematical inevitability independent of substrate. Despite the common, flawed fashion in which the concept of evolution by natural selection is popularly understood – as a biological process driven by the objective of self-preservation – it is incontestably a simple statistical phenomenon.

In the case of viruses, for example, it is the viral DNA that make it more effectively into the host nucleus and replicate faster that will dominate. This phenomenon is the basis of the mechanism of evolution by natural selection. By adding limited resources, such as the number of cells that can be infected, one can extrapolate from this simple mathematical fact to things other than viruses. Even though viruses are naturally perfect carriers of computing code, biological programming and reprogramming is, in general, very similar because we know that everything is encoded in the cell's genetic material.

9.4   A Computational Approach to Life and Disease

Equipped with an understanding of these mechanisms of evolution by natural selection and the way in which self-replicating cells can be hacked, cancer researchers have devised a hack to fight the disease. Traditional drug and radiation therapies have not been very successful so far but so-called *gene therapy* promises spectacular results. *Gene therapy* is a mechanism to deliver specific genes into tumours by using similar, if not exactly the same, strategies that viruses use to deliver their payload. Indeed, this approach relies heavily on actually using viruses to deliver anti-tumour genes into the target cancer cells. Various hacking approaches using viruses can be devised. If the foreign code enters the nucleus it will be replicated but make a buggy virus that only reaches the cytoplasm once it penetrates the cell membrane, and it can only interfere with protein transcription (by leading the enzymes in the cytoplasm to believe that the instructions are legitimate and come from other cells or are regular instructions coming from the cell nucleus). The fake instructions will then modify cell behaviour (function), with the nucleus remaining intact and replication of its original untouched code guaranteed. Indeed, one can see how genes are effectively subroutines, computer programs that regulate the type (shape) and number of building blocks (proteins) to be produced, that in turn fully determine the cell function(s). Let the instructions reach only outside the nucleus and they will interfere only with messages (RNA) indicating protein production, or let them reach inside the nucleus and they will modify the actual genome of the cell (DNA) to replicate the virus (which is what viruses do, integrating their DNA content into the host genome and undergoing a co-evolution).

There is therefore the potential to develop these ideas into a more systematic information-theoretic and software-engineering view of life, cancer, and immune-related diseases based on these amazing, purely computational mechanisms in biology. While it is clear that information processing is, in a very fundamental sense, key to essential aspects of the biology of different species, by applying ideas related to computability and algorithmic information theory we can advance toward viewing processes from fresh and different angles, so as to conceive new strategies for modifying the behaviour of cells in our fight against diseases.

Cancer, for example, is like a computer programming bug that does not serve the purpose of the multicellular organism. Cancer cells are ultimately cells that grow uncontrollably and do not fulfill their contract to die or stop proliferating at the rate at which they must if the multicellular host is to remain stable and healthy. From a software engineering perspective it is a cellular computer program that has gone wrong, resulting in an infinite loop with no halting condition, in effect being out of control. Normal cells have a programmed mechanism to stop their cell cycle at the point at which the population of older cells is replaced. When the cell cycle produces more than the number of cells necessary to replace the population, the result is a mass of abnormal cells that we call a tumour.

Cancer can be seen as a purely information-theoretic problem: the information dictating the way in which a cell replicates is compromised, either because it, as it were, reneges on a contract with the multicellular organism, resulting in the cell behaving selfishly and replicating with no controls, or else because noise in the environment in the form of external stimuli has broken the programming code that makes a cell's behaviour 'normal'. In other words, it is either a bug or a broken message. The question is therefore how a cell can transfer its information at replication time without generating instructions that produce cancer states when the said cell encounters noise.

Likewise, the immune system can be seen as an error-correcting code. One key aspect of the immune system is its diversity, which is largely owed by T and B lymphocytes, cell types of the adaptive immune system. By the rearrangement of segments in their antigen receptor genes, a highly complex repertoire of different receptors expressed by individual B or T cell clones is generated, which are specific for different antigens. The B cell antigen receptor, surface-bound or secreted as an antibody molecule by terminally differentiated B cells, recognises whole pathogens without any need for antigen processing, while the T cell antigen receptor is exclusively on the T cell surface and recognises processed antigens presented on so-called MHC molecules.

When B cells and T cells are activated and begin to replicate, some of their progeny become long-lived memory cells, remembering each specific pathogen encountered, and can mount a strong, faster response if the pathogen is detected again. This means that memory can take the form of either passive short-term memory or active long-term memory.

9.5   Algorithmic Information Dynamics in Biology

We have proposed ways to study and quantify the information content of biological networks based on the related concept of algorithmic probability, and we have found that it is possible to characterise and profile these networks in different ways and with a considerable degree of accuracy.

Important sources of information are epigenetic phenomena, an additional layer of complexity reversing the traditional molecular biology dogma that describes how information is transferred from the genome all the way to the upper levels. Epigenetics shows that information can flow downward from all upper layers to the lower layer (genome). This information coming back to the cell alters how genes are finally expressed, even if the cell's genome (the DNA code) remains exactly the same. Information flowing in this direction can be detected as information that cannot be explained by the underlying candidate models explaining the driving signals and major components of a system.

We have seen that uncomputability prescribes limits to what can be known about nature or models of nature, limits that are likely to apply to natural and biological systems or the models we build of them, and therefore we cannot help but develop an encompassing behavioural approach that can utilise ideas and tools from both theoretical computer science and software engineering. These ideas can then be further developed to yield new concepts, classifications, and tools with which to reprogram cells against diseases.

We will move towards biological systems by first applying our new tools and methods to synthetic dynamic systems (such as a cellular automaton associated with processes reminiscent of life) and artificial objects (such as integer sequences), then moving on to cognitive processes, genetics and actual regulatory networks, and biological evolution.

10   Algorithmic Dynamics in Artificial Environments

As we have introduced it, algorithmic information dynamics is a calculus with which to study changes in the causal content of a dynamical system's orbits when the complex system is perturbed or unfolds over time. We demonstrate the application and utility of these methods in characterising evolving emergent patterns and interactions (collisions) in a well-studied example of a dynamical (discrete) complex system that has been proven to be very expressive by virtue of being computationally universal [114].

Conway's Game of Life (GoL) [115] is a two-dimensional cellular automaton (see Fig. 10.1). A cellular automaton is a computer program that applies in parallel a global rule composed of local rules on a tape of cells with symbols (e.g., binary). The local rules governing GoL are traditionally written as follows:

1. A live cell with fewer than two live neighbours dies.
2. A live cell with more than three live neighbours dies.
3. A live cell with two or three live neighbours continues to live.
4. A dead cell with three live neighbours becomes a live cell.

Each of these is a local rule governing a special case, while the set of rules 1–4 constitute the global rule defining the GoL.

Following [114], we call a configuration in GoL that contains only a finite number of 'alive' cells and prevails a *pattern*. If such a pattern occurs with high frequency we call it a *motif*.

GoL is an example of a two-dimensional cellular automaton that is not only Turing-universal but also intrinsically universal [114]. This means that the GoL not only computes any computable function but can also emulate the behaviour of any other two-dimensional cellular automaton (under rescaling).

10.1   Emergent Patterns

So-called *gliders* are a (small) pattern that emerges in GoL with high frequency. The most frequent glider motif (see Fig. 10.2(d)) travels diagonally at a speed of 1/4 (the

Material adapted from Hector Zenil, Narsis A. Kiani, Jesper Tegnér. Algorithmic information dynamics: A computational approach to causality with applications to living systems. In Andrew Adamatzky, Selim G. Akl and Georgios Ch. Sirakoulis (eds.) *From Parallel to Emergent Computing*. CRC Press, 2019. Reproduced by permission of Taylor & Francis Group.

(a)   (b)   (c)

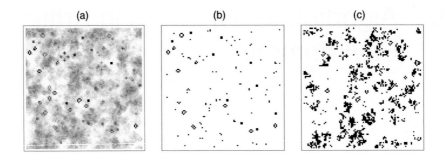

Figure 10.1 A typical run of the Game of Life (GoL). (a) Density plot with persistent motifs highlighted and vanishing ones in various lighter shades of grey. (b) Only prevalent motifs from the initial condition, as depicted in (c).

distance travelled in time t would be $t/4$) across the grid, and is the smallest and fastest motif in GoL, where t is the automaton runtime from initial condition $t = 0$.

Glider collisions and interactions can produce other particles, such as the so-called blocks, beehives, blinkers, traffic lights, and a less common pattern known as the 'eater'. Particle collisions in cellular automata, as in high particle physics supercolliders, have been studied before, demonstrating the computational capabilities of such interactions where both annihilation and new particle production is key [116]. Particle collision and interaction profiling may thus be key in controlling the way in which computation can happen within the cellular automaton, as shown in [117].

We have exhaustively demonstrated that Shannon entropy and popular lossless compression algorithms are not well equipped to deal with a series of possible challenges, such as algorithmic patterns that are not of a statistical nature and thus cannot be captured by computable measures like Shannon entropy, and small changes and patterns that cannot be profiled or characterised by popular lossless compression algorithms. Here, we will only show that the algorithmic tools enabling the area that we have called algorithmic information dynamics can be used and are actually known to fare better at these tasks than Shannon entropy and common compression [59, 66, 103].

Algorithmic information dynamics, or simply algorithmic dynamics, helps track the local dynamical changes of patterns and motifs in GoL that may shed light on the local but also the global behaviour of a discrete dynamical system, of which GoL is a well-known case.

In the interests of fairness, Shannon entropy (and, for that matter, popular lossless compression algorithms) and algorithmic complexity will be applied to exactly the same objects and exactly the same representations of these objects. For example, to binary matrices of size 4×4 as described in Chapter 6.

As illustrated in Fig. 10.3, an isolated observation window does not contain all the algorithmic information of an evolving system. In particular, it may not contain the complexity to enable inference of the set of local generating rules, and hence of the global rule of a deterministic system (Fig. 10.3a). Thus, in practice, the phenomena in the window appear to be driven by external processes that are random, while some

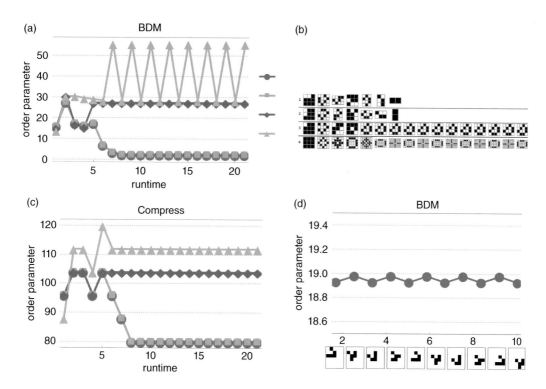

Figure 10.2 (a) Algorithmic probability approximation of local GoL orbits by BDM on evolving patterns of size 3×3 cells/pixels in GoL that remain 'alive', as shown in (b). (c) Same behavioural analysis using Compress (based on LZW), which under-performs (compared to BDM) in the characterisation of small changes in local emergent patterns. (d) The algorithmic dynamics of a free particle (the most popular local moving pattern in GoL, the glider), with BDM capturing its two oscillating shapes in a closed moving window of 4×4 cells running for 11 steps.

others can be explained in terms of interacting/evolving local patterns in space and time (Fig. 10.3c). This means that even though GoL is a fully deterministic system, its algorithmic complexity K can only grow by $\log(t)$ (Fig. 10.3b), because the shortest generating program is always the same at any time t and to reproduce the system at any time all that is needed is to encode the value of t, which can be done in binary in about $\log(t)$ bits. One can meaningfully estimate the $K(w)$ of a cross-section w (Fig. 10.3c) of an orbit of a deterministic system like GoL and study its algorithmic dynamics (the change of $K(w)$ over time).

It has been proven that there are quantitative connections between indicators of algorithmic information content (or algorithmic complexity) and the chaotic behaviour of dynamical systems that are related to their sensitivity to initial conditions. Some of these results and the relevant references are given in [118], for instance. Previous numerical approaches, such as the one used in [118] and others cited in that paper, including those proposed by the authors of the landmark textbook on Kolmogorov complexity [71], make use of computable measures, in particular measures based on

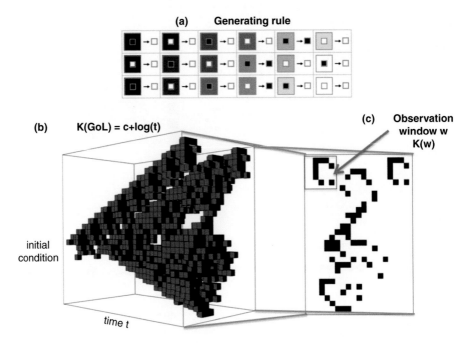

(a)   Generating rule

(b)   K(GoL) = c+log(t)

(c)   Observation window w
K(w)

initial condition

time t

Figure 10.3 The algorithmic complexity of an observation. (a) Generating rule of Conway's Game of Life (GoL), a two-dimensional cellular automaton whose global rule is composed of local rules that can be represented by the average of the values of the cells in the (Moore) neighbourhood (a property also referred to as 'totalistic' [35]). (b) 3D space-time representation of successive configurations of GoL after 30 steps. (c) Projected slice window w of an observation of the evolution of (b) the last step of GoL.

popular lossless compression algorithms, and suggest that non-computable approximations cannot be used in computer simulations or in the analysis of experimental results. One of the aims of this chapter is to illustrate how CTM and BDM [81, 82, 108] based on the concept of algorithmic probability, that has been shown to be more powerful [59, 103] than computable approximations [60] such as popular lossless compression algorithms (e.g., LZW), can overcome some previous limitations and difficulties in profiling orbit complexity, difficulties encountered particularly in the investigation of the behaviour of local observations and small perturbations typical of systems.

10.2   Algorithmic Probability of Emergent Patterns

The distribution of motifs (the 100 most frequent local persistent patterns, also referred to as *ash* as they are the debris of random interactions) of GoL are reported in http://wwwhomes.uni-bielefeld.de/achim/freq_top_life.html starting from 1,829,196 (likely different) random seeds (in a torus configuration) with initial density 0.375 black cells over a grid size of 2048×2048 and from which 50,158,095,316 objects were found.

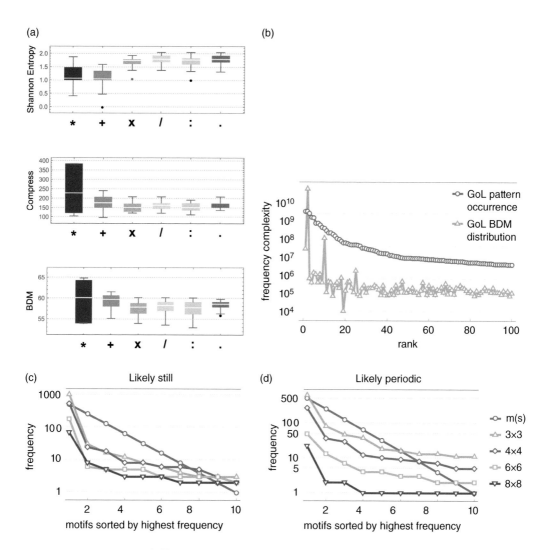

Figure 10.4 (a) Classical and algorithmic measures versus symmetries of the top 100 most frequent patterns (hence motifs) in Conway's Game of Life (GoL). The measures show diverse (and similar) abilities to separate patterns with the highest and lowest number of symmetries. Notation for the square dihedral group D_4: invariant to all possible rotations (*), to all reflections (+), to two rotations only (X), to two reflections (/), one rotation (:) and one reflection (.). (b) The heavily long-tail distribution of local persistent patterns in GoL (of less than 10×10 pixels) from the 100 most frequent emerging patterns and of (c, d) most likely still and periodic structures.

Figure 10.4(a) suggests that highly symmetric patterns/motifs that produce about the same number of black and white pixels and look similar (small standard variation) to entropy can actually have more complex shapes than those collapsed by entropy alone. Similar results were obtained before and after normalising by pattern size (length × width). Symmetries considered include the square dihedral group D_4, i.e., those invariant to rotations and reflections. Shannon entropy characterises the highest symmetry as having the lowest randomness, but both lossless compression and algorithmic probability (BDM) suggest that highly symmetric shapes can also reach higher complexity.

Given the structured nature of the output of GoL, taking larger blocks (square windows capturing a pattern in GoL from block size 3×3 to 8×8) reveals this structure (see Figure 10.4(b–d)). If the patterns were statistically random, the block decomposition would display high block entropy values, and the distributions of patterns would look more uniform for larger blocks. However, larger blocks remain highly non-uniform, indicating a heavy tail, as is consistent with a distribution corresponding to the algorithmic complexity of the patterns – that is, the simpler the more frequent. Indeed, the complexity of the patterns can explain 43% (according to a Spearman rank correlation test, p-value 8.38×10^{-6}) of the simplicity bias in the distribution of these motifs (see Fig. 10.4(b)).

Algorithmic probability may not account for a greater percentage of the deviation from uniform or normal distributions because patterns are filtered by persistence, i.e., only persistent patterns are retained after an arbitrary runtime step, and therefore no natural halting state exists, likely producing a difference in distribution as reported in [59], where distributions from halting and non-halting models of computation were studied and analysed. Values of algorithmic probability for some motifs (the top and bottom 20 motifs in GoL) are given in Fig. 10.5.

On the other hand, as plotted in Fig. 10.4(b–d), the frequency and algorithmic complexity of the patterns in GoL follow a rank distribution and are negatively correlated amongst each other, as the algorithmic coding theorem establishes. That is, the most frequent emergent patterns are also the most simple, while the most seldom occurring ones are more algorithmically random (and their algorithmic probability low). This is also illustrated by plotting the complexity of the distribution of patterns in GoL as they emerge with long tails for both still and periodic patterns and for all patterns of increasing square window size.

10.3   Algorithmic Dynamics of Evolving Patterns

While each pattern in GoL evolving in time t comes from the same generating global rule for which $K(GoL(t))$ is fixed (up to $\log(t)$, corresponding to the binary encoding of the runtime step), a pattern within an observational window (Fig. 10.3) that does not necessarily display the action of all the local rules of the global rule can be regarded as an (open) system separate from the larger system governed by the global rule. This is similar to what happens in the practice of understanding real-world complex systems to which we only have partial access and where a possible underlying global rule exists but is unknown.

An application to Conway's GoL evolving over time using BDM shows some advantages in the characterisation of emergent patterns, as seen in Figs. 10.2, 10.6, and 10.7.

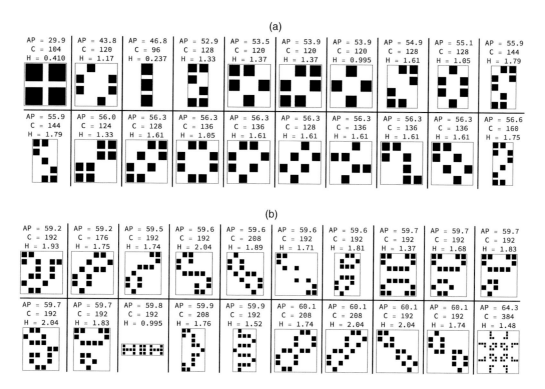

Figure 10.5 (a) Top 20 and (b) bottom 20 most and least algorithmically complex local persistent patterns in Conway's Game of Life (AP is the BDM estimation, C is lossless compression by Compress, and H is classical Shannon entropy).

We took a sliding window consisting of a small number of $n \times m$ cells from a 2D cross section of the 3D evolution of GoL as shown in Fig. 10.3. For most cases $n = m$. The size of n and m is determined by the size of the pattern of interest, with the sliding window following the unfolding pattern. The values of n or m may increase if the pattern grows but they can never decrease, even if the pattern disappears. Each line in all plots corresponds to the algorithmic dynamics (complexity change) of the orbit of a local pattern in GoL, unless otherwise established (e.g., such as in collapsed cases). Fig. 10.2, for example, demonstrates how the algorithmic probability approach implemented by BDM can capture dynamical changes even for small patterns where lossless compression may fail because they are limited to statistical regularities that are not always present. For example, in Fig. 10.2(a), BDM captures the periodic/oscillating behaviour (period 2) of a small pattern, something that compression, as an approximation of algorithmic complexity, was unable to capture for the same motifs in Fig. 10.2(b). Likewise, the BDM approximation of algorithmic complexity captures the periodic behaviour of the glider in Fig. 10.2(d) for 10 steps.

Figures 10.8(a) and (b) illustrate cases of diagonal particle (glider) collisions. In a slightly different position, the same two particles can produce a single still pattern as shown in Fig. 10.8(d), which reaches a maximum complexity when new particles are

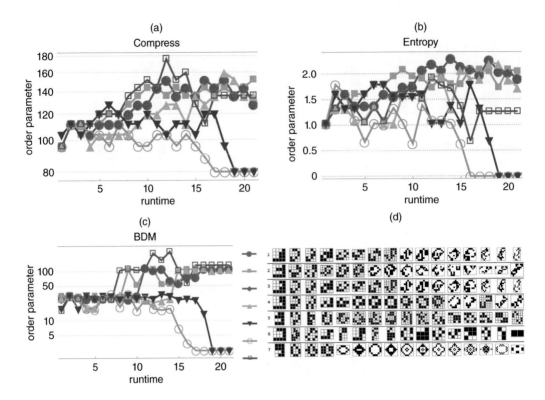

Figure 10.6 Orbit algorithmic dynamics of local emergent patterns in Conway's Game of Life (GoL). Compress (a) and Entropy (b) retrieve very noisy results compared to BDM (c), which converges faster and separates the dynamic behaviour of all emerging patterns in GoL of size 4×4 pixels. (d) Shows the dynamic behaviour of all emerging patterns in GoL of size 4×4 pixels.

produced, thereby profiling the collision as a transition between a dynamic and a still configuration. In Fig. 10.8(a), the particles annihilate each other after a short transition of different configurations. In Fig. 10.8(b) the collision of four gliders produces a stable non-empty configuration of still particles after a short transition of slightly more complicated interactions. We call this interaction a 'near-miss' because the particles seem to have missed each other even though there is an underlying interaction. In Fig. 10.8(c), an unstable collision characterised by the open-ended number of new patterns evolving over time in a growing window can also be characterised by their algorithmic dynamics using BDM, as shown in Fig. 10.8(d) and marked as an unstable collision.

More cases, both trivial and non-trivial, are shown in Figures 10.6 and 10.7(a, b). Figure 10.6 shows seven other cases of evolving motifs starting from different initial conditions in small grid sliding windows of size up to 4×4 displaying different evolutions captured by their algorithmic dynamics. Figure 10.7 shows all evolving patterns of size 3×3 in GoL and the algorithmic dynamics characterising each particle's behaviour, with BDM and entropy showing similar results, albeit with a better separation for BDM.

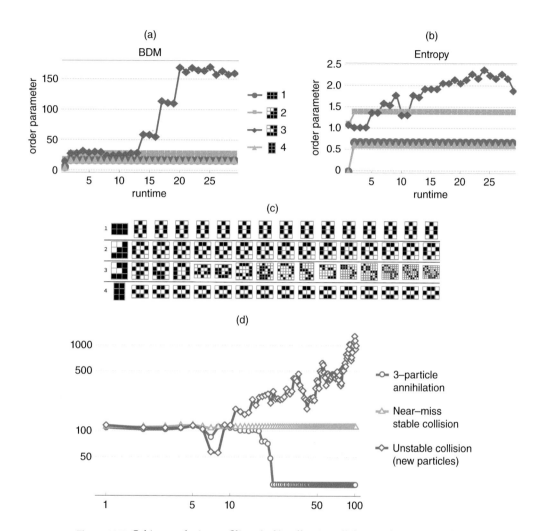

Figure 10.7 Orbit complexity profiling. (a, b) collapsing all the simplest cases (1, 2 and 4) to the bottom – closest to zero – values diverging from the only open-ended case (3). (a) The measure BDM returns the best separation compared to entropy (c) 16 steps corresponding to evolving steps of the four cases captured in (a) and (b). (d) The algorithmic information dynamics of three particle interactions/collisions. The unstable collision corresponds to Figure 10.8(d), the 3-particle annihilation is qualitatively similar to the 2-particle Figure 10.8(a) and the near-miss stable collision corresponds to Figure 10.8(b), where the four particles look as if they are about to collide but appear not to (hence a 'near miss'). Starting seeds are shown in Figure 10.9.

10.4 Algorithmic Dynamic Profiling of Particle Collisions

We traced the evolution of collisions of so-called gliders. Figure 10.8 shows concrete examples of particle collisions of gliders in GoL and the algorithmic dynamic characterisation of one such interaction, and Fig. 10.10(a) illustrates all cases for a sliding window of up to size 17×17, where all cases for up to four colliding gliders are reported, analysed, and classified using different information-theoretic indices,

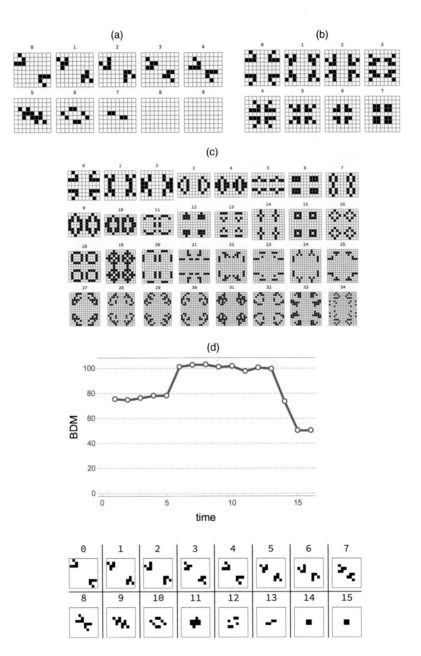

Figure 10.8 (a, b, c) Three possible collisions showing 2-particle annihilation (a), stability (b), and instability, i.e., production of new particles (d). (d) The algorithmic information dynamics of a 2-particle stable collision.

including compression as a typical estimator of algorithmic complexity, and BDM as an improvement over both Shannon entropy as a stand-alone measure and typical lossless compression algorithms. The results show that cases can be classified in a few

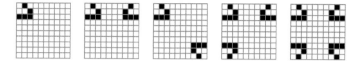

Figure 10.9 Set of initial conditions for particle (glider) collision. From left to right: free particle, 2-particle sideways collision, 2-particle frontal collision, 3-particle collision, and 4-particle collision.

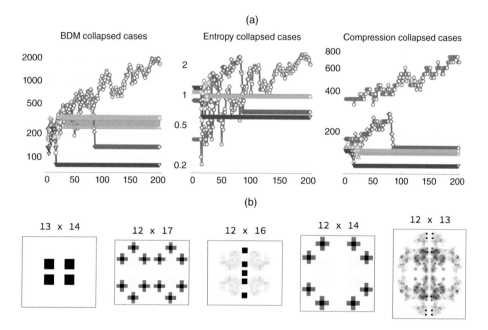

Figure 10.10 All possible collision cases of up to four glider particles in blocks (square arrays) of up to 17×17 bits/cells. (a) Clusters of dynamical system attractors of colliding gliders in Conway's Game of Life. (b) Density plot of all non-trivial (particles that are not entirely annihilated) qualitative interactions among four particles. The darker they are, the later, and more persistent over time.

categories corresponding to the qualitative behaviour of the possible outcomes of up to four particle collisions.

Figure 10.7(d) summarises the algorithmic dynamics of different collisions and for all cases with up to four gliders in Fig. 10.10(a) by numerically producing all collisions but collapsing cases into similar behaviour corresponding to qualitatively different cases, as shown in the density plots in Fig. 10.10(b). The interaction of colliding particles is characterised by their algorithmic dynamics, with the algorithmic probability estimated by BDM remaining constant in the case in which four particles prevail, the annihilation case collapsing to 0, and the unstable collision producing more particle divergence.

10.5   Chapter Summary

We have explained how observational windows can be regarded as apparently open systems, even if they come from a closed deterministic system $D(t)$ for which the algorithmic complexity $K(D)$ cannot differ by more than $\log_2(t)$ over time t – a (mostly) fixed algorithmic complexity value.

However, in local observations patterns seem to emerge and interact, generating algorithmic information as they unfold, requiring different local rules and revealing the underlying mechanisms of the larger closed system.

We have shown the different capabilities that both classical information and algorithmic complexity (the former represented by the lossless compression algorithm Compress, and the latter based on algorithmic probability) display in the characterisation of these objects and how they can be used to track changes and analyse their spatial dynamics.

We have also illustrated the way in which the method and tools of *algorithmic dynamics* can be used to measure the *algorithmic information dynamics* of discrete highly dynamical and evolving systems, in particular of emerging local patterns (particles) and interacting objects (such as colliding particles), as exemplified in a much studied two-dimensional cellular automaton. To help readers apply these tools to their own data we have developed (and continue to develop) free tools online at http://complexitycalculator.com/ that can be used, among other things, to estimate the algorithmic complexity of blocks by submitting local patterns and taking the estimation difference as the algorithmic information dynamics of a possible evolving system.

11 Applications to Integer and Behavioural Sequences

11.1 Introduction

Knowledge formalised in academic fields, including mathematics, is built on a foundation of inborn core knowledge [119, 120, 121], which follows a specific developmental trajectory across its lifespan [122]. Numerosity (our approximate implicit sense of quantity) has been a privileged target of recent research because numbers form one of the main pillars of elementary mathematical knowledge [123], but the study of randomness perception and statistical reasoning has also yielded striking results in the field of probability: adults with no formal education [124] as well as 8- to 12-month-old children [125, 126] have the wherewithal for simple implicit probabilistic reasoning. One of the toughest problems when it comes to Bayesian reasoning, however, is the detection of randomness, i.e., the ability to decide whether an observed sequence of events originates from a random source or has a deterministic origin [127].

Humans, both adults and children [128, 129], have a keen ability to detect structure, both of a statistical and an algorithmic nature, e.g., (0101... and 1234...) that only algorithmic complexity can capture (as opposed to, e.g., entropy rate).

Within the field of study devoted to our sense of complexity, the task of randomly arranging a set of alternatives is of special interest as it poses almost insurmountable problems to any cognitive system. The complexity of a subject-produced pseudorandom sequence may serve as a direct measure of cognitive functioning, one that is surprisingly resistant to practice effects [130] and largely independent of the kind of items to be randomised, e.g., dots [131], digits [132], words [133], tones [134], or heads-or-tails [135]. Although random item generation (RIG) tasks usually demand the vocalisation of selections, motor versions have comparable validity and reliability [136, 137]. RIG tasks are believed to tap into our approximate sense of complexity (ASC), while also drawing heavily on focused attention, sustained attention, updating, and inhibition [138, 139]. Indeed, to produce a random sequence of symbols one has to avoid any routine and inhibit prepotent responses. The ability to inhibit such responses is a sign of efficient cognitive processing, notably a flexibility assumed to be mediated by the prefrontal cortex.

Instructions may require responses at various speeds [140], or else the generation of responses may be unpaced [141]. Participants are sometimes asked to guess a forthcoming symbol in a series ('implicit randomisation' [142]), or vaguely instructed to 'create a random-looking string' [143]. The consensus is that, beyond their diversity, all RIG tasks rely heavily on an ASC, akin to a probabilistic core knowledge [144, 145].

Theoretical accounts of the reasons why RIG tasks are relevant tests of prefrontal functions are profuse, but pieces of experimental evidence are sparse. Sparse empirical research indirectly validates the status of RIG tasks as measures of controlled processing, such as the detrimental effect of cognitive load or sleep deprivation [146] or the fact that they have proved useful in the monitoring of several neuropsychological disorders [141, 147, 148, 149, 150].

As a rule, the development of cognitive abilities across a person's lifespan follows an inverse U-shaped curve, with differences in the age at which the peak is reached [122, 151]. The decrease rate following the peak also differs from one case to another, moving between two extremes. 'Fluid' components tend to decrease at a steady pace until stabilisation is reached, while 'crystalised' components tend to remain high after the peak, significantly decreasing only in those in their sixties or older [152]. Other evolutions may be thought of as a combination of these two extremes.

Two studies have addressed the evolution of complexity in adulthood, but with limited age ranges and, more importantly, limited 'complexity' measures. The first compared young and older adults' responses and found a slight decrease in several indices of randomness [140]. The second found a detrimental effect of aging on inhibition processes, but also an increase of the cycling bias (a tendency to postpone the re-use of an item until all possible items had been used once), which tends to make the participants' productions more uniform [133]. In both studies, authors used controversial indices of complexity that only capture particular statistical aspects, such as repetition rate or first-order entropy. Such measures have been proven to have some usefulness in gauging the diversity and type of long sequences (with, e.g., thousands of data points) such as those appearing in the study of physiological complexity in [153, 154, 155], but are inadequate when it comes to short strings (e.g., of less than a few tens of symbols), such as the strings typically examined in the study of behavioural complexity. Moreover, such indices are only capable of detecting statistical properties. Some researchers have proposed the use of algorithmic complexity as a way to overcome these difficulties [156, 157]. However, because algorithmic complexity is uncomputable, it was widely believed to have no practical interest or application. In the past few years, however, methods have been introduced related to algorithmic complexity that are particularly suitable for short strings [57, 58], and native n-dimensional data [65]. These methods are based on a massive computation to find short computer programs producing short strings and have been made publicly available [74] and successfully deployed in a range of different applications [65, 73, 81].

The main objective of the present study is to provide the first fine-grained description of the evolution over the lifespan of the (algorithmic) complexity of human pseudo-random production. Secondary objectives are to demonstrate that, across a variety of different tasks of random generation, the novel measure of behavioural complexity does not rely on the collection of tediously long response sequences as hitherto required. The playful instructions to produce brief response sequences by randomising a given set of alternatives are suitable for children and elderly people alike, can be applied in work with various patient groups, and are convenient for individual testing as well as internet-based data collection.

Participants with ages ranging from 4 to 91 performed a series of RIG tasks online. Completion time (CT) served as an index of speed in a repeated multiple choice framework. An estimate of the algorithmic complexity of (normalised) responses was used to assess randomisation performance (e.g., response quality). The hypothesis being tested was that the different RIG tasks are correlated since they all rely on similar core cognitive mechanisms, despite their differences. To ensure a broad range of RIG measurements, five different RIG tasks were selected from the ones most commonly used in psychology.

The experiment is a sort of reversed Turing test where humans are asked to produce configurations of high algorithmic randomness that are then compared to what computers can produce by chance according to the theory of algorithmic probability [57, 58].

11.2   Methods

The five tasks used, described in Table 11.1, are purposely different in ways that may affect the precise cognitive ability they estimate. For instance, some tasks draw on short-term memory because participants cannot see their previous choices (e.g., 'pointing to circles'), whereas in other tasks memory requirements are negligible, because the

Table 11.1 Description of the five RIG tasks used in the experiment. The order was fixed across participants.

Task	Description
Tossing a coin	Participants had to create a series of 12 heads-or-tails that would 'look random to somebody else' by clicking on one of the two sides of a coin appearing on the screen. The resulting series was not visible on the screen (the participant could only see the last choice made).
Guessing a card	Participants had to select one of 5 types of cards (Zener cards; see e.g., [158]), 10 times. In contrast to the other tasks, they were not asked to make the result look random. Instead, they were asked to guess which card would appear after a random shuffle.
Rolling a die	Participants had to generate a string of 10 numbers between 1 and 6, as random as possible ('the kind of sequence you'd get if you really rolled a die'). In contrast to the preceding cases, they could see all previous choices, but could not change any of them.
Pointing to circles	Participants had to point 10 times at 1 out of 9 circles displayed simultaneously on the screen. They could not see their previous choices. This task is an adaptation of the classical Mittenecker pointing test [131].
Filling a grid	Participants had to blacken cells in a 3×3 grid so that the result would look randomly patterned, starting from a white grid. In contrast to the other tasks, they could see their choices and click as many times as they wished. Clicking on a white cell made it black, and vice versa.

participant's previous choices remain visible ('rolling a die'). Completion times across the various tasks showed a satisfactory correlation (Cohen's $\alpha = .79$), suggesting that participants did not systematically differ in the cognitive effort they devoted to the different tasks. Any difference between task-related complexities is thus unlikely to be attributable to differences in time on task.

Complexities were weakly to moderately positively correlated across the different tasks (Cohen's $\alpha = .45$), mostly as a consequence of the 'filling the grid' task being almost uncorrelated with the other tasks [109]. Despite this moderate link, however, all trajectories showed a similar pattern across the lifespan, with a peak around age 25, a slow, steady decline between ages 25 and 60, followed by accelerated decline after age 60, as shown in Fig. 11.1.

The hypothesis being tested, i.e., that the different tasks were positively related to each other, was partially supported by the data, especially in view of the results obtained on the 'filling the grid' task. Furthermore, CTs showed correlation, also supporting the hypothesis, as did the fact that developmental complexity curves were in agreement. This suggests that all the tasks tap into our ASC as well as into other cognitive components with similar developmental trajectories, but that different tasks actually require different supplementary abilities, or else weight the components of these abilities differently.

The 'filling the grid' task appeared unique in that it was loosely correlated with all the other tasks. The fact that it required binary responses cannot account for this lack of association, since the 'tossing a coin' task yielded results uncorrelated with the 'filling the grid' responses. Bi-dimensionality could possibly have had an effect, but the 'pointing to circles' task was also unrelated to the grid task. On the other hand, one factor distinguished the grid task from all others in the set: the option offered to the participants to change their previous choices as many times as they wished. For that reason, the grid task may in fact have relied more on randomness perception, and less on inhibition and randomness production heuristics. Indeed, participants could change the grid until they felt it was random, relying more on their ASC than on any high order cognitive ability serving output structure. This hypothesis is supported by the fact that participants did indeed change their minds. There were only nine cells (that could turn white or black) on the grids and participants' end responses had a mean of 4.08 (SD = 1.8) selected (black) cells, thus generally favouring whiter configurations (possibly having to do with the all-white initial configuration). However, the number of clicks used by participants during this task was far larger ($M = 10.16$, SD = 9.86), with values ranging from 5 to 134 (with 134 clicks covering almost a fifth of all possible configurations). Thus the option to change previous choices in a RIG task may have been an important factor, and should accordingly be considered a novel variable in future explorations of randomisation tasks (and balanced with an all-black initial configuration). In this view, the 'filling the grid' task would reflect our ASC in a more reliable fashion than the other tasks, while being less dependent on inhibition processes.

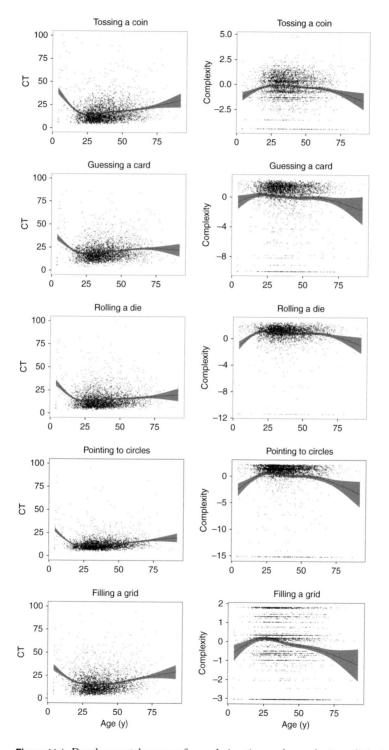

Figure 11.1 Developmental curves of completion time and complexity, split by task, with (red) trend curves and (shaded area) 95% confidence regions. From N. Gauvrit, H. Zenil, F. Soler-Toscano, J.-P. Delahaye, P. Brugger. Human Behavioral Complexity Peaks at Age 25. *PLoS Comput Biol* 13(4): e1005408, 2017.

11.2.1   Limitations

The present findings are based on data collected online.

Although direct control over a participant's behaviour online is certainly limited, as compared to the laboratory environment, there are an increasing number of studies demonstrating the convergence of laboratory-based and web-based data collection [122]. This is the case in very particular procedural situations, such as lateralised [159] or very brief [160] stimulation, presentation and the measurement of reaction times [161], and it also holds for the assessment of cognitive changes across a person's lifespan [162]. Compared to these special situations, our research procedure was simple, the tasks were entertaining, and response times did not have to be collected with millisecond precision [163]. We thus think that any disadvantages of online-testing were more than compensated for by the advantages of enrolling a large number of participants across a wide age range.

11.2.2   Modulating Factors

To investigate possible modulating factors (besides age), we used general linear models with complexity and CT as dependent variables (DV), and age (squared), sex, education, field of study, and paranormal beliefs as IV.

The variable *sex* was chosen in order to test in a large sample whether the absence of differences in laboratory RIG experiments could be replicated in an online test. Similarly, *education* was important to test for given previous claims in the RIG literature that human randomisation quality seemed independent of educational level [142]. Paradoxically, participants with a scientific background may perform worse at producing random sequences, thanks to a common belief among them that the occurrence of any string is as statistically likely as any other (a bias deriving from their knowledge of classical probability theory), which justifies further controlling for *field of education*, simplified as humanities versus science. Finally, the variable *paranormal belief* was included as it has been related to RIG performance in previous studies [164].

The variables *field* and *paranormal belief* were, however, only tested in a subset of the 3,313 participants that were above the age of 15, and we ignored the responses of younger participants as they were not considered to have a differentiated educational background or a fixed belief concerning paranormality. The analysis was performed on a task-wise basis. As we report, neither field nor educational level had a significant effect on any of the complexity or CT scores.

11.2.3   Experiment

A sample of 3,429 participants took part in the experiment (age range: 4–91y, M = 37.72, SD = 13.38). Participants were recruited through social networks, radio broadcasts and journal calls during a 10-month period. Basic demographic characteristics of the sample are displayed in Table 11.2, and the experiment is still available online for people wishing to test themselves (URL available in the next section). Each of the

Table 11.2 Sample descriptive statistics (*n*)

Sex	Male	2,333
	Female	1,085
	Unknown	11
Mother tongue	English	274
	French	1,448
	German	1,303
	Spanish	220
	Other	184
Education level	Kindergarten or below	38
	Primary school	83
	Secondary school	387
	High School	621
	Undergraduate	347
	Graduate	1,364
	Post graduate	538
	Unknown	51
Field	Humanities	609
	Science	1,684
	Other	550
	Irrelevant	586

five (self-paced) RIG tasks consisted in the production, as fast and as accurately as possible, of a short (with length range 9–12) pseudo-random series of symbols, with various numbers of possible symbols and variations among other instructional features (Table 11.1).

11.2.4   Procedure

A specific web application was designed to implement the experiment online. Participants freely logged on to the website. The experiment was available in English, French, German, and Spanish (written up by native speakers in each language). In the case of young children as yet unable to read or use a computer, an adult was instructed to read the instructions out loud, make sure they were understood, and enter the child's responses without giving any advice or feedback. Participants were informed that they would be taking part in an experiment on randomness. They then performed a series of tasks before entering demographic information such as sex, age, mother tongue, educational level, and main field of education (science, humanities, or other) if relevant.

One last item served as an index of paranormal beliefs and was included since probabilistic reasoning is among the factors associated with the formation of such beliefs [165, 166]. Participants had to rate on a 6-point Likert scale how well the following statement applied to them: 'Some "coincidences" I have experienced can best be explained by extrasensory perception or similar paranormal forces'.

11.2.5   Measures

For each task, CT (in seconds) was recorded. The sum of CTs (total CT) was also used in the analyses. An estimate of the algorithmic complexity of participants' responses was computed using the acss function included in the freely available acss R-package [74] that implements the complexity methods used in this project. Complexities were then normalised, using the mean and standard deviation of all possible strings with the given length and number of possible symbols, so that a complexity of 0 corresponds to the mean complexity of all possible strings. For each participant, the mean normalised complexity (averaged over the five tasks) was also computed, serving as a global measure of complexity.

11.3   **Results and Discussion**

Sex had no effect on any of the complexity scores, but a significant effect on two CT scores, with male participants performing faster in the first two tasks: 'tossing a coin' ($p = 6.26 \times 10^{-10}, \eta_p^2 = .012$) and 'guessing a card' ($p = 2.3 \times 10^{-10}, \eta_p^2 = .012$). A general linear model analysis of the total CT scores as a function of sex, age (squared), educational field, educational level, and paranormal belief was performed on the same subset and revealed a strongly significant effect of sex ($p = 9.41 \times 10^{-8}, \eta_p^2 = .009$), with male participants performing faster than female participants. A simpler model, including only sex and age (squared) as independent variables (IV), still showed an effect of sex ($p = 6.35 \times 10^{-13}, \eta_p^2 = .016$), with male participants needing less time. The sex difference in CT, mostly appearing in adulthood, was in line with previous findings that in adults choice CT is lower in men than in women [167].

Paranormal belief scores were unrelated to CTs for all tasks. However, they were negatively linked with the complexity score in the 'filling a grid' task ($p = .0006$, $\eta_p^2 = .004$), though not with any other task.

Paranormal beliefs have been previously reported to be negatively linked to various RIG performances [164, 165]. Our results replicated these findings but only on the 'filling a grid' task. One possible explanation is that the grid task actually measures randomness perception rather than randomness production, and that a paranormal bent is more strongly linked with a biased perception of randomness than with a set of biased procedures used by participants to mimic chance. This hypothesis is supported by the finding that believers in extrasensory perception are more prone to see 'meaningful' patterns in random dot displays [164]. A complementary hypothesis is that the type of biases linked to beliefs in the paranormal only usually appear over the long haul, and are here pre-empted by the fact that we asked participants to produce short sequences (of 12 items at most). Indeed, when it comes to investigating the effects of paranormal belief on pure randomness production, rather long strings are needed, as the critical measure is the number of repetitions produced [165, 168].

To get a better sense of the effect of paranormal belief, we performed a general linear model analysis of the mean complexity score as a function of sex, age (squared), educational field, educational level, and paranormal belief. For this analysis, we again used the subset of 3,313 participants over the age of 15. Paranormal belief no longer had an effect.

11.3.1 Mean Complexity and Total CT Trajectories

Our main objective was to describe the evolution over the lifespan of mean complexity, which is achieved here using an approximation of algorithmic complexity for short strings. In line with Craik and Bialystok's [152] view, the developmental curve of complexity found in Fig. 11.2 suggests that RIG tasks measure a combination of fluid mechanics (reflected in a dramatic performance drop with aging) and more crystallised processes (represented by a stable performance from 25 to 65 years of age). This trajectory indirectly confirms a previous hypothesis [133]: attention and inhibition decrease in adulthood, but an increased sense of complexity based on crystallised efficient heuristics counters the overall decline in performance.

Plotting complexity and CT trends on a single two-dimensional diagram allowed a finer representation of these developmental changes (Fig. 11.3). It confirmed the entanglement of complexity (accuracy) and CT (speed). In the first period of life (< 25), accuracy and speed increased together in a linear manner. The adult years were remarkable in that complexity remained at a high level for a protracted period, in spite of a slow decrease of speed during the same period. This suggests that in adulthood people tend to invest more and more computational time to achieve a stable level of output complexity. Later in life (>70), however, speed stabilises, while complexity drops in a dramatic way.

These speed-accuracy trade-offs were evident in the adult years, including the turn toward old age. During childhood, however, no similar pattern is discernible. This sug-

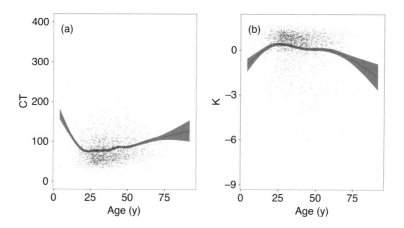

Figure 11.2 (a) Total completion time (CT) and (b) mean complexity as a function of age, with (red) trend curve and (shaded area) 95% confidence region. From N. Gauvrit, H. Zenil, F. Soler-Toscano, J.-P. Delahaye, P. Brugger. Human Behavioral Complexity Peaks at Age 25. *PLoS Comput Biol* 13(4): e1005408, 2017.

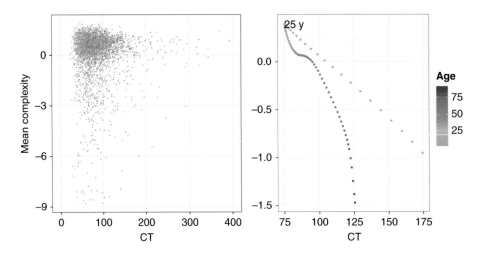

Figure 11.3 Scatterplot and developmental change trend of completion time (CT) and complexity combined. The trend is obtained by use of smooth splines of CT and complexity (df = 7). From N. Gauvrit, H. Zenil, F. Soler-Toscano, J.-P. Delahaye, P. Brugger. Human Behavioral Complexity Peaks at Age 25. *PLoS Comput Biol* 13(4): e1005408, 2017.

gests that aging cannot simply be considered a 'regression', and that CT and complexity provide different, complementary information. This is again supported by the fact that in the 25–60 year range, where the effect of age is reduced, CT and complexity are uncorrelated ($r = -.012, p = .53$). These findings add to a rapidly growing literature that views RIG tasks as good measures of complex cognitive abilities (see [139] for a review).

We have gone further here than any previous studies in several respects. First, we present a set of data collected in RIG tasks with a broad variety of instructions as to what and how to randomise. Our participants playfully solved binary randomisation tasks along with multiple alternative variants. They explicitly attempted to generate random sequences, but also distributed their responses in a guessing task, typically considered 'implicit random generation'. The expected outcome was unidimensional in some tasks and two-dimensional in others. Constraints imposed by working memory capacity were high in some tasks, but almost absent in others. In the cognitive science literature such diverse tasks have never been compared directly. We do not deny that the various tasks we used may tap into slightly different subcategories of prefrontal functioning, with some relying more on working memory and others on inhibitory control. Yet, we set out to illustrate the commonalities among the different tasks, leaving a more fine-grained analysis to future studies.

Cross-sectional studies should try to relate behavioural complexity to the degree of maturation or degeneration of specific prefrontal cortex regions. Neuropsychological investigations could use the tasks and measures employed here with selected patient groups to perform lesion-symptom mappings, as has been done recently [169], but preferably in patients with focal vascular lesions. In parallel with such investigations, internet-based work such as the project presented here may still play a powerful role. They may complement RIG tasks with brief behavioural tasks having a known neurocognitive basis and well-studied developmental trajectories. Thus, laboratory testing and web-based approaches may conjointly help pinpoint the cognitive mechanisms underlying the age-related modulation of behavioural complexity.

A second extension of the existing literature on subject-generated random sequences is the number of participants tested and their age range. To date, only two studies have investigated age-related changes in RIG tasks in an age range comparable to the one investigated here [133, 140]. They both compared groups of young adults and older adults and were thus unable to describe the continuous evolution of complexity across lifespan.

Finally, one of the most exciting and novel aspects of this research is that we have presented an estimate of algorithmic complexity that relies on sequences shorter than any that research on RIG reported in the psychological literature would have dared to use, given the limitations of other complexity indices.

12 Applications to Evolutionary Biology

Central to the modern synthesis and general evolutionary theory is the understanding that evolution is gradual and is explained by small genetic changes in populations over time [170]. Genetic variation in populations can arise by chance through mutation, with these small changes leading to major evolutionary changes over time. Of interest in connection to the possible links between the theory of biological evolution and the theory of information is the place and role of randomness in the process that guarantees the variety necessary to allow organisms to change and adapt over time.

On the one hand, while there are known sources of non-uniform random mutations in plants and animals, mutations as functions of environment, gender, and age, for example, when all conditions are the same, traditionally considered to be uniformly distributed across coding and non-coding regions. Non-coding DNA regions are subject to different mutation rates throughout the genome because they are subject to less selective pressure than coding regions. The same is true for the so-called microsatellites, repetitive DNA segments that are mostly non-coding, where the mutation rate increases as a function of the number of repetitions. However, beyond physical properties, where the probability of a given nucleotide mutating also depends on their weaker or stronger chemo- and thermodynamic bonds, other departures from non-uniformity are less well understood, and seem to be the result of a process rather than of direct, discrete physical or chemical interactions.

On the other hand, random mutation has no role for a directing force, and in artificial genetic algorithms mutation has traditionally been taken to be uniform, even though other strategies are subject to continuous investigation and have been introduced as a function of, for example, time or data size.

We have found that the introduction of computation into a model of random mutation, as we first did in [171], has interesting ramifications and may help explain some genetic and evolutionary processes. The introduction of computation in the form of algorithmic probability (as opposed to randomness) entails a bias toward simplicity, and transforms our model from a space in which evolution can explore every corner to a space constrained by informational laws, with the result that most random space regions are less likely to be explored. Physicists can relate to this restriction because the laws of physics impose a similar one, limiting, to various degrees, the freedom of regions removed from randomness. For example, on earth, macro-molecules and everything made out of them do not move randomly as in Brownian motion; they tend to fall, pulled by gravity toward earth's centre, which effectively limits the space available for these macro-molecules, and everything made of them, to move. The set of all laws applied to all things over time can be seen as the imposition of regions of information more difficult to access or simply unavailable to general processes.

12.1 Methodology

12.1.1 Chaitin's Evolutionary Model

In the context of these ideas, we can consider a model introduced by Gregory Chaitin, a founder of the theory of algorithmic information. In this model, 'organisms' can 'evolve' relative to their environment considerably faster than by *classical random mutation* [172, 173, 174]. Algorithmic information dynamics enabled this model to be approximated and tested. This represented the first step toward advancing a sound algorithmic framework for biological evolution.

Starting with an empty binary string, Chaitin's example approximates his Ω number, defined as $\Omega = \sum_{p \in HP} 2^{-|p|}$, where HP is the set of all halting programs [30], in an expected time of $O(t^2 (\log t)^{(1+O(1))})$, which is significantly faster than the exponential time that the process would take if random mutations from a uniform distribution were applied. This *speed-up* is obtained by drawing mutations according to the universal distribution [50, 175], a distribution that results from the operation of computer programs that we will explain in detail in the following section.

In a previous result [176], we showed that Chaitin's model exhibits open-ended evolution (OEE) [177], according to a formal definition of OEE as set forth in [176] – in line with the general intuition about OEE – and that no decidable system with computable dynamics can achieve OEE under such a computational definition. In [171], we introduced a system that, by following the universal distribution, optimally approaches OEE.

In Chaitin's evolutionary model [172, 173, 174], a successful mutation is defined as a computable function μ, chosen according to the probabilities given by the universal distribution [50, 175], that changes the current state of the system (as an input of the function) to a better approximation of the constant Ω [178]. In order to be able to simulate this system we would need to compute the universal distribution and the fitness function. However, both the universal distribution and the fitness function of the system require the solution of the halting problem [30], which is uncomputable. Nevertheless, as with Ω itself, this solution can be approximated [179, 103]. In [171] we proposed a model that, to the best of our knowledge, was the first computable approximation of Chaitin's proposal.

For this first approximation we made four important initial concessions: one with respect to the *real* computing time of the system, and three with respect to Chaitin's model:

- We assumed that building the probability distributions for each instance of the evolution would take no computational time, while in the *real computation* this is the single most resource-intensive step.

- The goal of our system was to approximate objects of bounded information content: binary matrices of a set size.
- We used BDM and Shannon's entropy as approximations of the algorithmic information complexity K.
- We did not approximate the algorithmic probability of the mutation functions, but rather that of their outputs.

We justified the first concession in a similar fashion to Chaitin: if we assume that the interactions and mechanics of the natural world are computable, then the probability of a decidable event[1] occurring is given by the universal distribution. The third concession is a necessity, as the algorithmic probability of an object is uncomputable (it requires a solution for *HP* too). In Section 12.3 we will show that Shannon's entropy is not as good as BDM for our purposes. Finally, note that given the universal distribution and a fixed input, the probability of a mutation is in inverse proportion to the descriptive complexity of its output, up to a constant error. In other words, it is highly probable that a mutation may reduce the information content of the input but improbable that it may increase it. Therefore, the last concession yields an adequate approximation, since a low information mutation can reduce the descriptive complexity of the input but not increase it in a meaningful way.

## 12.1.2	Our Expectations

It is important to note that the goal of our system is not identical to that of Chaitin's metabiology model [173]. Having changed the goal of our system, we must also change our expectations as regards its behaviour.

Chaitin's evolution model [173] is faster than *regular* random models, despite targeting a highly random object, thanks to the fact that positive mutations have low algorithmic information complexity and hence a (relatively) high probability of being stochastically chosen under the universal distribution. The universally low algorithmic complexity of these positive mutations relies on the fact that, when assuming an oracle for *HP*, we are also implying a constant algorithmic complexity for its evaluation function and target since we can write a program that verifies that a change on a given approximation of Ω is a positive one without needing a codification of Ω itself.

In contrast, we expected our model to be sensitive with respect to the algorithmic complexity of the target matrix, obtaining high speed-up for structured target matrices that decreases as the algorithmic complexity of the target grows. However, this change of behaviour remains congruent with the main argument of metabiology [173] and our assertion that, contrary to *regular random* mutations, algorithmic probability-driven evolution tends to produce structured novelty at a faster rate, which we hope to prove in the upcoming set of experiments.

---

[1] An event is decidable if it can be *decided* by a Turing machine.

In summary, we expected that when using an approximation of the universal distribution:

- Convergence would be reached in fewer total mutations than when using the uniform distribution for structured target matrices
- The stated difference would decrease in relation to the algorithmic complexity of the target matrix

We also aimed to explore the effect of the number of allowed shifts (mutations) on the expected behaviour.

12.1.3 Evolutionary Model

We set up an experiment in which we could test some ideas related to the fact that, as some researchers believe, life can be seen as evolving in a sort of software space. This is the idea that life does not evolve randomly but as a dynamical system. What we did was to take this idea seriously and work with binary representations of organisms, as is done in areas such as genetic algorithms that help solve problems of all sorts. These organisms are obviously oversimplified versions of real organisms, but in the end they are not completely different from them. We know that all life on earth is actually based on a very basic dictionary of only four letters, so simplifying things to two is not insane, and nothing would have changed if we had used all four letters representing the DNA alphabet.

Broadly speaking, our evolutionary model is a tuple $\langle S, \mathbb{S}, M_0, f, t, \alpha \rangle$, where:

- $\mathbb{S}$ is the *state space* (last sections of Chapter 2)
- M_0, with $M_0 \in \mathbb{S}$, is the *initial state* of the system
- $f: \mathbb{S} \mapsto \mathbb{R}^+$ is a function, called the *fitness or aptitude function*, which goes from the state space to the positive real numbers
- t is a positive integer called the *extinction threshold*
- α is a real number called the convergence parameter
- $S: \mathbb{S} \mapsto \mathbb{S} \times (\mathbb{Z}^+ \cup \{\bot, \top\})$ is a non-deterministic evolution dynamic such that if $S(M, f, t) = (M', t')$ then $f(M') < f(M)$ and $t' \leq t$, where t' is the number of *steps or mutations* it took S to produce M', $S(M, f, t) = (\bot, t')$ if it was unable to find M' with a *better fitness* in the given time, and $S(M, f, t) = (\top, t')$ if it finds M' such that $f(M') \leq \alpha$

Specifically, the function S receives an individual M and returns an *evolved individual* M' in the time specified by t, which is an improvement on the value of the fitness function f and the time it took to do so; $\bot$ if it was unable to do so and $\top$ if it reached the convergence value.

A *successful evolution* is the sequence $M_0, (M_1, t_1) = S(M_0, f, t), ..., (\top, t_n)$ and $\sum t_i$ is the total evolution time. We say that the evolution failed or that we got an *extinction* if we finish the process by $(\bot, t_n)$ instead, with $\sum t_i$ being the extinction time. The evolution is undetermined otherwise. Finally, we will call each element (M_i, t_i) an *instance* of the evolution.

12.2   Results

In [171], we showed that what we call algorithmic evolution converges faster toward structured solutions or goals.

We know that we live in environments that are far from random [180] despite having to deal with degrees of randomness, and that biological systems are highly structured systems, among the most structured.

In our approach to evolutionary strategies, we introduced a bias toward simplicity in the kind of mutations possible. But this is not an arbitrary bias. We know that processes that are not completely random and that are subject to physical laws will produce this same type of bias, so it is actually a more apt type of bias than simply assuming, let's say, uniform distribution, which assigns equal probability to all possible mutations. And in fact, we already know this in biology; we know, for example, that different human populations are subject to different kinds of mutations. Multiple sclerosis, for example, is more common among white people. And mutation rates vary across the genome, so different regions of the DNA have different rates and are more or less prone to certain mutations. In fact, mutations vary over the human lifespan, and all this is related to laws and principles of physics and biology. So rather than being assumed to be random, mutations have traditionally been assumed to be anything but.

What we found was that our strategies were slower if we tried to approach random objects but much faster when they had to evolve to highly structured or organised objects.

This is a perfect formal example of the way in which questions related to, e.g., intelligent design, can be addressed. The common argument is that reaching the exceptional complexity of living systems would require longer than the age of the universe if they came about as a result of pure chance in the form of completely random mutations, but we know this is not the case. First, because the universe is governed by laws and rules. And second, because once a useful structure is found by natural selection, this structure will evolve towards other types of structures in a very modular fashion.

12.2.1   Emergence of Persistent Structures and Modularity

We also find the emergence and evolution of persistent regular structures. A persistent structure is a substructure that is very simple to begin with, and so has a high probability of persisting over time because it is very unlikely that mutations biased towards simplicity affect that structure, allowing it to be carried forward through future populations. This means that in our simple model we can explain the kind of modularity we were talking about. These substructures are like genes, with which nature can produce a large diversity of organisms. Thus, our simple model may explain genetic memory in a very natural way. And modularity has some consequences.

For example, genetic memory can help build other structures faster because persistent structures, simply by persisting across populations, have proven themselves useful in adaptation. And they can also generate greater diversity, like Lego pieces used as building blocks. This may explain the outbreaks of diversity we have witnessed in the past, such as the Cambrian explosion, which gave rise to thousands or millions of species in a relatively short period of time.

However, this same phenomenon can also account for disasters such as massive extinctions, because if a substructure such as a gene becomes a persistent structure, and such a structure is no longer useful in, for example, a different environment, then all species carrying such a structure will move toward extinction. And such an extinction does not even need to be caused by a sudden change in the environment. All it takes is the presence in a population of a certain number of structures that are not able to adapt even to small changes. This is because when a persistent structure is retained because it is simple and has no evolutionary disadvantage, it is very hard to evolve away from it.

This also means that evolution happens at different scales, at a local level but also at a functional level, both in these sorts of structures and in genes, where evolution is not confined to the level of the single nucleotide. Genes can be caused to mutate by multiple factors, even at the transcription stage, by regulation of the ways in which the gene can express, or by external factors such as epigenetic ones. We simulated these scenarios by applying mutations to structures rather than bits.

12.2.2 Applications to Biological Research

How can all this really help in a pragmatic sense? Well, if you consider that not all mutations are equally likely, and that you can identify the operative biases, you can start exploring genetic and biological functioning in this new light and with new methods. As in fact we did. We explored a signalling network, that is, a network of signals sent by cells and cellular functions by way of proteins and chemicals that are exchanged among cells for communication purposes. We calculated which paths in such a network would be more prone to mutation according to our framework. What we found was that our algorithm exactly pinpointed the pathways that have been reported to be associated with diseases like cancer, whereas assuming that all pathways are equally likely to suffer mutations is not informative at all. So we think our approach may have great potential.

12.2.3 Application to Genetic Algorithms and Evolutionary Strategies

We also demonstrated that a classical benchmark problem in the area of artificial genetic algorithms for optimisation problems could benefit from the same speed-ups that we found when applying mutation based on the universal distribution.

For the experiment, our phase state was the set of all binary matrices of sizes $n \times n$, our fitness function was defined as the Hamming distance $f(M) = H(M_t, M)$, where M_t is the *target matrix*, and our convergence parameter was $\alpha = 0$. In other words, the evolution converges when we produce the target matrix, guided only by the Hamming distance to it, which is defined as the number of different bits between the *input matrix* and the target matrix.

This setup was chosen because it allows us to easily define and control the descriptive complexity of the fitness function by controlling the target matrix, and therefore also control the complexity of the evolutionary system itself. It is important to note that our setup can be seen as a generalisation of the *One-Max problem* [181], where the initial state is a binary 'initial gene' and the target matrix is the 'target gene'; when we obtain a Hamming distance of 0 we have obtained the gene equality.

Each *evolution instance* was computed by iterating over the same dynamic. We started by defining the set of possible mutations as those that were within a fixed n number of bits from the input matrix. In other words, for a given input matrix M, the set of possible mutations in a single instance was defined as the set

$$\mathbb{M}(M) = \{M' | H(M',M) \leq n\}.$$

Then, for each matrix in $\mathbb{M}$, we computed the probability $P(M')$ defined as:

- $P(M) = \dfrac{1}{|\mathbb{M}|}$ in the case of the uniform distribution.
- $P(M) = \dfrac{\beta}{2^{BDM(M)}}$ for the BDM distribution, and
- $P(M) = \dfrac{\beta'}{h(M)}$ or $P(M) = \dfrac{\beta''}{2^{h(M)}}$ for Shannon entropy (for an uninformed observer with no access to the possible deterministic or stochastic source),

where β, β', and β'' are normalisation factors such that the sum of the respective probabilities is 1.

For implementation purposes, we chose a minor variation of the entropy probability distribution, to be used and compared to BDM. The probability distributions for the set of possible mutations using entropy were built using two heuristics: Let M' be a possible mutation of M, then the probability of obtaining M' as a mutation is defined as either $\dfrac{\beta'}{h(M')+\epsilon}$ or $\dfrac{\beta''}{2^{h(M')}}$. The first definition assigns a linearly higher probability to mutations with lower entropy. The second definition is consistent with our use of BDM in the rest of the experiments. The constant ϵ is an arbitrary small value that was included to avoid undefined (infinite) probabilities. For the experiments presented, ϵ was set at 1^{-10}.

Once the probability distribution was computed, we set the number of steps at 0, and then, using a (pseudo)random number generator (RNG), we proceeded to stochastically draw a matrix from the stated probability distributions and evaluate its fitness with the function f, adding 1 to the number of steps. If the resultant matrix did not show an improvement in fitness, we drew another matrix and added another 1 to the number of steps, not stopping the process until we obtained a matrix with superior fitness or we reached the extinction threshold. We can either replace the matrix drawn or leave it out of the pool for subsequent iterations. A visualisation of the stated work flow for a 2×2 matrix is available in Fig. 12.1.

To produce a complete evolution sequence, we iterated the stated process until either convergence or extinction was reached. As stated before, we can choose to not replace an evaluated matrix from the set of possible mutations in each instance, but we chose to not keep track of evaluated matrices after an instance was complete. This was done in order to keep open the possibility of dynamic fitness functions in future experiments.

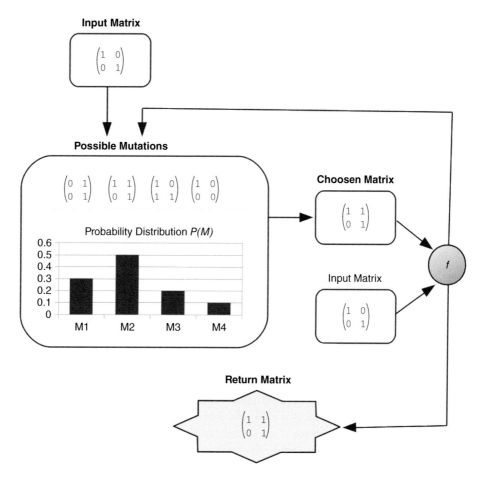

Figure 12.1 An evolution instance. The instances are repeated by updating the input matrix until convergence or extinction is reached.

In this case, the *evolution time* is defined as the sum of the number of *steps* (or draws) it took the initial matrix to reach equality with the target matrix. When computing the evolutionary dynamics by one of the different probability distribution schemes we will denote it by *uniform strategy*, *BDM strategy*, or *h strategy*, respectively. That is, the uniform distribution, the distribution for the algorithmic probability estimation by BDM, and the distribution by Shannon entropy.

12.2.4   The Speed-Up Quotient

We will measure how fast (or slow) a strategy is compared to the uniform strategy by the speed-up quotient, which we will define as follows:

Definition 12.1   The speed-up quotient, or simply *speed-up*, between the uniform strategy and a given strategy f is defined as

$$\delta = \frac{S_u}{S_f},$$

where S_u is the average number of steps it takes a sample (a set of initial state matrices) to reach convergence under the uniform strategy and S_f is the average number of steps it takes under the f strategy.

12.3 Further Results

12.3.1 Cases of Negative Speed-Up

In order to better explain the choices we have made in our experimental setup, first we will present a series of cases where we obtained no speed-up or slow-down. Although these cases were expected, they shed important light on the behaviour of the system.

12.3.2 Entropy versus Uniform on Random Matrices

For the following experiments, we generated 200 random matrices separated into two sets: *initial matrices* and *target matrices*. After pairing them based on their order of generation we evolved them using 10 strategies: the uniform distribution, *block* Shannon's entropy for blocks of size 4×4, denoted below by h_b, entropy for single bits denoted by h, and their variants where we divide by h^2 and h_b^2, respectively. The strategies were repeated for 1- and 2-bit shifts (mutations).

The results obtained are summarised in Table 12.1, which lays out the strategy used for each experiment, the number of shifts/mutations allowed, and the average number of steps it took to reach convergence, as well as the standard error of the sample mean. As we were able to see, the differences in the number of steps required to reach convergence were not statistically significant, validating our assertion that, for random matrices, entropy evolution is not much different from the uniform evolution.

Table 12.1 Results obtained for the 'Random Graphs'

Strategy	Shifts	Average	SE
Uniform	1	214.74	3.55
h_b	1	214.74	3.55
h	1	215.53	3.43
h_b^2	1	214.74	3.55
h^2	1	213.28	3.33
Uniform	2	1,867.10	78.94
h_b	2	1,904.52	79.88
h	2	2,036.13	83.38
h_b^2	2	1,882.46	78.63
h^2	2	1,776.25	81.93

Because, in general, the algorithmic complexity of a network makes sense only with respect to its unlabelled version, in [81, 82, 103] we showed, both theoretically and numerically, that approximations of the algorithmic complexity of adjacency matrices of labelled graphs are a good approximation (up to a logarithmic term or the numerical precision of the algorithm) of the algorithmic complexity of the unlabelled graphs. This means that we can consider any adjacency matrix of a network a good representation of the network, disregarding graph isomorphisms.

12.3.3   Entropy versus Uniform on a Highly Structured Matrix

For this set of experiments, we took the same set of 100 8×8 initial matrices and evolved them into a highly structured matrix, which is the adjacency matrix of the star with eight nodes. For this matrix, we expected entropy to be unable to capture its structure, and the results obtained accorded with our expectations. These results are shown in Table 12.2.

As we can see from the results, entropy was unable to show a statistically significant speed-up compared to the uniform distribution. Over the next sections we show that we have obtained a statistically significant speed-up by using the BDM approximation to algorithmic probability distributions, which is expected because *BDM manages to better capture the algorithmic structures of a matrix rather than just the distribution of the bits, which is what entropy measures*. Based on the previous experiments, we conclude that entropy is not a good approximation of K, and we will omit its use in the rest of the chapter.

12.3.4   Randomly Generated Graphs

For this set of experiments, we generated 200 random 8×8 matrices and 600 16×16 matrices, both sets separated into initial and target matrices. We then proceeded to evolve the initial matrix into the corresponding target using the following strategies: uniform and BDM within 2-bit and 3-bit shifts (mutations) for the 8×8 matrices and

Table 12.2 Results obtained for the 'Star'

Strategy	Shifts	Average	SE
Uniform	1	216.24	3.48
h_b	1	216.71	3.54
h	1	212.74	3.41
h_b^2	1	216.71	3.54
h^2	1	211.74	3.69
Uniform	2	1,811.84	85.41
h_b	2	1,766.69	88.18
h	2	1,859.11	75.73
h_b^2	2	1,764.03	84.52
h^2	2	1,853.04	74.48

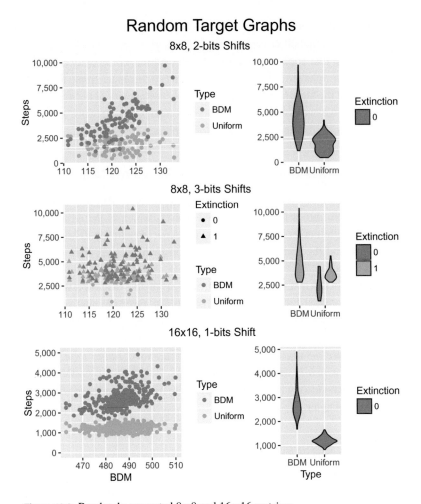

Figure 12.2 Randomly generated 8×8 and 16×16 matrices.

only 2-bit shifts for the 16×16 matrices due to computing time constraints. The results obtained are shown in Fig. 12.2. In all cases, we do not replace drawn matrices and the extinction threshold was set at 2,500.

From the results we can see two important behaviours for the 8×8 matrices. The matrices generated are of high BDM complexity, and evolving the system using the uniform strategy tends to be faster than using BDM for these highly random matrices. Secondly, although increasing the number of possible shifts by 1 seems, at first glance, a small change in our setup, it has a big impact on our results: the number of extinctions has gone from 0 for both methods to 92 for the uniform strategy and 100 for BDM. This means that most evolutions will rise above our threshold of 2,500 drafts for a single successful evolutionary step, leading to an extinction. As for the 16×16 matrices, we can see a formation of two easily separable clusters that coincide perfectly with the uniform and BDM distributions, respectively.

12.3.5   The Causes of Extinction

For the uniform distribution, the reason is simple: the number of 3-bit shifts on 8×8 matrices gives a space of possible mutations of $\binom{8 \times 8}{3} = 41,664$ matrices, which is much larger than the number of possible mutations present within 2-shifts and 1-shift (mutation), which are $\binom{8 \times 8}{2} = 2,016$ and $8 \times 8 = 64$, respectively. Therefore, as we get close to convergence, the probability of getting the right evolution, if the needed number of shifts is two or one, is about 0.04%, and removing repeated matrices does not help in a significant way to avoid extinction, since 41,664 is much larger than 2,500.

Given the values discussed, we have chosen to set the extinction threshold at 2,500 and the number of shifts at 2 for 8×8 matrices, as allowing just 64 possible mutations for each stage is a number too small to show a significant difference in the evolutionary time between the uniform and BDM strategies, while requiring evolutionary steps of $\sim 41,664$ for an evolutionary stage is too computationally costly. The threshold of 2,500 is close to the number of possible mutations and has been shown to consume significant computational resources. For 16×16 matrices, we performed 1-bit shifts, and occasionally 2-bit shifts, when computationally possible.

12.3.6   The BDM Strategy, Extinctions, and Persistent Structures

The interesting case is the BDM strategy. As we can see clearly in Fig. 12.3 for the 8×8 3-bit case, the overall number of steps needed to reach each extinction is often significantly higher than 2,500 under the BDM strategy. This behaviour cannot be explained by the analysis done for the uniform distribution, which predicts the sharp drop observed in the blue curve.

After analysing the set of matrices drawn during failed mutations (all the matrices drawn during a single failed evolutionary stage), we found that most of these matrices have in common highly regular structures. We will call these structures *persistent structures*. Formally, regular structures can be defined as follows:

Definition 12.2   Let M be the description used for an organism or population and Γ a substructure of M in a *computable position* such that $K(M) = K(\Gamma) + K(M - \Gamma) - \epsilon$, where ϵ is a *small* number and $M - \Gamma$ is the codification of M without the contents of Γ. We will call Γ a *persistent or regular structure* of degree γ if the probability of choosing a mutation M' with the subsequence Γ is $1 - 2^{-(\gamma - \epsilon)}$.

Now, note that γ grows in inverse proportion to $K(\Gamma)$ and the difference in algorithmic complexity of the mutation candidates and $K(\Gamma)$: Let M contain Γ in a computable position. Then the probability of choosing M' as an evolution of M is

$$\frac{1}{2^{K(M')}} \geq \frac{1}{2^{K(M' - \Gamma) + K(\Gamma) + O(1)}}.$$

Furthermore, if the possible mutations of M can only mutate a bounded number of bits, and there exists C such that, for every other subsequence of Γ' that can replace Γ we have it that $K(\Gamma') \geq K(\Gamma) + C$, then:

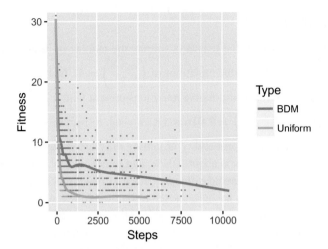

Figure 12.3 Fitness graph for random 8×8 matrices with 3-bit shifts (mutations). Evolution convergence is reached at a fitness of 0. The curves are polynomial approximations computed to serve as visual aids.

$$P(M' \text{ contains } \Gamma) \geq 1 - O(2^{-C}).$$

The previous inequality is a consequence of the fact that the possible mutations are finite and only a small number of them, if any, can have a smaller algorithmic complexity than the mutations that contain Γ; otherwise we controvert the existence of C. In other words, as Γ has relatively *low complexity*, the structures that contain Γ tend to also have low algorithmic complexity, and hence a higher probability of being chosen.

Finally, as shown in Section 12.3.5, *we can expect the number of mutations with persistent structures to increase in factorial order with the number of possible mutations, and in polynomial order with respect to the size of the matrices that compose the state space.*

Proposition 12.3 *As a direct consequence of the last statement, we have that, for systems evolving as described in Section 2 under the universal distribution:*

- *Once a structure with low descriptive complexity is developed, it is* exponentially hard *to get rid of it.*
- *The probability of finding a mutation without the structure decreases* in factorial order *with respect to the set of possible mutations.*
- *Evolving toward random matrices is* hard (improbable).
- *Evolving from and to* unrelated *regular structures is also* hard.

Given the fourth point, we will always choose random initial matrices from now on, as the probability of drawing a mutation other than an empty matrix (of zeroes), when one is present in the set of possible mutations, is extremely low (below 9×10^{-6} for 8×8 matrices with two shifts).

12.3.7   Positive Speed-Up Instances

In the previous section, we established that the BDM strategy yields a negative speed-up when targeting randomly generated matrices, which are expected to be of high algorithmic information content or *unstructured*. However, as stated in Section 12.1.2, that behaviour is within our expectations. In the next section we will show instances of positive speed-up, including cases where, previously, entropy failed to show statistically significant speed-up or was outperformed by BDM.

12.3.8   Synthetic Matrices

For the following set of experiments we manually built three 8×8 matrices that encode the adjacency matrices of three undirected non-random graphs with eight nodes that are intuitively *structured*: the *complete graph*, the *star graph*, and a *grid*. The matrices used are shown in Fig. 12.4.

After evolving the same set of 100 randomly generated matrices for the three graphs, we reported that we found varying degrees of positive speed-up, which correspond to their respective descriptive complexities as approximated by their BDM values. The complete graph, along with the empty graph, is the graph that has the lowest approximated descriptive complexity with a BDM value of just 24.01. As expected, we get the best speed-up quotient in this case. After the complete graph, the star intuitively seems to be one of the less complex graphs we can draw. However, its BDM value (105.434) is higher than the grid (83.503). Accordingly, the speed-up obtained is lower. The results are shown in Fig. 12.5.

As we can see from Fig. 12.5, a positive speed-up quotient was consistently found within 2-bit shifts without replacements. We have one instance of negative speed-up with one shift with replacements for the grid, and negative speed-up for all but the complete graph with two shifts.

However, it is important to say that almost all the instances of negative speed-up are not statistically significant as we have a very high extinction rate of over 90%, and the difference between the averages is lower than two standard errors of the mean. The one exception is the grid with 1-bit shift, which had 45 extinctions for the BDM strategy.

12.3.9   Mutation Memory

The cause of the extinctions found in the grid are what we will call *maladaptive* persistent structures (Definition 12.2), as they occur at a significantly higher rate under

Figure 12.4 Adjacency matrices for the labelled complete, star, and grid graphs.

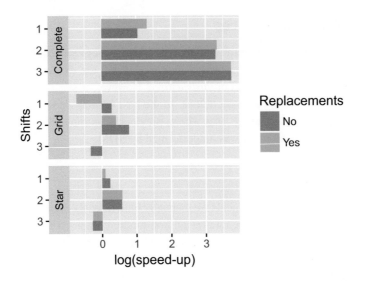

Figure 12.5 The logarithm (natural) of the speed-up obtained for the matrices in Fig. 12.4.

the BDM distribution. Also, as the results suggest, a strategy to avoid this problem is adding *memory* to the evolution. In our case, we will not replace matrices already drawn from the set of possible mutations.

We do not believe this change to be at odds with the stated goals, since another way to see this behaviour is that the universal distribution *dooms* (with very high probability) *populations with certain mutations to extinction*, and *evolution must find strategies* to eliminate these mutations fast from the population. This argument also implies that extinction is faster under the universal distribution than regular random evolution when a persistent maladaptive mutation is present, which can be seen as a form of mutation memory. This requirement has the potential to explain evolutionary phenomena such as the Cambrian explosion, as well as mass extinctions: once a positively structured mutation is developed, further algorithmic mutations will keep it (with a high probability), and the same applies to negatively structured mutations. This can also explain the recurring structures found in the natural world. Degradation of a structure is still possible, but will be relatively slow. In other words, evolution will remember positive and negative mutations (up to a point) when they are structured.

From now on, we will assume that our system has memory and that mutations are not replaced when drawn from the distribution.

12.3.10   The Speed-Up Distribution

Having explored various cases, and found several conditions where negative and positive speed-up are present, the aim of the following experiment was to offer a broader view of the distribution of speed-up instances as functions of their algorithmic complexity.

For the 8×8 case, we generated 28 matrices by starting with the undirected complete graph with eight nodes, represented by its adjacency matrix, and then we removed one edge at a time until the empty graph (the diagonal matrix) was left, obtaining our *target matrix set*. It is important to note that the resultant matrices are always symmetrical. The process was repeated for the 16×16 matrices, obtaining a total of 120 target matrices.

For each target matrix in the first *target matrix* set, we generated 50 random initial matrices and evolved the population until convergence was reached using the two stated strategies: uniform and BDM, both without replacements. We saved the number of steps it took for each of the 2,800 evolutions to reach convergence and computed the average speed-up quotient for each target matrix. The process was repeated for the second target matrix set, but by generating 20 random matrices for each of the 120 target matrices, to conserve computational resources. The experiment was repeated for shifts of 1 and 2 bits and the extinction thresholds used were 2,500 for 8×8 and 10,000 for 16×16 matrices.

As we can see from the results in Fig. 12.6, the average number of steps required to *reach convergence* is lower when using the BDM distribution for matrices with low algorithmic complexity, and this difference drops along with the complexity of the

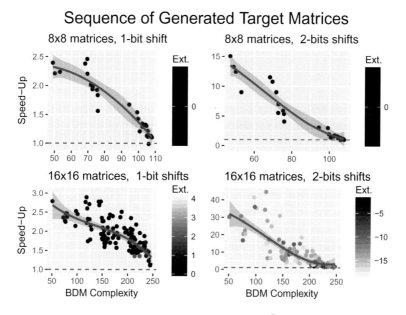

Figure 12.6 The *speed-up quotient* is defined as $\delta = \frac{S_u}{S_{BDM}}$, where S_u is the average number of steps it took to reach convergence under the uniform strategy and S_{BDM} for BDM, and 'Ext.' is the difference $E_u - E_{BDM}$, where each factor is the number of extinctions obtained for the universal and BDM distribution, respectively. In the case of an extinction, the sample was not used to compute the average number of steps. The red (dashed) line designates the *speed-up threshold at* $y = 1$: above this line we have positive speed-up and below it we have negative speed-up. The blue (continuous) line represents a cubic fit by regression over the data points.

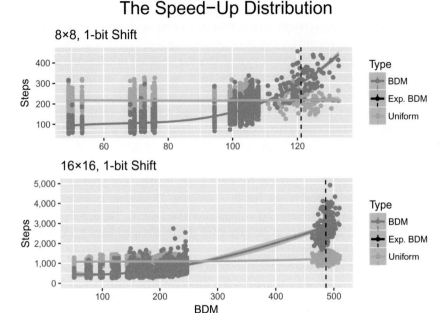

Figure 12.7 The positive speed-up instances are those where the coral curve, computed as a cubic linear regression to all the evolution times for the BDM strategy, is below the teal line, which is a cubic approximation of the evolution times for the uniform strategy. The vertical black dashed line is the *expected* BDM value for a randomly chosen matrix. The large gap in the data reflects the fact that it is hard to find structured (non-random) objects.

matrices but never crosses the *extinction threshold*. This suggests that symmetry over the diagonal is enough to guarantee a degree of structure that can be captured by BDM. It is important to report that we found no extinction case for the 8×8 matrices, 13 in the 16×16 matrices with 1-bit shifts, all for the BDM distribution, and 1,794 with 2-bit shifts, mostly for the uniform distribution.

This last experiment was computationally very expensive. Computing the data required for the 16×16, 2-bit shifts sequence took 12 days, 6 hours, and 22 minutes on a single core of an i5-4570 PC with 8GB of RAM. Repeating this experiment for 3-bit shifts is infeasible with our current setup, as it would take us roughly two months shy of 3 years.

Now, by combining the data obtained for the previous sequence and the random matrices used in Section 12.3.4, we can approximate the positive speed-up distribution. Given the nature of the data, this approximation (Fig. 12.7) is given as two curves, each representing the expected evolution time from a random initial matrix as a function of the algorithmic information complexity of the target matrix for both strategies, uniform and BDM respectively. The positive speed-up instances are those where the BDM curve is below the uniform curve.

The first result we get from Fig. 12.7 is an expected one: unlike the uniform strategy, the BDM strategy is highly sensitive to the algorithmic information content of the target matrix. In other words, *it makes no difference for a uniform probability mutation space whether the solution is structured or not, while an algorithmic probability-driven mutation will naturally converge faster to structured solutions.*

The results obtained allow us to expand upon the theoretical development presented in Section 12.3.6. As the set of possible mutations grows, so do the instances of persistent structure and the slow-down itself. This behaviour is evident from the fact that when we increase the dimension of the matrices we obtain a wider gap within the intersection point of the two curves and the expected BDM value, which corresponds to the expected algorithmic complexity of randomly generated matrices. However, we also increase the number of structured matrices, ultimately producing a richer and more interesting evolution space.

12.4   Chasing Biological and Synthetic Dynamic Networks

12.4.1   A Biological Case

We now set as target the adjacency matrix of a biological network corresponding to the topology of an ERBB signalling network [182]. The network is involved in responses ranging from cell division, death, motility, and adhesion, and when dysregulated it has been found to be strongly related to cancer [183, 184].

As one of our main hypotheses is that algorithmic probability is a better model for explaining biological diversity, it is important to explore whether naturally occurring structures are more likely to be produced under the BDM strategy than the uniform strategy, which is equivalent to showing them evolving faster.

The binary target matrix is shown in Fig. 12.8 and it has a BDM of 349.91 bits. For the first experiment, we generated 50 random matrices that were evolved using 1-bit shift mutations for the uniform and BDM distributions, without repetitions. The BDM of the matrix is at the right of the intersection point inferred from the cubic models shown in Fig. 12.7. Therefore, we predict a slow-down. The results obtained are shown in the Table 12.3.

As the results show, we obtained a slow-down of 0.71, without extinctions. However, as mentioned above, the BDM of the target matrix is relatively high, so this result is consistent with our previous experiments. However, the strategy can be improved.

12.4.2   Evolutionary Networks

An *evolutionary network N* is a tensor of dimension 4 of nodes M_i that are networks themselves, with edges drawn if M_k evolves into $M_{k'}$ and weight corresponding to the number of times that a network M_k has evolved into $M_{k'}$. Figure 12.9 shows a subnetwork of the full network for each evolutionary strategy from 50 (pseudo-)randomly generated networks, with the biological ERBB signalling network as target.

Mutations and overexpression of ERB receptors (ERBB2 and ERBB3 in this network) have been strongly associated with more than 10 types of tissue-specific cancers, and they can be seen at the highest level regulating most of the acyclic network.

Table 12.3 Results obtained for the ERBB Network

Strategy	Shifts	Average	SE	Extinctions
Uniform	1	1222.62	23.22	0
BDM	1	1721.86	56.88	0

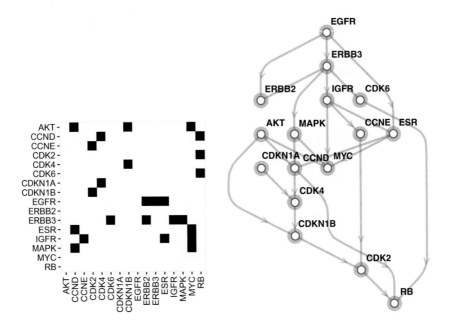

Figure 12.8 (Left) Adjacency matrix of (right) an ERBB signalling network.

We designate as *forward mutations* mutations that led to the target network, and as *backward mutations*, mutations that got away from the target network, through the same evolutionary paths induced by forward mutations. The forward mutations in the neighbourhood of the target (the evolved ERBB network) for each strategy are as follows. For the uniform distribution, each of the following forward mutations (regulating links) had equal probability (1/5, assuming independence, even though unlikely): ESR→ERBB2, ERBB2→CDK4, CDK2 →AKT, CCND→EGFR, CCND→CDKN1A, as shown in Fig. 12.9.

For the BDM strategy, the forward mutations in the top five most likely immediate neighbourhood followed and sorted by their occurring probability are: CCND→ CDK4, 0.176471; ESR→ CCND, 0.137255; CDK6→ RB, 0.137255; CDKN1B→ CDK2, 0.0784314; IGFR→ ESR, 0.0588235.

One of the mutations by BDM involves the breaking of the only network cycle of size 6: EGFR→ERBB3, ERBB3→IGFR, IGFR→ESR, ESR→MYC, MYC→EGFR by deletion of the interaction MYC→EGFR, with probability 0.05 among the possible mutations in the BDM immediate neighbourhood of the target. In the cycle is ERBB3, which has been found to be related to many types of cancer when overexpressed [183].

BDM-based mutation

(approximation to universal distribution)

Total number of edges:6309

> 1 - graph:

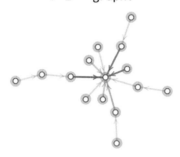

local BDM-based mutation

(modularity induced)

Total number of edges:6327

> 1 - graph:

Uniform random mutation

Total number of edges:6339

> 1 - graph:

Figure 12.9 Evolutionary convergence in evolutionary subnetworks closest to the target matrix, with edges shown only if they were used more than once. The darker the edge colour, the more times that an evolutionary path was used to reach the target. (Top) The highest number is 7 for the BDM-based mutation (top) and the lowest 2 (e.g., all those in the uniform random distribution). The BDM-based one is the only disconnected graph, meaning that it produced a case of convergent evolution even before reaching the target.

For the local BDM strategy, the following were the top five forward mutations: EGFR→ERBB2, 0.32; EGFR→ ERBB3, 0.107; IGFR→ CCNE, 0.0714; ERBB3→ ERBB2, 0.0714; EGFR→ ESR, 0.0714; with ERBB2 and ERBB3 heavily involved in three of the top five possible mutations with added probability 0.49, and thus more likely than any other pair and interaction of proteins in the network.

Under a hypothesis that mutations can be reversals of states in past evolutionary pathways, then mutations of such interactions may be the most likely backward mutations to occur.

12.4.3   The Case for Localised Mutations and Modularity

As previously mentioned in Proposition 12.3, the main causes of slow-down under the BDM distribution are maladaptive persistent structures. These structures will negatively impact the evolution speed in factorial order relative to the size of the state space. One direct way to reduce the size set of possible mutations is to reduce the size of the matrices we are evolving. However, doing so will also reduce the number of interesting objects we can evolve towards. Another way to accomplish the objective while using the same heuristic is to rely on *localised* (or *modular*) mutations. That is, we force the mutation to take place on a submatrix of the input matrix.

The way we implement the stated change is by adding a single step in our evolution dynamics: at each iteration, we will randomly draw, with uniform probability, one submatrix of size 4×4 out of the set of adjacent submatrices that compose the input matrix, with no overlap, and force the mutation to be there by computing the probability distribution over all the matrices that contain the bit-shift only at the chosen place. We will call this method the *local BDM* method.

It is important to note that, within 1-bit shifts (point mutations), the space of total possible mutations remains the same when we compare the uniform and BDM strategies. Furthermore, the behaviour of the uniform strategy would remain unchanged if the extra step is applied using the uniform distribution.

We repeated the experiment shown in Table 12.3 with the addition of the local BDM strategy and the same 50 random initial matrices. The results are shown in Table 12.4. As we can see from the results obtained, local BDM obtains a statistically significant speed-up of 1.25 when compared to the uniform strategy.

One potential explanation of why we failed to obtain speed-up for the network with the BDM strategy is that, as an approximation of K, the model depends on finding global algorithmic structures, while the sample is based on a substructure that might not have enough information about the underlying structures that we hypothesise govern the natural world and allow scientific models and predictions.

However, biology evolves modular systems [185], such as genes and cells, that in turn produce building blocks such as proteins and tissues. Therefore, local algorithmic

Table 12.4 Results obtained for the ERBB Network

Strategy	Shifts	Average	SE	Extinctions
Uniform	1	1,222.62	23.22	0
BDM	1	1,721.86	56.88	0
Local BDM	1	979	25.94	0

mutation is a better model. This is a good place to recall that local BDM was devised as a natural solution to the problem presented by maladaptive persistent structures in global algorithmic mutation. Which also means that this type of modularity can be evolved by itself, given that it provides an evolutionary advantage, as our results demonstrate. This is compatible with the biological phenomenon of non-point mutations, in contrast to point mutations, which affect only a single nucleotide. For example, in microsatellites, mutations may lead to the gain or loss of the entire repeated unit, and sometimes several repeats simultaneously.

We will further explore the relationship between BDM and local BDM within the context of global structures in the next section. Our current setup is not optimal for further experimentation in *biological* and local structured matrices, as the computational resources required to build the probability distribution for each instance grows in quadratic order relative to matrix size, though these computational resources are not needed in the real world (see Section 12.5).

12.4.4 Chasing Synthetic Evolving Networks

The aim of the next set of experiments was to follow, or *chase*, the evolution of a moving target using our evolutionary strategies. In this case, we chased four different *dynamical networks*: the ZK graphs [66], K-ary trees, an evolving N-star graph, and a star-to-path graph dynamic transition artificially created for this project. These dynamical networks are families of directed labelled graphs that evolve over time using a deterministic algorithm, some of which display interesting graph-theoretic and entropy-fooling properties [66]. As the evolution dynamics of these graphs are fully deterministic, we expected BDM to be (statistically) significantly faster than the other two evolutionary strategies, uniform probability and local BDM.

We chased these dynamics in the following way: We let $S_0, S_1, \ldots, S_n, \ldots$ be the stages of the system we are chasing. Then the initial state S_0 was represented by a random matrix and, for each evolution $S_i \mapsto S_{i+1}$, the input was defined as the adjacency matrix corresponding to S_i, while the target was set as the adjacency matrix for S_{i+1}. In order to normalise the matrix size, we defined the networks as always containing the same number of nodes (16 for 16×16 matrices). We followed each dynamic until the corresponding stage could not be defined in 16 nodes.

The results that were obtained, starting from 100 random graphs and 100 different evolution paths at each stage, are shown in Fig. 12.10. It is important to note that, since the graphs were directed, the matrices used were non-symmetrical.

From the results we can see that local BDM consistently outperformed the uniform probability evolution, but the BDM strategy was faster by a significant margin. The results are as expected and confirm our hypothesis: *uniform evolution cannot detect any underlying algorithmic cause of evolution, while BDM can, inducing a faster overall evolution.* Local BDM can only detect local regularities, which is good enough to outrun uniform evolution in these cases. However, as the algorithmic regularities are global, local BDM is slower than (global) BDM.

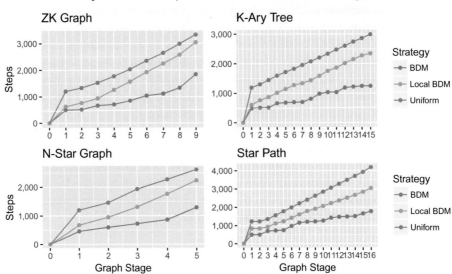

Figure 12.10 Each point of the graph is composed of the average accumulated time from 100 samples. All evolutions were carried out with 1-bit shifts (mutations).

12.5   Discussion and Conclusions

The results of our numeric experiments are statically significant and, as shown in Figs. 12.6 and 12.7, the speed-up quotient increases in relationship to the ratio between the algorithmic complexity of the target matrix and the *expected* random matrix, confirming our theoretical expectations. The speed-up obtained can be considered low when the stated quotient is sufficiently close to 1, but on a *rich* evolution space we expect this difference to be significant: as a rough estimate, the human genome can potentially store 700 megabytes of data, while the biggest matrix used in our experiments represents a space limited to objects of $16 \times 16 = 256$ bits. Therefore, we expect the effects of speed-up (and slow-down) to be significantly higher in natural evolution than in these experiments.

On the one hand, classical mechanics establishes that random events are only apparent and not fundamental. This means that mutations are not truly random but the result of interacting deterministic systems that may distribute differently than in a random fashion. A distribution representing causal determinism is that suggested by algorithmic probability and the universal distribution, because of its theoretical stability under changes of formalism and description language. Its relevance, even for non-Turing universal models of computation, has also been proven [59], being able to explain a bias toward simplicity of up to more than 50%.

On the other hand, the mathematical mechanisms of biological information, from Mendelian inheritance to Darwin's evolution and the discovery of the digital nature of the genetic code together with the mechanistic nature of the mechanisms of translation, transcription, and other inter cellular processes, suggests a strong algorithmic basis to fundamental biological processes. By taking it to the next step, these ideas indicate that evolution by natural selection may not be (very) different from, and can thus be regarded and studied as, evolving programs in software space, as suggested by Chaitin [172, 174].

Our findings demonstrate that computation can thus be a powerful driver of evolution, and better able to explain key aspects of life. Effectively, algorithmic probability reduces the space of possible mutations. By abandoning the uniform distribution assumption, matters ranging from the apparition of sudden major stages of evolution, the emergence of 'subroutines' in the form of modular persistent structures, and the need for an evolving memory carrying information organised in such modules that drive evolution by selection, may be explained.

The algorithmic distribution emerges naturally from the interaction of deterministic systems [59, 50]. In other words, we are simulating the conditions of an algorithmic/procedural world and there is no reason to believe that it requires greater real-world (thus highly parallel) computation than is required by the assumption of the uniform distribution given the highly parallel computing nature of physical laws. The algorithmic distribution can thus be considered natural, and in a sense even more natural than the uniform distribution.

The interplay of the evolvability of organisms from the persistence of such structures also explains two opposed phenomena: recurrent explosions of diversity and mass extinctions, events which have occurred during the history of life on earth that have not been satisfactorily explained under the uniform mutation assumption. The results suggest that extinction may be an intrinsic mechanism of biological evolution.

In summary, taking the informational and computational aspects of life based on modern synthesis to their natural conclusions, the present approach based on weak assumptions of deterministic dynamic systems offers a novel framework of algorithmic evolution within which to study both biological and artificial evolution.

Postface

We have shown that computability and algorithmic complexity are key to addressing the challenge of causality in science, and that though these are more difficult to evaluate numerically, methods are now available for estimating universal measures of complexity. Only universal measures of complexity can characterise any (computable) feature while remaining invariant relative to object description. These methods confer an advantage when it comes to finding indications of causal content in a system, which is key to the development of algorithmic calculi for studying causal mechanisms in artificial and biological systems, with a view to reprogramming them.

Appendix: Mutual and Conditional BDM

A.1   Conditional BDM

As introduced and specified in [186], let X and Y be tensors of the same dimension and $\{\alpha_i\}$ a computable *partition strategy* for aggregating objects for which the CTM values are known. The BDM value for X with respect to $\{\alpha_i\}$ is defined as

$$\mathrm{BDM}(X) = \sum_{(r_i, n_i) \in \mathrm{Adj}(X)} \mathrm{CTM}(r_i, n_i) + \log(n_i),$$

where Adj is the set of pairs (r_i, n_i) resulting from applying the partition $\{\alpha_i\}$ to X, r_i is a subtensor for which the CTM value is known, and n_i are the respective multiplicities within the tensor.

Definition A.1   *Given the definition stated above, we can define the* conditional BDM *of X with respect to the tensor Y with respect to $\{\alpha_i\}$ as*

$$\mathrm{BDM}(X|Y) = \sum_{(r_i, n_i) \in \mathrm{Adj}(X) - \mathrm{Adj}(Y)} (\mathrm{CTM}(r_i, n_i) + \log(n_i)) + \sum_{\mathrm{Adj}(X) \cup \mathrm{Adj}(Y)} f\left(n_j^x, n_j^y\right),$$

where n_j^x and n_j^y are the multiplicity of the subtensor within X and Y, respectively, and f is the function defined as

$$f\left(n_j^x, n_j^y\right) = \begin{cases} 0 & \text{if } n_j^x = n_j^y, \\ \log(n_j^x) & \text{otherwise.} \end{cases}$$

    Intuitively, Definition A.1 approximates the algorithmic information within the tensor X that is not present in Y: If we assume knowledge of Y, then to describe the elements of the decomposition of X using the partition strategy $\{\alpha_i\}$ we need the description of the subtensors that are not Y. In the case of common subtensors, if the multiplicity is the same, then we can assume that X does not contain additional information, but that it does if the multiplicity differs.

A.2   On the Second Term

As stated in the previous section, the term

$$\sum_{\text{Adj}(X)\cup\text{Adj}(Y)} f\left(n_j^x, n_j^y\right)$$

quantifies the additional information contained within X when the multiplicity of the subtensors varies between X and Y. This term is important in cases where such multiplicity dominates the complexity of the objects. An alternative definition for the term is

$$\sum_{\text{Adj}(X)\cup\text{Adj}(Y)} \left|n_j^x - n_j^y\right|.$$

However, we believe this would introduce ambiguity, given that we do not have information about whose multiplicity is higher. Moreover, the stated definition will be shown to be *well behaved* in Section A.4.

A.3   Joint and Mutual BDM

Given that we are not working within a Shannon information framework, we found the concept of joint BDM harder to extend from classical information theory. In classical information theory, we can think of joint entropy as the information contained in two or more events occurring concurrently, and the joint entropy of the channels as the average over all combinations of events. Within AIT, we interpret this concept as the '*amount of information contained within two or more objects*'.

In contrast to classical information theory, we started by defining conditional BDM. Therefore, we think that the best way to define joint BDM is by way of the *chain rule*.

Definition A.2   *The* joint BDM *of X and Y with respect to $\{\alpha_i\}$ is defined as*

$$\text{JointBDM}(X, Y) = \text{BDM}(Y|X) + \text{BDM}(X).$$

Following the same path, we could define *mutual BDM*:

Definition A.3   *The* mutual BDM *of X and Y with respect to $\{\alpha_i\}$ is defined as*

$$\text{MutualBDM}(X, Y) = \text{BDM}(X) - \text{BDM}(X|Y).$$

A.4   The Relationship between Conditional, Joint, and Mutual Information

The results shown in this section are evidence that our definition of conditional BDM is *well behaved*, as they are analogous to important properties for conditional, joint, and mutual entropy.

Proposition A.4   *If $X = Y$ then* $\text{BDM}(X|Y) = 0$.

Proof   This is a direct consequence of Definition A.1. □

It is important to note that $\mathrm{BDM}(X|Y) = 0$ does not imply that $X = Y$. However, it does imply that $\mathrm{Adj}(X) = \mathrm{Adj}(Y)$. This is a consequence of the fact that BDM does not measure the information encoded in the position of the subtensors.

Proposition A.5 $\mathrm{BDM}(X) \geq \mathrm{BDM}(X|Y)$.

Proof   As we consider subsets of $\mathrm{Adj}(X)$, this is a direct consequence of Definition A.1. □

Proposition A.6   *If X and Y are independent with respect to the partition $\{\alpha_i\}$, this is equivalent to* $\mathrm{Adj}(X) \cap \mathrm{Adj}(Y) = \emptyset$*, then* $\mathrm{BDM}(X|Y) = \mathrm{BDM}(X)$.

Proof   This is a direct consequence of Definition A.1, given that

$$\mathrm{Adj}(X) - \mathrm{Adj}(Y) = \mathrm{Adj}(X).$$

□

Proposition A.7 $\mathrm{MutualBDM}(X,Y) = \mathrm{MutualBDM}(Y,X)$.

Proof   First, consider the equation

$$
\begin{aligned}
\mathrm{MutualBDM}(X,Y) = {} & \mathrm{BDM}(X) - \mathrm{BDM}(X|Y) \\
= {} & \sum_{(r_i,n_i)\in\mathrm{Adj}(X)} \mathrm{CTM}(r_i,n_i) + \log(n_i) \\
& - \sum_{(r_i,n_i)\in\mathrm{Adj}(X)-\mathrm{Adj}(Y)} (\mathrm{CTM}(r_i,n_i) + \log(n_i)) \\
& - \sum_{\mathrm{Adj}(X)\cup\mathrm{Adj}(Y)} f\left(n_k^x, n_k^y\right).
\end{aligned}
$$

While, on the other hand, we have that

$$
\begin{aligned}
\mathrm{MutualBDM}(Y,X) = {} & \mathrm{BDM}(Y) - \mathrm{BDM}(Y|X) \\
= {} & \sum_{(r_j,n_j)\in\mathrm{Adj}(Y)} \mathrm{CTM}(r_j,n_j) + \log(n_j) \\
& - \sum_{(r_j,n_j)\in\mathrm{Adj}(Y)-\mathrm{Adj}(X)} \left(\mathrm{CTM}(r_j,n_j) + \log(n_j)\right) \\
& - \sum_{\mathrm{Adj}(Y)\cup\mathrm{Adj}(X)} f\left(n_k^y, n_k^x\right).
\end{aligned}
$$

Notice that in both equations we have the sum over all the pairs that are in both sets, $\mathrm{Adj}(X)$ and $\mathrm{Adj}(Y)$, with the difference being in the terms corresponding to the *multiplicity*. Now we have to consider two cases. If $n_i^x = n_i^y$, we have equality. Otherwise, in the first equation we have terms of the form $\log(n_j^x) - f(n_k^x, n_k^y)$, which, by definition of f, is 0; likewise for the second equation. Therefore, we have equality. □

Proposition A.8 $\mathrm{MutualBDM}(X,Y) = \mathrm{BDM}(X) + \mathrm{BDM}(Y) - \mathrm{JointBDM}(X,Y)$.

Proof

$$\text{MutualBDM}(X, Y) = \text{MutualBDM}(Y, X)$$
$$= \text{BDM}(Y) - \text{BDM}(Y|X)$$
$$= \text{BDM}(Y) + \text{BDM}(X) - (\text{BDM}(Y|X) + \text{BDM}(X))$$
$$= \text{BDM}(X) + \text{BDM}(Y) - \text{JointBDM}(X, Y). \qquad \square$$

A.5   Properties and Relationship with Entropy

As mentioned before, the goal behind the definition of conditional BDM, $\text{BDM}(X|Y)$, is to measure the amount of information contained in X not present in Y. Ideally, this is measured by the conditional algorithmic information $K(X|Y)$. However, this value is not computable, and approximating it by means of CTM would require extensive computational resources, in an exponential relationship with the ones required for the values used in the current iterations of BDM. The definition of conditional BDM presented in this text relies heavily on the entropy-like behaviour of BDM, and perhaps it has a closer relationship to conditional entropy than to BDM itself in most use cases. However, unlike conditional entropy, conditional BDM extends *naturally* to tensors of larger size.

A.6   The Impact of the Partition Strategy

As shown in previous results, BDM better approximates the universal measure $K(X)$ as the number of elements resulting from applying the partition strategy $\{\alpha_i\}$ to X. This is not the case for conditional BDM. Instead $\text{BDM}(X|Y)$ is a good approximation of $K(X|Y)$ when the $\text{Adj}(X)$ and $\text{Adj}(Y)$ share a high number of *base tensors* r_i, and the probability of this occurring is lower in inverse proportion to the number of elements of the partition. It is for this reason that we have emphasised throughout this text that conditional BDM is dependent on the chosen partition strategy $\{\alpha_i\}$.

For a simple example, think of the binary string $X = 11110000$ and consider its inverse $Y = 00001111$. Since we have the CTM approximation for strings of size 8, the best BDM value for each string is found when $\text{Adj}(X) = \{(11110000, 1)\}$ and $\text{Adj}(Y) = \{(00001111, 1)\}$. However, given that the elements of the partitions are different, we have

$$\text{BDM}(11110000|00001111) = \text{BDM}(11110000) = 25.1899,$$

even though we intuitively know that, algorithmic information wise, they should be very close. However, conditional BDM is able to capture this with partitions of size 1 to 4 with no overlapping, assigning a value of 0 to $\text{BDM}(X|Y)$.

In general, it seems that there is no strategy for finding a *best partition strategy*. This is an issue shared with conditional block entropy, and like the original BDM

definition, at its worst, conditional BDM will behave like conditional entropy in comparable situations, while in the best cases approaching the ideal of conditional algorithmic
complexity.

A.7   Conditional BDM as an Extension of Conditional Entropy

Conditional entropy $H(X|Y)$ is defined as

$$H(X|Y) = \sum_{y \in Y} p(y) \, H(X|Y = y)$$

$$= -\sum_{y \in Y} p(y) \sum_{x \in X} p(x|y) \log p(x|y),$$

which can be interpreted as '*the amount of information introduced by X when Y is
assumed to be known*'[1] and is defined for two variables over the same probability space.
In other words, where the probability $p(x, y)$ is defined. In this section, we will show
how restrictions create problems when we try to use this mathematical construct over
the probability distribution suggested by finite binary strings.

Two important properties for conditional entropy are that $H(X|Y) \leq H(X)$ and
$H(X|Y) = H(Y, X) - H(Y)$, where $H(X, Y)$ can be considered a 'vector-valued variable' [45]. Formally,

$$H(X, Y) = -\sum_{x \in X} \sum_{y \in Y} p(x, y) \log p(x, y).$$

Now, let's consider two binary strings $X = 1111$ and $Y = 1010$. The probability
suggested by X is $P(X = 1) = 1$ while that suggested by Y is $P(Y = 1) = 0.5$.
Therefore, in terms of the second relation we have that

$$H(X, Y) = -\sum_{x \in \{1\}} \sum_{y \in \{0,1\}} p(x, y) \log p(x, y)$$

$$= -p(1, 0) \log p(1, 0) - p(1, 1) \log p(1, 1).$$

Now, given that $H(1010) = 1$ and $H(1111) = 0$, if we assume that the distributions
are independent, and therefore $p(x, y) = p(x)p(y)$, we have that $H(X|Y) = H(X, Y) -
H(Y) = 1 - 1 = H(X)$. But if the distributions are always assumed to be independent,
then we will always have the case of $H(X|Y) = H(X)$, given that Y will have nothing
to say about X and $\sum_{x \in X} p(x|y) \log p(x|y)$, and therefore $H(X|Y)$ will always be equal
to $H(X)$. However, if we assume that the variables are not independent, then we face
an impossible task, given that $H(Y, X) - H(Y)$ must be equal to or less than zero, the
value of 1 can only be reached when both probabilities are $\frac{1}{2}$, and entropy cannot be
a negative value. This issue is heightened when using block entropy, since with very
high probability the resulting *alphabets* will be unequal.

---

[1] http://www.scholarpedia.org/article/Entropy

The problem encountered here arises from the fact that Shannon's entropy was originally defined for communication channels for which a finite string could only give us a glimpse of the underlying distribution. For instance, if we assume that there is an underlying common distribution for the strings 1010 and 1111, then conditional entropy will behave as expected.

In contrast, conditional BDM implicitly assumes the universal distribution as the underlying probability distribution for the elements of finite strings, and uses algorithmic information techniques to compute an approximation to the information content of a string as a whole with respect to other strings. For this reason, as shown in Section A.4, conditional BDM is always well behaved across different finite strings **when assuming the same partition strategy**. Furthermore, as we will show empirically in the next section, conditional BDM is able to capture the conditional information content of strings that come from the same distribution as well as conditional entropy.

A.8   Numerical Results

In this section we explore the behaviour of conditional BDM and contrast it with conditional entropy over a series of numerical experiments.

A.8.1   Conditional BDM Compared to Conditional Entropy over Biased Distributions

For the first experiment we generated a sample of 19,000 pseudo-random binary strings of length 20 that are pairwise related by virtue of being drawn from one of 19 *biased* distributions where the expected number of 1s varies from 1 to 19. For each pair, we computed the conditional BDM with partitions of size 1, and divided it by the conditional BDM of the first string with respect to a random string from a uniform distribution. To both the divisor and the dividend, we added 1 to avoid divisions by zero. We repeated the experiment for conditional entropy. Both results where normalised by dividing the quotients obtained by the maximum value obtained for each distribution. In Fig. A.1 we show the average obtained for each biased distribution.

From Fig. A.1, we can see that as the underlying distribution associated with strings is increasingly biased, the expected shared information content of two related strings is higher (conditional BDM is lower) when compared to the conditional BDM of two unrelated strings. This behaviour is in line with what we expect and observe for conditional entropy. That the area under the normalised cube is smaller is expected, given that BDM is a finer-graded information content measure than entropy and is not perfectly symmetric, as BDM and CTM are computational approximations of an uncomputable function and are also inherently more sensitive to the fundamental limits of computable random number generators.

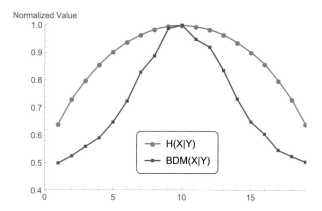

Figure A.1 Each point represents the normalised average of the conditional BDM (BDM($X|Y$)) and conditional entropy (H($X|Y$)), respectively, corresponding to 5,000 pairs of strings randomly chosen from a distribution where the expected number of 1s is the value shown on the *x axis* divided between the conditional BDM or conditional entropy of the first element of the pair and an unrelated randomly chosen binary string. All strings are of length 20. The partition strategy used for BDM is that of sets of size 1. From this plot, we can see that conditional BDM manages to capture the statistical relationship of finite strings generated from the same distribution.

A.8.2   Conditional BDM over Different Partition Sizes

The experiment shown in the previous section used BDM with partitions of size 1, since it is under such a partition strategy that we expect the behaviour to be closer to entropy. In this section we will briefly explore the effect that different partition sizes have on the conditional BDM of random strings.

For this experiment we generated 2,400,000 random binary strings of size 20, with groups of 600,000 strings belonging to one of four different distributions: *uniform* (ten 1s expected), *biased 3/20* (three 1s expected), *biased 1/4* (five 1s expected), and *biased 7/20* (seven 1s expected). Then, we formed pairs of strings belonging to the same distribution and computed the conditional BDM using different partition sizes, from 1 to 20, for a total of 30,000 pairs per data point, normalising the result by dividing it by the partition size to avoid this factor being the dominant one. In Fig. A.2 we show the average obtained for each data point.

From Fig. A.2 we can observe two main behaviours. The first is that as the partition size increases so does the conditional BDM value. This is because bigger partitions take into account more information about the position of each bit, and we do not expect randomly generated strings to share positional information. The drop observed after partitions of size 12 is the result of CTM values being available up to strings of size 12, the point where the program begins to rely on BDM for the computation. Additionally, the partition strategy ignores smaller partitions than the ones stated, thus reducing the overall amount of information taken into account.

The second behaviour observed is that not only is conditional BDM able to capture the discrepancies expected from the different distributions for partition sizes where

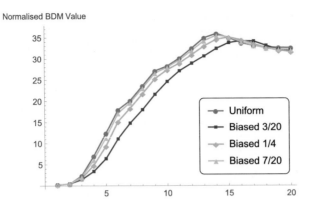

Figure A.2 Each point represents the average of the conditional BDM from 30,000 pairs of binary strings of size 20 randomly generated from four different distributions: *uniform* (ten 1s expected), *biased 3/20* (three 1s expected), *biased 1/4* (five 1s expected), and *biased 7/20* (seven 1s expected). The *y-axis* indicates the partition size used to compute the respective conditional BDM values, normalised by dividing by the partition size.

there is no loss of statistical information (this being from size 1 to 10), but it seems to improve on its ability to do so with larger partition sizes up to 10, thus improving upon the results presented in Section A.8.1.

It is important to note that a significant reduction of accuracy for partitions of size larger than 10 was expected, given that the partition strategy used discards substrings of smaller sizes than the ones stated. For instance, the partition of size 3 of the string 10111 is just {101}, thus losing information. Furthermore, for big partition sizes with respect to the string length, the statistical similarity vanishes, given that now each substring is considered a *different symbol of an alphabet*. Therefore, the abrupt change of behaviour observed beyond partitions of size 15 is expected, and is a product of causality.

A.8.3   Distance to Conditional Kolmogorov Complexity

The ideal and universal measure of conditional information content is the conditional algorithmic information complexity $K(X|Y)$, which is defined as

$$K(X|Y) = \min\{|p| : p(Y) = X\},$$

where p is a program of length $|p|$ for a reference universal Turing Machine U that receives Y as an input. This conditional complexity function measures the total amount of information of X that can be recovered from Y using any computable method.

Although theoretically sound, conditional information content is an uncomputable function. Therefore, it represents a theoretical ideal that cannot be attained in practice.

Conditional BDM is designed to be an approximation of this measure. However, it differs in not taking into account two information sources: the information content shared between base blocks and the position of each block.

As an example of the first limitation, consider the string 1010 and its *negation* 0101. We know that they are algorithmically close, but for a partition strategy of size 2 with no overlapping, the Adj sets {('10', 2)} and {('01', 2)} are disjoint. Therefore, conditional BDM assigns the maximum BDM value to the shared information content. Within this limitation, conditional BDM represents a better approximation of K in comparison to entropy, mainly because BDM uses the CTM approximation value for each block rather than just its distribution, and the information content of its multiplicity, thus representing a more accurate approximation of the overall information content of the non-shared base blocks.

As for the second limitation, it can only become significant when the size of the base blocks is *small* when compared to that of the objects we are analysing, given that the positional information can become the dominant factor of the information content within an object. This is an issue shared with entropy. In these kinds of cases BDM, and therefore conditional BDM, approaches entropy. However, conditional BDM has the added benefit that it is defined for finite tensors generated from different distributions.

Glossary

Algorithmic coding theorem (not to be confused with Shannon's coding theorem): A theorem that formally establishes an inversely proportional relationship between Kolmogorov–Chaitin complexity and algorithmic probability.

Algorithmic information theory: The literature based on the concept of Kolmogorov–Chaitin complexity and related concepts such as algorithmic probability, compression, optimal inference, the universal distribution, and Levin's semi-measure.

Algorithmic probability: The probability of producing an object from a random digital computer program whose binary digits are chosen by chance. The calculation of algorithmic probability is a lower semi-computable problem.

Algorithmic randomness: How removed the length of the shortest generating program is from the size of the uncompressed data that such a program generates.

Block decomposition method (BDM): A divide and conquer method that chunks a string or multidimensional object into smaller pieces for which better lower bounds of algorithmic probability can be produced.

Causal gaps: The gap between objects that Shannon entropy can identify as non-random, and that estimations of algorithmic (Kolmogorov–Chaitin) complexity can identify as non-random, the latter being a much larger set.

Coding theorem method (CTM): A method for estimating Kolmogorov–Chaitin complexity by way of algorithmic probability based on the relationship established by the algorithmic coding theorem.

Computability: The theory that studies the problems that can be solved by algorithmic/mechanistic means using, for example, digital computers.

Halting problem: The strongest problem of predictability in algorithms, related to whether a computer program will or will not halt. Turing proved that it is an undecidable problem and is equivalent to the undecidability results by Gödel.

Invariance theorem: A foundational theorem that establishes that the shortest programs in different computer languages differ in length by at most a constant value, and thus guarantees that the Kolmogorov–Chaitin complexity of an object converges up to a constant, making this measure robust in the face of changes of model (such as a reference universal Turing machine).

Kolmogorov–Chaitin complexity: Also known as the algorithmic complexity of a string, it is the length of the shortest computer program that generates said string. The calculation of the Kolmogorov–Chaitin complexity is an upper semi-computable problem, meaning upper bounds found are short representations, if not necessarily the shortest, computer programs.

Semi-computability: A semi-computable problem is one that allows approximations from above or below. If from above, then it is considered upper semi-computable, and if from below, it is considered lower semi-computable.

Semi-measure: A measure of probability whose sum never equals 1 because some events are undetermined, for example, by the halting problem in the context of computability theory.

Shannon entropy: A measure of combinatorial and statistical complexity based on the diversity of symbols used to define an object according to a probability distribution.

The universal distribution: The distribution produced by the output of random computer programs whose instruction bits are the result of, say, tossing a coin. It is considered universal not only because it is the limit distribution of the output of all halting programs running on a universal Turing machine, but also because it is independent of the choice of particular universal Turing machine or computer language, up to a constant.

Turing machine: An abstraction of a general-purpose computer introduced by Alan M. Turing.

Undecidable problem: A problem that cannot be decided by traditional algorithms such as, for example, all those implemented in a universal Turing machine.

Universal Turing machine: A Turing machine that can emulate any other Turing machine by reprogramming the machine using proper inputs.

References

[1] Hector Zenil, Carlos Gershenson, James A. R. Marshall, and David A. Rosenblueth. Life as thermodynamic evidence of algorithmic structure in natural environments. *Entropy*, 14 (11):2173–2191, 2012.

[2] Christopher Hitchcock. Probabilistic causation. In Edward N. Zalta, ed., *The Stanford Encyclopedia of Philosophy*. Spring ed. Metaphysics Research Lab, Stanford University, 2021.

[3] Justin Matejka and George Fitzmaurice. Same stats, different graphs: Generating Datasets with varied appearance and identical statistics through simulated annealing. In *Proceedings of the 2017 CHI Conference on Human Factors in Computing Systems* (CHI '17). Association for Computing Machinery, New York, 2017, pp. 1290–1294.

[4] Duncan J. Watts and Steven H. Strogatz. Collective dynamics of 'small-world' networks. *Nature*, 393(6684):440–442, 1998.

[5] Dapeng Hao, Cong Ren, and Chuanxing Li. Revisiting the variation of clustering coefficient of biological networks suggests new modular structure. *BMC Systems Biology*, 6(1):34, 2012.

[6] Patrick Aloy and Robert B. Russell. Taking the mystery out of biological networks. *EMBO reports*, 5(4):349–350, 2004.

[7] Haiyuan Yu, Dov Greenbaum, Hao Xin Lu, Xiaowei Zhu, and Mark Gerstein. Genomic analysis of essentiality within protein networks. *Trends in Genetics*, 20(6):227–231, 2004.

[8] Attila Gursoy, Ozlem Keskin, and Ruth Nussinov. Topological properties of protein interaction networks from a structural perspective. *Biochemical Society Transactions*, 36(6):1398–1403, 2008.

[9] Gert Sabidussi. The centrality index of a graph. *Psychometrika*, 31(4):581–603, 1966.

[10] Linton C. Freeman. A set of measures of centrality based on betweenness. *Sociometry*, 40(1):35–41, 1977.

[11] Linton C. Freeman. Centrality in social networks conceptual clarification. *Social Networks*, 1(3):215–239, 1978.

[12] R. Milo, S. Shen-Orr, S. Itzkovitz, N. Kashtan, D. Chklovskii, and U. Alon. Network motifs: Simple building blocks of complex networks. *Science*, 298(5594):824–827, 2002.

[13] Paul Erdős and Alfréd Rényi. On random graphs I. *Publicationes Mathematicae Debrecen*, 6:290–297, 1959.

[14] Paul Erdős and Alfréd Rényi. On the evolution of random graphs. *Bulletin of the International Statistical Institute*, 38:343–347, 1961.

[15] Derek J. de Solla Price. Networks of scientific papers. *Science*, 149(3683):510–515, 1965.

[16] Derek De Solla Price. A general theory of bibliometric and other cumulative advantage processes. *Journal of the American Society for Information Science*, 27(5):292–306, 1976.

[17] Albert-László Barabási and Réka Albert. Emergence of scaling in random networks. *Science*, 286(5439):509–512, 1999.

[18] Stanley Milgram. The small world problem. *Psychology Today*, 1(1):61–67, 1967.

[19] Ray Solomonoff and A. Rapoport. Connectivity of random nets. *Bulletin of Mathematical Biophysics*, 13:107–17, 1951.

[20] Albert László Barabási and Zoltán N. Oltvai. Network biology: Understanding the cell's functional organization. *Nature Reviews Genetics*, 5(2):101–113, 2004.

[21] Pooja Sharma, Hasin A. Ahmed, Swarup Roy, and Dhruba K. Bhattacharyya. Unsupervised methods for finding protein complexes from PPI networks. *Network Modeling Analysis in Health Informatics and Bioinformatics*, 4(1):1–15, 2015.

[22] Henri Poincaré. Sur le problème des trois corps et les équations de dynamique. *Acta Mathematica*, 13:1–270, 1890.

[23] Alfred J. Lotka. Contribution to the theory of periodic reactions. *Journal of Physical Chemistry*, 14(3):271–274, 1910.

[24] Vito Volterra. Variazioni e fluttuazioni del numero d'individui in specie animali conviventi. *Memoria della Reale Accademia Nazionale dei Lincei*, 2:31–113, 1926.

[25] S. A. Kauffman. Metabolic stability and epigenesis in randomly constructed genetic nets. *Journal of Theoretical Biology*, 22(3):437–467, 1969.

[26] Itziar Irurzun-Arana, José Martín Pastor, Iñaki F. Trocóniz, and José David Gómez-Mantilla. Advanced Boolean modeling of biological networks applied to systems pharmacology. *Bioinformatics*, 33(7):1040–1048, 2017.

[27] Brenton Prettejohn, Matthew Berryman, and Mark McDonnell. Methods for generating complex networks with selected structural properties for simulations: A review and tutorial for neuroscientists. *Frontiers in Computational Neuroscience*, 5:11, 2011.

[28] Carlos Gershenson, Diederik Aerts, and Bruce Edmonds. *Worldviews, Science and Us: Philosophy and Complexity*. World Scientific, 2007.

[29] Judea Pearl. *Causality: Models, Reasoning, and Inference*. Cambridge University Press, 2000.

[30] A. M. Turing. On computable numbers, with an application to the Entscheidungsproblem. *Proceedings of the London Mathematical Society*, 42(2):230–265, 1936.

[31] The Experimental Algorithmic Information Project. The Shortest Universal Turing Machine Implementation Contest. www.mathrix.org/experimentalAIT/TuringMachine.html.

[32] Claude E. Shannon. A universal Turing machine with two internal states. In C. E. Shannon and J. McCarthy (eds.), *Automata Studies*, vol. 34, Princeton University Press, 1956, pp. 157–165.

[33] Hector Zenil. On the dynamic qualitative behavior of universal computation. *Complex Systems*, 20(3):265–278, 2011.

[34] Jürgen Riedel and Hector Zenil. Cross-boundary behavioural reprogrammability reveals evidence of pervasive universality. *International Journal of Unconventional Computing*, 13:309–357, 2018.

[35] Stephen Wolfram. *A New Kind of Science*. Wolfram Media, 2002.

[36] T. Rado. On non-computable functions. *Bell System Technical Journal*, 41(3):877–884, 1962.

[37] RANDOM.ORG. True Random Number Service. www.random.org/.

[38] A. N. Kolmogorov. Three approaches to the quantitative definition of information. *International Journal of Computer Mathematics*, 2(1–4):157–168, 1968.

[39] Gregory J. Chaitin. On the length of programs for computing finite binary sequences. *Journal of the ACM (JACM)*, 13(4):547–569, 1966.

[40] Rodney G. Downey and D. R. Hirschfeldt. *Algorithmic Randomness and Complexity*. Springer, 2010.

[41] Per Martin-Löf. The definition of random sequences. *Information and Control*, 9:602–619, 1966.

[42] C. P. Schnorr. A unified approach to the definition of random sequences. *Mathematical Systems Theory*, 5(3):246–258, 1971.

[43] Ray J. Solomonoff. A formal theory of inductive inference. Part I. *Information and Control*, 7(1):1–22, 1964.

[44] L. A. Levin. Laws of information conservation (non-growth) and aspects of the foundation of probability theory. *Problems of Information Transmission*, 10:206–210, 1974.

[45] Thomas M. Cover and Joy A. Thomas. *Elements of Information Theory*. Wiley-Interscience, 2nd ed., 2006.

[46] Cristian S. Calude. *Information and Randomness: An Algorithmic Perspective*. Springer-Verlag, 2nd ed. 2002.

[47] R. J. Solomonoff. Complexity-based induction systems: Comparisons and convergence theorems. *IEEE Transactions on Information Theory*, 24(4):422–432, 1978.

[48] Ray Solomonoff. The application of algorithmic probability to problems in artificial intelligence. In L. N. Kanal and J. F. Lemmer (eds.) *Machine Intelligence and Pattern Recognition*, Vol. 4. North-Holland, 1986, pp. 473–491.

[49] Ray J. Solomonoff. A system for incremental learning based on algorithmic probability. *Proceedings of the Sixth Israeli Conference on Artificial Intelligence, Computer Vision and Pattern Recognition*, 1989.

[50] Walter Kirchherr, Ming Li, and Paul Vitányi. The miraculous universal distribution. *The Mathematical Intelligencer*, 19(4):7–15, 1997.

[51] Gregory J. Chaitin. A theory of program size formally identical to information theory. *Journal of the ACM*, 22(3):329–340, 1975.

[52] Jean-Paul Delahaye and Hector Zenil. Towards a stable definition of Kolmogorov–Chaitin complexity. ArXiv abs/0804.3459, 2008.

[53] C. S. Calude, E. Calude, and M. J. Dinneen. A new measure of the difficulty of problems. *Journal of Multiple-Valued Logic and Soft Computing*, 12(3):285–307, 2006.

[54] G. J. Chaitin. Algorithmic information theory. *IBM Journal of Research and Development*, 21(4):350–359, 1977.

[55] C. H. Bennett. Logical depth and physical complexity. In R. Herken (ed.) The Universal Turing Machine: *A Half-Century Survey*. Oxford University Press, 1988, pp. 227–257.

[56] Jacob Ziv and Abraham Lempel. Compression of individual sequences via variable-rate coding. *IEEE Transactions on Information Theory*, 24(5):530–536, 1978.

[57] Jean-Paul Delahaye and Hector Zenil. Numerical evaluation of algorithmic complexity for short strings: A glance into the innermost structure of randomness. *Applied Mathematics and Computation*, 219(1):63–77, 2012.

[58] Fernando Soler-Toscano, Hector Zenil, Jean-Paul Delahaye, and Nicolas Gauvrit. Calculating Kolmogorov complexity from the output frequency distributions of small turing machines. *PLoS ONE*, 9(5):e96223, 2014.

[59] Hector Zenil, Liliana Badillo, Santiago Hernández-Orozco, and Francisco Hernández-Quiroz. Coding-theorem like behaviour and emergence of the universal distribution from resource-bounded algorithmic probability. *International Journal of Parallel, Emergent and Distributed Systems*, 34(2):161–180, 2019.

[60] Hector Zenil. Algorithmic data analytics, small data matters and correlation versus causation. In Wolfgang Pietsch, Jörg Wernecke, and Maximilian Ott, eds., *Berechenbarkeit der Welt? Philosophie und Wissenschaft im Zeitalter von Big Data*. Springer Fachmedien, 2017, pp. 453–475.

[61] C. S. Calude and M. A. Stay. Most programs stop quickly or never halt. *Advances in Applied Mathematics*, 40:295–308, 2008.

[62] H. Zenil. From computer runtimes to the length of proofs: With an algorithmic probabilistic application to waiting times in automatic theorem proving. In B. Khousainov Dinneen and A. Nies, eds., *Computation, Physics and Beyond*. Springer, 2012, pp. 223–240.

[63] A. H. Brady. The determination of the value of Rado's noncomputable function $\Sigma(k)$ for four-state Turing machines. *Mathematics of Computation*, 40(162):647–665, 1983.

[64] Fernando Soler-Toscano, Hector Zenil, Jean-Paul Delahaye, and Nicolas Gauvrit. Correspondence and independence of numerical evaluations of algorithmic information measures. *Computability*, 2(2):125–140, 2013.

[65] Hector Zenil, Fernando Soler-Toscano, Jean Paul Delahaye, and Nicolas Gauvrit. Two-dimensional Kolmogorov complexity and an empirical validation of the Coding theorem method by compressibility. *PeerJ Computer Science*, 1:e23, 2015.

[66] Hector Zenil, Narsis A. Kiani, and Jesper Tegnér. Low-algorithmic-complexity entropy-deceiving graphs. *Physical Review E*, 96(1):12308, 2017.

[67] L. A. Levin. Universal sequential search problems. *Rossiiskaya Akademiya Nauk. Problemy Peredachi Informatsii*, 9(3):115–116, 1973.

[68] Robert P. Daley. An example of information and computation resource trade-off. *Journal of the ACM*, 20(4):687–695, 1973.

[69] R. P. Daley. Minimal-program complexity of sequences with restricted resources. *Information and Control*, 23(4):301–312, 1973.

[70] Jürgen Schmidhuber. The speed prior: A new simplicity measure yielding near-optimal computable predictions. In J. Kivinen and R. H. Sloan (eds.), Computational Learning Theory. COLT 2002. *Lecture Notes in Computer Science*, Vol. 2375, Springer, 2002.

[71] M. Li and P. Vitányi. *An Introduction to Kolmogorov Complexity and Its Applications*. Springer Verlag, 2008.

[72] R. Cilibrasi and P. Vitanyi. Clustering by compression. *IEEE Transactions on Information Theory*, 51(4):1523–1545, 2005.

[73] Nicolas Gauvrit, Fernando Soler-Toscano, and Hector Zenil. Natural scene statistics mediate the perception of image complexity. *Visual Cognition*, 22(8):1084–1091, 2014.

[74] Nicolas Gauvrit, Henrik Singmann, Fernando Soler-Toscano, and Hector Zenil. Algorithmic complexity for psychology: A user-friendly implementation of the coding theorem method. *Behavior Research Methods*, 48(1):314–329, 2016.

[75] Vera Kempe, Nicolas Gauvrit, and Douglas Forsyth. Structure emerges faster during cultural transmission in children than in adults. *Cognition*, 136:247–254, 2015.

[76] Andreas Holzinger, Bernhard Ofner, Christof Stocker, et al. On graph entropy measures for knowledge discovery from publication network data. In *Lecture Notes in Computer Science (Including Subseries Lecture Notes in Artificial Intelligence and Lecture Notes in Bioinformatics)*, volume 8127 LNCS, pages 354–362. Springer, 2013.

[77] Frank Emmert-Streib and Matthias Dehmer. Exploring statistical and population aspects of network complexity. *PLoS ONE*, 7(5):e34523, 2012.

[78] Matthias Dehmer. A novel method for measuring the structural information content of networks. *Cybernetics and Systems*, 39(8):825–842, 2008.

[79] Matthias Dehmer and Abbe Mowshowitz. A history of graph entropy measures. *Information Sciences*, 181(1):57–78, 2011.

[80] Abbe Mowshowitz and Matthias Dehmer. Entropy and the complexity of graphs revisited. *Entropy*, 14(3):559–570, 2012.

[81] H. Zenil, F. Soler-Toscano, K. Dingle, and A. A. Louis. Correlation of automorphism group size and topological properties with program-size complexity evaluations of graphs and complex networks. *Physica A: Statistical Mechanics and its Applications*, 404:341–358, 2014.

[82] Hector Zenil, Narsis A. Kiani, and Jesper Tegnér. Methods of information theory and algorithmic complexity for network biology. *Seminars in Cell and Developmental Biology*, 51:32–43, 2016.

[83] Cristian S. Calude, Kai Salomaa, and Tania K. Roblot. Finite state complexity. *Theoretical Computer Science*, 412(41):5668–5677, 2011.

[84] Christopher G. Langton. Studying artificial life with cellular automata. *Physica D: Nonlinear Phenomena*, 22(1):120–149, 1986.

[85] Gustav A. Hedlund and Marston Morse. Unending chess, symbolic dynamics and a problem in semigroups. *Duke Mathematical Journal*, 11(1):1–7, 1944.

[86] David Bailey, Peter Borwein, and Simon Plouffe. On the rapid computation of various polylogarithmic constants. *Mathematics of Computation*, 66(218):903–913, 1997.

[87] Wolfram. GraphData – Wolfram Language Documentation. https://reference .wolfram.com/language/ref/GraphData.html?view=all.

[88] Jacob Ziv and Abraham Lempel. A universal algorithm for sequential data compression. *IEEE Transactions on Information Theory*, 23(3):337–343, 1977.

[89] Laszlo Babai and Eugene M. Luks. Canonical labeling of graphs. In *Conference Proceedings of the Annual ACM Symposium on Theory of Computing* (STOC '83). Association for Computing Machinery, New York, 1983, pp. 171–183.

[90] E. N. Gilbert. Random graphs. *The Annals of Mathematical Statistics*, 30(4):1141–1144, 1959.

[91] S. Boccaletti, G. Bianconi, R. Criado, et al. The structure and dynamics of multilayer networks. *Physics Reports*, 544(1):1–122, 2014.

[92] Zengqiang Chen, Matthias Dehmer, Frank Emmert-Streib, and Yongtang Shi. Entropy bounds for dendrimers. *Applied Mathematics and Computation*, 242:462–472, 2014.

[93] Chiara Orsini, Marija M Dankulov, Pol Colomer-De-Simon, et al. Quantifying randomness in real networks. *Nature Communications*, 6(1):1–10, 2015.

[94] Hector Zenil, Narsis A. Kiani, and Jesper Tegnér. The thermodynamics of network coding, and an algorithmic refinement of the principle of maximum entropy. *Entropy*, 21(6):560, 2019.

[95] Ginestra Bianconi. The entropy of randomized network ensembles. *Europhysics Letters*, 81(2):28005, 2008.

[96] Yilun Shang. Bounding extremal degrees of edge-independent random graphs using relative entropy. *Entropy*, 18(2):53, 2016.

[97] Ernesto Estrada, José A. de la Peña, and Naomichi Hatano. Walk entropies in graphs. *Linear Algebra and Its Applications*, 443:235–244, 2014.

[98] Dipendra C. Sengupta and Jharna D. Sengupta. Application of graph entropy in CRISPR and repeats detection in DNA sequences. *Computational Molecular Bioscience*, 06(03):41–51, 2016.

[99] Yilun Shang. The Estrada index of evolving graphs. *Applied Mathematics and Computation*, 250:415–423, 2015.

[100] J. Korner and K. Marton. Random access communication and graph entropy. *IEEE Transactions on Information Theory*, 34(2):312–314, 1988.

[101] Matthias Dehmer, Stephan Borgert, and Frank Emmert-Streib. Entropy bounds for hierarchical molecular networks. *PLoS ONE*, 3(8):e3079, 2008.

[102] Mikołaj Morzy, Tomasz Kajdanowicz, and Przemysław Kazienko. On measuring the complexity of networks: Kolmogorov complexity versus entropy. *Complexity*, 2017(1):1–12, 2017.

[103] Hector Zenil, Fernando Soler-Toscano, Narsis A. Kiani, Santiago Hernández-Orozco, and Antonio Rueda-Toicen. A decomposition method for global evaluation of Shannon entropy and local estimations of algorithmic complexity. *Entropy*, 20(8):605, 2018.

[104] Réka Albert and Albert-László Barabási. Statistical mechanics of complex networks. *Reviews of Modern Physics*, 74(1):47, 2002.

[105] Harry Buhrman, Ming Li, John Tromp, and Paul Vitányi. Kolmogorov random graphs and the incompressibility method. *SIAM Journal on Computing*, 29(2):590–599, 1999.

[106] Uri Alon. Network motifs: Theory and experimental approaches. *Nature Reviews Genetics*, 8(6):450–461, 2007.

[107] Hector Zenil, Narsis A. Kiani, and Jesper Tegnér. Algorithmic complexity of motifs clusters superfamilies of networks. In *Proceedings of the 2013 IEEE International Conference on Bioinformatics and Biomedicine (BIBM '13)*. IEEE, 2013.

[108] Hector Zenil, Narsis A. Kiani, Francesco Marabita, et al. An algorithmic information calculus for causal discovery and reprogramming systems. *iScience*, 19:1160–1172, 2019.

[109] Judea Pearl and Dana Mackenzie. *The Book of Why: The New Science of Cause and Effect*. Basic Books, 2018.

[110] Nicolas Gauvrit, Hector Zenil, Fernando Soler-Toscano, Jean-Paul Delahaye, and Peter Brugger. Human behavioral complexity peaks at age 25. *PLoS Computational Biology*, 13(4):e1005408, 2017.

[111] Hector Zenil, Narsis A. Kiani, Felipe S. Abrahão, Antonio Rueda-Toicen, Allan A. Zea, and Jesper Tegnér. Minimal algorithmic information loss methods for dimension reduction, feature selection and network sparsification. *ArXiv:1802.05843 [physics]*, 2020.

[112] Hector Zenil. A review of methods for estimating algorithmic complexity: Options, challenges, and new directions. *Entropy*, 22(6):612, 2020.

[113] Hector Zenil, Narsis A. Kiani, Allan A. Zea, and Jesper Tegnér. Causal deconvolution by algorithmic generative models. *Nature Machine Intelligence*, 1(1):58–66, 2019.

[114] Bruno Durand and Zsuzsanna Róka. The game of life: Universality revisited. Technical report, Ecole Normale Supérieure de Lyon, Laboratoire de l'Informatique du Parallélisme, 1998.

[115] Martin Gardner. Mathematical Games: The fantastic combinations of John Conway's new solitaire game 'life'. *Scientific American*, 223:120–123, 1970.

[116] Genaro J. Martínez, Andrew Adamatzky, and Harold V. McIntosh. A computation in a cellular automaton collider rule 110. In A. Adamatzky (ed.) *Advances in Unconventional Computing*. Springer, 2016, pp. 391–428.

[117] Melanie Mitchell. Computation in cellular automata: A selected review. In *Non-Standard Computation*. John Wiley & Sons, Ltd, 1998, pp. 95–140.

[118] V. Benci, C. Bonanno, S. Galatolo, G. Menconi, and M. Virgilio. Dynamical systems and computable information. *Discrete and Continuous Dynamical Systems - Series B*, 4(4):935, 2004.

[119] Elizabeth S. Spelke and Katherine D. Kinzler. Core knowledge. *Developmental science*, 10(1):89–96, 2007.

[120] Vé Ronique Izard, Coralie Sann, Elizabeth S. Spelke, and Arlette Streri. Newborn infants perceive abstract numbers. *Proceedings of the National Academy of Sciences*, 106(25):10382–10385, 2009.

[121] Rosa Rugani, Giorgio Vallortigara, Konstantinos Priftis, and Lucia Regolin. Number-space mapping in the newborn chick resembles humans' mental number line. *Science*, 347(6221):534–536, 2015.

[122] Justin Halberda, Ryan Ly, Jeremy B. Wilmer, Daniel Q. Naiman, and Laura Germine. Number sense across the lifespan as revealed by a massive internet-based sample. *Proceedings of the National Academy of Sciences*, 109(28):11116–11120, 2012.

[123] Stanislas Dehaene. *The Number Sense: How the Mind Creates Mathematics*, rev. ed. Oxford University Press, 2011.

[124] Laura Fontanari, Michel Gonzalez, Giorgio Vallortigara, and Vittorio Girotto. Probabilistic cognition in two indigenous Mayan groups. *Proceedings of the National Academy of Sciences*, 111(48):17075–17080, 2014.

[125] Edward Téglás, Ernand Vul, Vittorio Girotto, Michel Gonzalez, Joshua B. Tenenbaum, and Luca L. Bonatti. Pure reasoning in 12-month-old infants as probabilistic inference. *Science*, 332(6033):1054–1059, 2011.

[126] Fei Xu and Vashti Garcia. Intuitive statistics by 8-month-old infants. *Proceedings of the National Academy of Sciences*, 105(13):5012–5015, 2008.

[127] Joseph J. Williams and Thomas L. Griffiths. Why are people bad at detecting randomness? A statistical argument. *Journal of Experimental Psychology: Learning, Memory, and Cognition*, 39(5):1473, 2013.

[128] Nicolas Gauvrit, Hector Zenil, Jean-Paul Delahaye, and Fernando Soler-Toscano. Algorithmic complexity for short binary strings applied to psychology: A primer. *Behavior Research Methods*, 46(3):732–744, 2014.

[129] Lili Ma and Fei Xu. Preverbal infants infer intentional agents from the perception of regularity. *Developmental Psychology*, 49(7):1330, 2013.

[130] Marjan Jahanshahi, T. Saleem, Aileen K. Ho, Georg Dirnberger, and R. Fuller. Random number generation as an index of controlled processing. *Neuropsychology*, 20(4):391–399, 2006.

[131] E. Mittenecker. Die Analyse zuflliger Reaktionsfolgen [The analysis of 'random' action sequences]. *Zeitschrift fr Experimentelle und Angewandte Psychologie*, 5:45–60, 1958.

[132] T. Loetscher, C. J. Bockisch, and P. Brugger. Looking for the answer: The mind's eye in number space. *Neuroscience*, 151(3):725–729, 2008.

[133] H. Heuer, M. Janczyk, and W. Kunde. Random noun generation in younger and older adults. *The Quarterly Journal of Experimental Psychology*, 63(3):465–478, 2010.

[134] S. Wiegersma and W. A. J. M. Van Den Brink. Repetition of tones in vocal music. *Perceptual and Motor Skills*, 55(1):167–170, 1982.

[135] U. Hahn and P. Warren. Perceptions of randomness: Why three heads are better than four. *Psychological Review*, 116(2):454–461, 2009.

[136] Günter Schulter, Erich Mittenecker, and Ilona Papousek. A computer program for testing and analyzing random generation behavior in normal and clinical samples: The Mittenecker Pointing Test. *Behavior Research Methods*, 42(1):333–341, 2010.

[137] Joseph H. R. Maes, Paul A. T. M. Eling, Miriam F. Reelick, and Roy P. C. Kessels. Assessing executive functioning: On the validity, reliability, and sensitivity of a click/point random number generation task in healthy adults and patients with cognitive decline. *Journal of Clinical and Experimental Neuropsychology*, 33(3):366–378, 2011.

[138] J. N. Towse. On random generation and the central executive of working memory. *British Journal of Psychology*, 89(1):77–101, 1998.

[139] A. Miyake, N. P. Friedman, M. J. Emerson, A. H. Witzki, A. Howerter, and T. D. Wager. The unity and diversity of executive functions and their contributions to complex 'Frontal Lobe' tasks: A latent variable analysis. *Cognitive psychology*, 41(1):49–100, 2000.

[140] Martial Van der Linden, Annick Beerten, and Mauro Pesenti. Age-related differences in random generation. *Brain and Cognition*, 38(1):1–16, 1998.

[141] R. G. Brown, P. Soliveri, and M. Jahanshahi. Executive processes in Parkinson's disease – Random number generation and response suppression. *Neuropsychologia*, 36(12):1355–1362, 1998.

[142] Peter Brugger. Variables that influence the generation of random sequences: An update. *Perceptual and Motor Skills*, 84(2):627–661, 1997.

[143] Raymond S. Nickerson. The production and perception of randomness. *Psychological Review*, 109(2):330, 2002.

[144] Alan Baddeley, Hazel Emslie, Jonathan Kolodny, and John Duncan. Random generation and the executive control of working memory. *The Quarterly Journal of Experimental Psychology A: Human Experimental Psychology*, 51A(4):819–852, 1998.

[145] Maarten Peters, Timo Giesbrecht, Marko Jelicic, and Harald Merckelbach. The random number generation task: Psychometric properties and normative data of an executive function task in a mixed sample. *Journal of the International Neuropsychological Society*, 13(4):626–634, 2007.

[146] Patricia Sagaspe, André Charles, Jacques Taillard, Bernard Bioulac, and Pierre Philip. Inhibition et mémoire de travail: Effet d'une privation aiguë de sommeil sur une tâche de génération aléatoire. *Canadian Journal of Experimental Psychology/Revue canadienne de psychologie expérimentale*, 57(4):265–273, 2003.

[147] P. Brugger, A. U. Monsch, D. P. Salmon, and N. Butters. Random number generation in dementia of the Alzheimer type: A test of frontal executive functions. *Neuropsychologia*, 34(2):97–103, 1996.

[148] Kevin Ka-Shing Chan, Christy Lai-Ming Hui, Jennifer Yee-Man Tang, et al. Random number generation deficit in early schizophrenia. *Perceptual and Motor Skills*, 112(1):91–103, 2011.

[149] Nicole J. Rinehart, John L. Bradshaw, Simon A. Moss, Avril V. Brereton, and Bruce J. Tonge. Pseudo-random number generation in children with high-functioning autism and Asperger's disorder: Further evidence for a dissociation in executive functioning? *Autism*, 10(1):70–85, 2006.

[150] H. Proios, S. S. Asaridou, and P. Brugger. Random number generation in patients with aphasia: A test of executive functions. *Acta Neuropsychologica*, 6:157–168, 2008.

[151] James R. Brockmole and Robert H. Logie. Age-related change in visual working memory: A study of 55,753 participants aged 8–75. *Frontiers in Psychology*, 4:12, 2013.

[152] Fergus I. M. Craik and Ellen Bialystok. Cognition through the lifespan: Mechanisms of change. *Trends in Cognitive Sciences*, 10(3):131–138, 2006.

[153] Ary L. Goldberger, C. K. Peng, and Lewis A. Lipsitz. What is physiologic complexity and how does it change with aging and disease? *Neurobiology of Aging*, 23(1):23–26, 2002.

[154] Brad Manor and Lewis A. Lipsitz. Physiologic complexity and aging: Implications for physical function and rehabilitation. *Progress in Neuro-Psychopharmacology and Biological Psychiatry*, 45:287–293, 2013.

[155] Lewis A. Lipsitz and Ary L. Goldberger. Loss of 'complexity' and aging: Potential applications of fractals and chaos theory to senescence. *JAMA: The Journal of the American Medical Association*, 267(13):1806–1809, 1992.

[156] Nick Chater and Paul Vitányi. Simplicity: A unifying principle in cognitive science? *Trends in Cognitive Sciences*, 7(1):19–22, 2003.

[157] Mónica Tamariz and Simon Kirby. Culture: Copying, compression, and conventionality. *Cognitive Science*, 39(1):171–183, 2015.

[158] Lance Storm, Patrizio E. Tressoldi, and Lorenzo Di Risio. Meta-analysis of ESP studies, 1987–2010:Assessing the success of the forced-choice design in parapsychology. *Journal of Parapsychology*, 76(2):243–273, 2012.

[159] Kenneth O. McGraw, Mark D. Tew, and John E. Williams. The integrity of web-delivered experiments: Can you trust the data? *Psychological Science*, 11(6):502–506, 2000.

[160] Matthew J. C. Crump, John V. McDonnell, and Todd M. Gureckis. Evaluating Amazon's mechanical turk as a tool for experimental behavioral research. *PLoS ONE*, 8(3):e57410, 2013.

[161] Andrey Chetverikov and Philipp Upravitelev. Online versus offline: The Web as a medium for response time data collection. *Behavior Research Methods*, 48(3):1086–1099, 2016.

[162] Laura T. Germine, Bradley Duchaine, and Ken Nakayama. Where cognitive development and aging meet: Face learning ability peaks after age 30. *Cognition*, 118(2):201–210, 2011.

[163] Pablo Garaizar and Ulf Dietrich Reips. Visual DMDX: A web-based authoring tool for DMDX, a Windows display program with millisecond accuracy. *Behavior Research Methods*, 47(3):620–631, 2015.

[164] Peter Brugger, Marianne Regard, Theodor Landis, Norman Cook, Denise Krebs, and Joseph Niederberger. 'Meaningful' patterns in visual noise: Effects of lateral stimulation and the observer's belief in ESP. *Psychopathology*, 26(5–6):261–265, 1993.

[165] P. Brugger, T. Landis, and M. Regard. A 'sheep-goat effect' in repetition avoidance: Extra-sensory perception as an effect of subjective probability? *British Journal of Psychology*, 81:455–468, 1990.

[166] Richard Wiseman and Caroline Watt. Belief in psychic ability and the misattribution hypothesis: A qualitative review. *British Journal of Psychology*, 97(3):323–338, 2006.

[167] Geoff Der and Ian J. Deary. Age and sex differences in reaction time in adulthood: Results from the United Kingdom health and lifestyle survey. *Psychology and Aging*, 21(1):62–73, 2006.

[168] Jochen Musch and Katja Ehrenberg. Probability misjudgment, cognitive ability, and belief in the paranormal. *British Journal of Psychology*, 93(2):169–177, 2002.

[169] Olivia Geisseler, Tobias Pflugshaupt, Andreas Buchmann, et al. Random number generation deficits in patients with multiple sclerosis: Characteristics and neural correlates. *Cortex*, 82:237–243, 2016.

[170] Daniel L. Hartl, and Andrew G. Clark. *Principles of Population Genetics*, 3rd ed. Sinauer Associates, 1997.

[171] S. Hernández-Orozco, N. Kiani, and H. Zenil. Algorithmically probable mutations reproduce aspects of evolution, such as convergence rate, genetic memory, and modularity. *Royal Society Open Science*, 5:180399, 2018.

[172] Gregory J. Chaitin. Evolution of mutating software. *Bulletin of the EATCS*, 97:157–164, 2009.

[173] Gregory J. Chaitin. *Proving Darwin: Making Biology Mathematical*. Vintage, 2013.

[174] Gregory J. Chaitin. Life as evolving software. In Hector Zenil (ed.), *A Computable Universe: Understanding and Exploring Nature as Computation*. World Scientific Publishing Company, 2012, pp. 277–302.

[175] Ray J. Solomonoff. The Kolmogorov lecture: The universal distribution and machine learning. *Computer Journal*, 46(6):598–601, 2003.

[176] Santiago Hernández-Orozco, Francisco Hernández-Quiroz, and Hector Zenil. Undecidability and irreducibility conditions for open-ended evolution and emergence. *Artificial Life*, 24(1):56–70, 2018.

[177] Mark A. Bedau. Four puzzles about life. *The Nature of Life: Classical and Contemporary Perspectives from Philosophy and Science*, 4:392–404, 2010.

[178] Gregory J. Chaitin. Information-theoretic limitations of formal systems. *Journal of the ACM*, 21(3):403–424, 1974.

[179] Cristian S. Calude, Michael J. Dinneen, Chi-Kou Shu, et al. Computing a glimpse of randomness. *Experimental Mathematics*, 11(3):361–370, 2002.

[180] Hector Zenil, Carlos Gershenson, James A. R. Marshall, and David A. Rosenblueth. Life as thermodynamic evidence of algorithmic structure in natural environments. *Entropy*, 14(11):2173–2191, 2012.

[181] J. D. Schaffer and L. J. Eshelman. On crossover as an evolutionary viable strategy. In R. K. Belew and L. B. Booker (eds), *Proceedings of the 4th International Conference on Genetic Algorithms*. Morgan Kaufmann, 1991, pp. 61–68.

[182] Narsis A. Kiani and Lars Kaderali. Dynamic probabilistic threshold networks to infer signaling pathways from time-course perturbation data. *BMC Bioinformatics*, 15(1):250, 2014.

[183] Y. Yarden and M. Sliwkowski. Untangling the ErbB signalling network. *Nature Reviews Molecular Cell Biology*, 2(2):127–137, 2001.

[184] M. A. Olayioye. The ErbB signaling network: Receptor heterodimerization in development and cancer. *The EMBO Journal*, 19(13):3159–3167, 2000.

[185] Koyel Mitra, Anne-Ruxandra Carvunis, Sanath Kumar Ramesh, and Trey Ideker. Integrative approaches for finding modular structure in biological networks. *Nature Reviews. Genetics*, 14(10):719–732, 2013.

[186] S. Hernández-Orozco, H. Zenil, J. Riedel, A. Uccello, N. A. Kiani, and J. Tegnér. Algorithmic probability-guided machine learning on non-differentiable spaces. *Frontiers in Artificial Intelligence*, 25(3):567356, 2021.

Index